Corrosion and Protection of Reinforced Concrete

Corrosion and Protection of Reinforced Concrete

Brian Cherry and Warren Green

CRC Press
Taylor & Francis Group
Boca Raton London New York

CRC Press is an imprint of the
Taylor & Francis Group, an **informa** business

First edition published 2021
by CRC Press
2 Park Square, Milton Park, Abingdon, Oxon, OX14 4RN

and by
CRC Press
6000 Broken Sound Parkway NW, Suite 300, Boca Raton, FL
33487-2742

CRC Press is an imprint of Taylor & Francis Group, LLC

ISBN: 978-0-367-51760-1 (hbk)
ISBN: 978-0-367-51761-8 (pbk)
ISBN: 978-1-003-08130-2 (ebk)

Typeset in Sabon
by SPi Global, India

Contents

Preface

The late Professor Brian Cherry was many things; a gentleman, a sailor, a scholar, and one of the most significant figures in the history of Australian engineering and engineering education. He was a man who had received the highest level of respect from all who knew him, and from all who had interacted with him professionally and socially. He inspired countless students, co-authors, practitioners, scientists, engineers, and researchers young and old.

Brian had a significant impact on the field of corrosion science and technology over many decades, and one of his passions in that regard was the 'Corrosion and Protection of Reinforced Concrete'. Some time ago he began work on a monograph on this subject. I have had the honour and delight of working with him on it when he was alive and progressing it since his passing in 2018.

Professor Brian Cherry was always a firm believer in first understanding the fundamentals of any aspects of corrosion science, then the mechanisms, before embarking on engineering solutions to the management of corrosion including in steel reinforced concrete.

This monograph has been developed to provide a sound understanding of the mechanisms of the corrosion and protection of reinforced concrete. It is particularly designed for asset managers, port engineers, bridge maintenance managers, building managers, heritage structure engineers, plant engineers, consulting engineers, architects, specialist contractors, and construction material suppliers, who have the task of resolving problems of corrosion of steel reinforced, prestressed, and post-tensioned concrete elements. It is also considered a most useful reference for students at postgraduate level.

Since its development in the mid-19th century, reinforced concrete has become the most widely used construction material in the world. Extended performance is often expected of our reinforced concrete structures. Some like marine structures are in aggressive environments, they may be many decades old, of critical importance in terms of function or location and be irreplaceable, and as such repair, and protection is necessary during their service lives.

This monograph provides readers with not only a comprehensive general knowledge, necessary for an understanding of corrosion prevention and protection in reinforced concrete, but also information on specific problems of corrosion in concrete learnt from the practical experience of both authors. To achieve this, the Chapters in this monograph are structured along the following lines:

- In Chapter 1, the characteristics of steel reinforced concrete are discussed. This includes its structure, different cements (binders), important aggregate, and mixing water issues, the more common admixtures that are necessary, and the different forms of reinforcing steel including conventional, prestressed, and post-tensioned. Fibre-reinforced concrete is also considered.
- The environments in which we place our reinforced concrete structures mean that various deterioration processes affect their in-service durability, loss of functionality, unplanned maintenance/remediation/ replacement, and in the worst cases a loss of structural integrity and a resultant safety risk. Degradation of concrete may involve one or more of mechanical, physical, structural, chemical, and biological causes. Cracks in concrete are routes for the ingress of aggressive environmental agents. The penetrability of hardened concrete is also relevant. Chapters 2 and 3 examine all these key factors.
- The most common form of damage to a reinforced concrete structure is caused by corrosion of the steel reinforcement. Key aspects of reinforcement corrosion in concrete are considered in Chapters 4 and 5. The high levels of protection afforded to steel reinforcement by suitably designed, constructed, and maintained concrete are discussed in these chapters. These include the physical protection provided to steel reinforcement by concrete, and the electrochemical protection provided by the passive iron oxide film produced on the steel surface by the concrete.

    The passivity provided to steel reinforcement by the alkaline environment of concrete may however be lost if the pH of the concrete pore solution falls because of carbonation or if aggressive ions such as chlorides penetrate in sufficient concentration to the steel reinforcement surface. Carbonation of concrete occurs as a result of atmospheric CO_2 gas (and atmospheric SO_x and NO_x gases) neutralising the concrete pore water (lowering its pH to 9) and thereby affecting the stability and continuity of the passive iron oxide film. Chloride ions in sufficient concentration can locally compromise the passivity of carbon-steel, prestressed, and post-tensioned steel reinforcement in concrete leading to pitting corrosion. Leaching of $Ca(OH)_2$ (and NaOH and KOH) from concrete also lowers the pH, which can allow corrosion initiation of the steel reinforcement. Stray electrical currents, most commonly from electrified traction systems and interference currents from cathodic protection systems can also breakdown the passive film, and cause corrosion of steel reinforced and prestressed concrete elements. Key issues relating to the mechanisms of corrosion of steel under these conditions are discussed. Fundamentals relating to the thermodynamics and kinetics of reinforcement corrosion are elucidated, together with the modelling of chloride and carbonation-induced corrosion initiation and propagation.

- It is essential that the mechanism(s) and extent of corrosion or concrete deterioration is known before appropriate remedial and/or protection measures can be developed for a reinforced concrete structure. Survey and diagnosis by onsite measurements as well as laboratory-based measurements are therefore important, and these are considered in Chapters 6 and 7.

- The methods that are employed to repair deteriorating reinforced concrete structures, or to protect reinforced concrete structures from deterioration depend on the nature and extent of deterioration occurring or likely to occur. In general repair techniques may be 'mechanical' or 'electrochemical', though often a combination of the two is necessary to ensure a long-lasting solution to the problem. Protection of the concrete is possible with coatings, penetrants, and membranes for example. Structural strengthening of concrete using fibre-reinforced polymers is also possible. Chapter 8 addresses repair and protection of concrete by mechanical methods. Cathodic protection (impressed current and galvanic) of reinforced concrete is examined in Chapter 9. Chapter 10 considers relevant aspects of electrochemical treatment methods available for reinforced concrete structures.

- Many of the problems that arise in the life of a reinforced concrete structure could be avoided if the appropriate precautions are taken at the design and construction stage. Chapter 11 looks at some options that can improve concrete durability such as concrete technology aspects, construction considerations, coatings and penetrants, coated and alternate reinforcement, and permanent corrosion monitoring.

- All engineering materials deteriorate with time, at rates dependent upon the type of material, the severity of the environment and the deterioration mechanisms involved. In engineering terms, the objective is to select the most cost-effective combination of materials to achieve the required design life. In doing so, it is critical to realise that the nature and rate of deterioration of materials is a function of their environment. Accordingly, the environment is a 'load' on a material as a force is a 'load' on a structural component. It is the synergistic combination of the structural load and environmental load which determines the performance of the structural component. Durability Planning is a system which formalises the process of achieving durability through appropriate design, construction, and maintenance. Chapter 12 closes the monograph by examining some key aspects of Durability Planning.

To Brian's wife Miriam, children James and Liz, and grandchildren Nicholas, Gretel, and William, this monograph is another contribution by your husband, dad, and grandfather to corrosion science and engineering.

To my wife Louise, and children Joshua, and Nicholas, I am indebted for your support and inspiration as well as that from my parents, brothers, educators, colleagues (current and previous but particularly in relation to this monograph Chelsea Derksen, Andrew Haberfield and Jack Katen) and mentors (most particularly Bruce Hinton, Brian Kinsella, Greg Moore, and Mark Byerley Sr).

Imagine the world without steel reinforced concrete structures and buildings. Imagine our primitive existence without the wonder and delight of steel reinforced concrete. Also, one cannot imagine such without the wonder and sheer delight of electrochemistry. Give that reinforced concrete element a hug next time rather than take it for granted.

Warren Green,
September 2020

Authors

Brian Cherry (deceased) was a Professor within the Department of Materials Science and Engineering at Monash University, Melbourne, Australia. He was the inaugural Associate Dean of Research at Monash University, and was also instrumental in the establishment of postgraduate degrees at Monash. Brian's legacy includes a stream of undergraduate students trained in corrosion, in addition to countless postgraduate students. These students have permeated academia and industry, nationally, and internationally, filling positions at large companies, consultancies, one-person companies, and everything in between.

Warren Green is a Director and a Principal Corrosion Engineer/Materials Scientist of Vinsi Partners, Sydney, Australia. He has experience in corrosion engineering and materials technology covering marine, civil, industrial, and building structures. Warren is also an Adjunct Associate Professor in the Institute for Frontier Materials at Deakin University, Melbourne, and a Visiting Adjunct Associate Professor within the Corrosion Centre of Curtin University, Perth, Western Australia.

Chapter I

Steel-reinforced concrete characteristics

1.1 CONCRETE AND REINFORCED CONCRETE

Concrete is the most widely used material of construction in the world (and the second most used material by mankind after water) (Miller, 2018). Concrete is strong in compression but weak in tension and so much of the concrete is reinforced, usually with steel. The steel reinforcement can take the form of conventional carbon steel (black steel), prestressing steel, post-tensioned steel, and steel fibres, and its widespread utility is primarily due to the fact that it combines the best features of concrete and steel. The properties of these two materials may be compared below (Table 1.1).

The properties of the materials thus complement one another and so by combining them together a composite that has good tensile strength, shear strength, and compressive strength combined with durability and fire resistance can be formed. Typical properties of reinforcing steel and concrete might be (Table 1.2).

So that the strains at failure (if the stress strain curves were linear) are approximately 2×10^{-3} for the steel and 8×10^{-6} for the concrete. Consequently, when the steel is operating at or near its yield point the concrete must be cracked and so different forms of prestressing have been developed to ensure that the cracking is controlled to an appropriate level.

Table 1.1 Properties of steel and concrete

Property	Concrete	Steel
Strength in tension	Poor	Good
Strength in compression	Good	Buckling can occur
Strength in shear	Fair	Good
Durability	Good	Corrodes if unprotected
Fire resistance	Good	Poor, low strength at high temperatures

Table 1.2 Properties of steel and concrete

Property	Steel	Concrete
Tensile strength	400 MPa	2.5 MPa
Modulus	200×10^3 MPa	320×10^3 MPa

The properties of the materials also complement each other in that the properties of the steel can be modified by alloying and working to vary its strength and corrosion resistance, the properties of the concrete can be modified to facilitate the building of the structure by changing the ease with which it can be formed into the required shape and the penetrability of the final structure to aggressive agents that might attack the steel. In this chapter attention, will be focussed on the structure of concrete and the way in which its properties are controlled by its components, the cement, the aggregate, the mixing water, and admixtures.

1.2 THE STRUCTURE OF CONCRETE

Basically, concrete consists of mineral aggregate held together by a cement paste, so that if we consider a normal mix it will consist of cement, sand (fine aggregate), coarse aggregate and water (and often other admixtures) which are mixed together to form eventually a hard, strong material. The primary binding agent in concrete is cement. The water reacts with the cement to make a cement paste which then hardens and binds the aggregate together to make the solid concrete and the cured composition of the concrete may be represented by the proportions shown in Figure 1.1.

Its structure may be as shown in Figure 1.2.

1.3 CEMENTS AND THE CEMENTING ACTION

1.3.1 General

The aggregate is bound together by hydrated cement. The hydraulic binder of concrete commonly consists of Portland cement or of mixtures of Portland cement and one or more of fly ash, ground granulated iron blast furnace slag or silica fume.

Portland cement takes its name from a cement manufactured in England in the late eighteenth century which bears a similarity to stone quarried near Portland in Dorset, England. It is defined in ASTM C150 (2016) as 'hydraulic

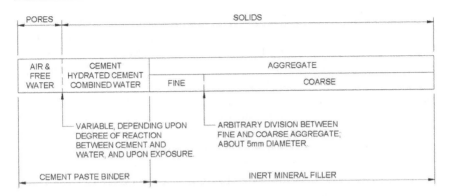

Figure 1.1 The components of a hardened concrete. (Courtesy of Troxell & Davies, 1956, p.4)

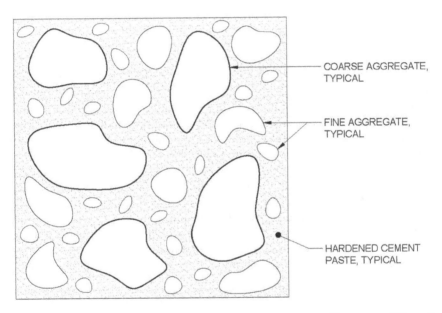

Figure 1.2 Diagrammatic representation of a concrete mix. (Courtesy of Troxell & Davies, 1956, p.63)

cement (cement that not only hardens by reacting with water but also forms a water-resistant product) produced by pulverising clinker which consists essentially of hydraulic calcium silicates, usually containing one, or more of the forms of calcium sulphate as an inter-ground addition'.

Mixtures of Portland cement with fly ash, ground granulated iron blast furnace slag or silica fume are referred to as blended cements and will be discussed later.

Portland cement, or ordinary Portland cement (OPC) or general purpose (GP) cement consists (nominally) of a mixture of oxides: CaO, SiO_2, Al_2O_3, Fe_2O_3, MgO, Na_2O, and K_2O. These oxides are, however, combined together as a series of cement compounds, which are themselves combinations of the principle oxides. The manufacture of Portland cement involves the grinding of limestone or chalk and clay to a suitable degree of fineness, followed by mixing of appropriate proportions of the ground raw materials and then burning them in a large rotary kiln at a temperature around 1480°C. The kiln is inclined and the raw materials are fed into the upper end while the heat source is at the lower end. Hence, there is a temperature gradient along the kiln. Firstly, water is evaporated from the clay and CO_2 is evolved from the limestone, as follows:

$$LIMESTONE \rightarrow CaO(s) + CO_2(g)$$

$$CLAY \rightarrow SiO_2(s) + Al_2O_3(s) + Fe_2O_3(s) + H_2O(g)$$

The main reaction products, the CaO, SiO_2, Al_2O_3, and Fe_2O_3 in the kiln fuse together and form a glassy mass of which the main constituents are the cement compounds: tricalcium silicate ($3CaO.SiO_2$) often abbreviated to C_3S; dicalcium silicate ($2CaO.SiO_2$) often abbreviated to C_2S; tricalcium aluminate ($3CaO.Al_2O_3$) abbreviated to C_3A; and tetracalcium aluminoferrite ($4CaO.Al_2O_3.Fe_2O_3$) abbreviated to C_4AF. Minor components include K_2O and Na_2O. The products are termed clinker at this stage and are cooled before grinding with about 5% gypsum ($CaSO_4$). Gypsum is added to regulate the setting time of cement. Different Portland cement types have different proportions and fineness of the final products.

The reactions of the Portland cement compounds with water result in the setting and hardening of cement paste so that it binds the aggregate together. The hydration reactions of the four principle components of Portland cement are as follows:

$$2(3CaO.SiO_2) + 6H_2O \rightarrow 3CaO.2SiO_2.3H_2O + 3Ca(OH)_2$$

$$2(2CaO.SiO_2) + 4H_2O \rightarrow 3CaO.2SiO_2.3H_2O + 3Ca(OH)_2$$

$$3CaO.Al_2O_3 + 12H_2O + Ca(OH)_2 \rightarrow 3CaO.Al_2O_3.Ca(OH)_2.12H_2O$$

$$4CaO.Al_2O_3.Fe_2O_3 + 22H_2O + 4Ca(OH)_2$$
$$\rightarrow 4CaO.Al_2O_3.13H_2O + 4CaO.Fe_2O_3.13H_2O$$

Table 1.3 Composition of a typical Portland cement

C_3S	C_2S	C_3A	C_4AF	K_2O, Na_2O etc
45%	27%	11%	10%	7%

A Portland cement may typically consist of a mixture of cement compounds with the approximate composition given in Table 1.3 and from the above equations it can be calculated that the amount of water required theoretically to hydrate the cement would be 38%.

In the process of hydration, however, the surfaces of the cement compounds dissolve in the water and then, as the hydrates are less soluble than the cement compounds, they precipitate out as a fine mesh of interlocking crystals which bind themselves and the aggregate together. This hydration process must take place on the surface of the cement particles and so the curing process thus involves the swelling of the particles of cement as they are hydrated leaving a core of unhydrated cement compounds. The volume of the hydrates is approximately 54% greater than that of the dry unhydrated cement compounds and so the particles swell trapping water between themselves, with the result that the cement paste has a porosity of ~28% with all the pores filled with water. The amount of water trapped in the pores can easily be calculated. The density of dry cement is 3.15 g/cm³ and so the volume of 100 g of the cement is 31.8 ml, which swells on hydration to 48.9 mls. If the volume of water trapped in the pores is v_g then $v_g/(v_g + 48.9) = 0.28$ or $v_g = 19.0$ ml. The amount of combined water in the cement paste can similarly be determined as the amount of non–evaporable water under given drying conditions and is put at 23% of the mass of the dry cement and so the total amount of water needed to hydrate 100 g of the cement is (23 + 19 = 42) g. The gel pores gradually diminish in size as further hydration causes the particles to swell further but in the fully hydrated state the water contained within them contributes 19% of the mass of the gel. However, since the hydration of the cement compounds can only take place by the cement swelling into the water and since the water already trapped within the gel pores is immobilised and therefore unable to migrate to where it is required for hydration, a greater water/cement ratio is necessary than the theoretical 38% to ensure complete hydration. Complete hydration of the cement compounds, leaving the gel pores full of water, requires a minimum theoretical water/cement ratio (by weight) of 0.42.

The additional water available above that which is necessary for complete hydration means that the hydrated cement compounds are unable to completely fill the volume contained by the plastic concrete and so gives rise to the presence of capillary pores.

The gel pores are perhaps 1.5–2 nm in diameter.

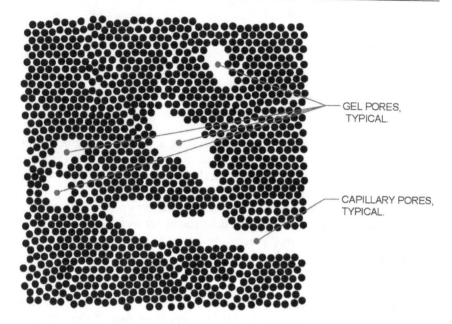

GEL PORES,
TYPICAL.

CAPILLARY PORES,
TYPICAL.

Figure 1.3 Structure of cement gel. (Courtesy of Neville, 1981, p.23)

The capillary pores are much bigger, possibly of the order of a micron or more and they may or may not be interconnected.

Interconnected pores are responsible for the penetrability of the concrete to air or water. As the hydration proceeds these pores diminish in size. The structure of the cement paste then has the appearance of Figure 1.3. Interconnected capillary pores are responsible for penetrability of concrete. Continued hydration may, however, produce a sufficient volume of material to produce blocks in the pores and turn them into discrete capillaries which do not provide a continuous path through the concrete. When the pores become segmented, the penetrability of the concrete is considerably reduced and its ability to inhibit access of aggressive agents that could accelerate corrosion of the reinforcement or degradation (e.g. chemical attack) of the cement paste is much enhanced. The time for this segmentation to take place is dependent upon the amount of water present initially, as can be seen in Table 1.4.

1.3.2 Heat of hydration

A Portland cement may typically consist of a mixture of cement compounds with the approximate composition given in Table 1.3. The heat of hydration varies considerably between the different cement compounds as given in Table 1.5.

Table 1.4 Approximate time required for capillaries to become segmented

Water/cement ratio	Time required for capillary segmentation
0.40	3 days
0.45	7 days
0.5	14 days
0.6	6 months
0.7	1 year
>0.7	impossible

Source: Neville (1975, p.30)

Table 1.5 Heat of hydration of pure cement compounds

Compound	Heat of hydration J/g
C_3S	502
C_2S	260
C_3A	867
C_4AF	419

Source: Neville (1975, p.38)

The very high heat of hydration of tricalcium aluminate (C_3A) can give rise to the phenomenon of 'flash set'. When the cement and water are mixed together the first thing that can happen is that C_3A reacts quite quickly with the water. The heat of hydration is considerable and causes a considerable rise in temperature. This can cause accelerated hydration of the whole mix so that under these circumstances it can stiffen up within minutes. This is called flash set. In order to eliminate the flash set, gypsum ($CaSO_4$) is added to the mix. This reacts with the C_3A as follows to produce an insoluble compound, $3CaO.Al_2O_3.3CaSO_4.31H_2O$, ettringite:

$$3CaO.Al_2O_3 + 3CaSO_4 + 31H_2O \rightarrow 3CaO.Al_2O_3.3CaSO_4.31H_2O$$

This is insoluble and eventually hydrates to the hydrated C_3A but because the C_3A reacts with the gypsum before its hydration reaction it can liberate enough heat to bring about flash set, the initial formation of the ettringite prevents the flash set.

1.3.3 Rate of strength development

The rate of development of strength varies considerably between the various cement compounds as can be seen in Figure 1.4. It was shown above that

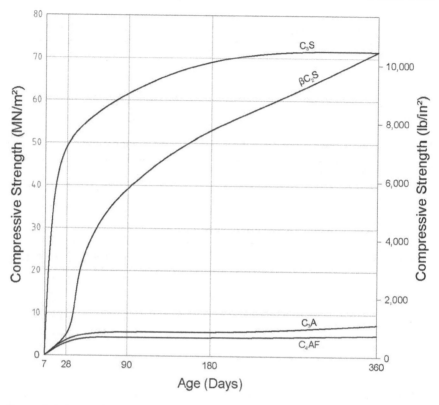

Figure 1.4 Rate of development of strength of different cement compounds. (Courtesy of Neville, 1975, p.42)

C_3A reacts with sulphate ions to produce ettringite. The same compound is formed by the reaction of the hydrated C_3A with sulphate ions:

$$2(3CaO.Al_2O_3.12H_2O) + 3(Na_2SO_4.10H_2O)$$
$$\rightarrow 3CaO.Al_2O_3.3CaSO_4.31H_2O + 2Al(OH)_3 + 6NaOH + 17H_2O$$

As the volume of the ettringite which is formed is about three times the volume of the reactants and as this reaction takes place after the concrete has cured, such a process leads to cracking and spalling (fretting) of the concrete. For a concrete that is to be exposed to sulphate bearing waters it is therefore necessary to restrict the amount of C_3A in the Portland cement.

On the basis of these different cement reactions a number of specific Portland cements for specific situations have been developed. Typical Portland cement compound compositions of the various types of Portland cement are listed in Table 1.6.

Table 1.6 Typical compound compositions for Portland cements (%)

Compound	High early strength	Low heat	Ordinary	Sulphate resisting
C_3S	55	30	45	45
C_2S	17	46	27	35
C_3A	11	5	11	4
C_4AF	9	13	10	10
Miscellaneous	8	6	7	6

Source: Taylor (1967, p.8)

Table 1.7 Cement types in accordance with AS 3972 (2010) or NZS 3122 (2009)

Cement type	Applications
General purpose	
Type GP	All types of construction
General purpose	
Type GB	All types of construction; curing and strength
General purpose blended	development differ from Type GP
Special purpose	
Type HE	Early strength requirements, e.g. cold
High early strength	weather and formwork removal
Type LH	Low temperature requirements, e.g. hot
Low heat	weather and mass concreting
Type SR	High resistance to sulphates, e.g. aggressive
Sulphate resistant	soils
Type SL	Controlled shrinkage applications
Shrinkage limited	

Table 1.7 shows a classification of binder systems as specified by Australian Standard AS 3972 (2010).

1.4 SUPPLEMENTARY CEMENTITIOUS MATERIALS – BLENDED CEMENTS

1.4.1 General

Concrete mix design/selection can involve the specification of the relative amounts of cement, coarse aggregate, fine aggregate, and water, and has four major objectives which apply to nearly all structures. The wet freshly mixed concrete must be sufficiently workable that it can be placed in position, and this often involves the ability to penetrate the small gaps between reinforcing bars. The cured concrete must have a required strength. The cured concrete must crack where it is permissible and for the cracks to be of a surface width sufficient to not compromise durability. The cured concrete must be sufficiently impenetrable (impermeable) to the ingress of aggressive agents that its durability is assured. These qualities of the concrete mix, in their different ways, are affected by the concrete mix design/selection and also by the conditions under which it is placed in position, by the curing conditions, and by additions that may be incorporated into the mix to modify its properties. Particularly effective in the modification of the concrete properties are the so called 'Supplementary Cementitious Materials' (SCMs).

During the manufacture of Portland cement, the oxides of calcium, aluminium, silicon, iron etc. are fused together to form the cement compounds and cooled sufficiently quickly that they form an amorphous glass which is ground up to form cement. There are a number of other major industrial processes that produce metallic oxides at a very high temperature and which because of a very high rate of cooling form glassy non-crystalline compounds in a similar fashion to a Portland cement. Coal fired electricity generation of pulverised-fuel black coal in which the molten residues of the burning process are recovered from the exhaust gases by electrostatic precipitation to yield classified fly ash (FA). The blast furnace production of iron produces a slag which, if subjected to rapid cooling, can be ground up to provide ground granulated iron blast furnace slag (BFS). Silicon and ferro-silicon alloys are produced in electric furnaces where SiO_2 is reduced by carbon at very high temperatures. SiO_2 vapours condense in the form of very tiny spheres ($\sim 0.1\mu$) of amorphous silica, silica fume (SF). SF is also known as 'condensed silica fume' and 'microsilica'.

These materials may be blended with Portland cement to improve its qualities or simply to benefit the environment and are termed SCMs. They are generally divided into two classes; pozzolanic and hydraulic. Pozzolanic SCMs (the name comes from the Italian village of Pozzuoli near Naples where vast amounts of Volcanic ash from Mt. Vesuvius were used for the manufacture of a primitive cement) such as SF, or Class F FA SCMs are low in CaO and are not themselves cementitious, but when they react with a solution of $Ca(OH)_2$ formed by the hydration of Portland cement, they dissolve, and form a cementitious compound. Hydraulic SCMs such as Class C FA or BFS, contain sufficient CaO that when mixed with water, they form a cementitious compound. The characteristic of these materials is that they have a very fine particle size; FA of the order of 1μ–10μ, SF of the order of 0.1μ and that they react quite slowly. Because of this, blended cements can develop superior qualities in the concrete in which they are used.

1.4.2 Fly ash

FA for use in cement in Australia, for example, needs to comply with AS 3582.1 (Standards Australia, 2016a). The fly ash content of a FA blended cement based concrete varies but is commonly in the range 20–35%. For example, some Road Authorities in Australia stipulate in their structural concrete specifications a minimum 25% FA content (Roads & Maritime Services, 2019; VicRoads, 2020) so as to achieve increased durability.

1.4.3 Slag

Ground granulated iron BFS for use in cement in Australia, for example, needs to comply with AS 3582.2 (Standards Australia, 2016b). The slag content of a BFS blended cement based concrete varies but is commonly in the range 50–70%. For example, some Road Authorities in Australia stipulate in their structural concrete specifications a minimum 50–65% BFS content (Roads & Maritime Services, 2019; VicRoads, 2020) so as to achieve increased durability.

1.4.4 Silica fume

Amorphous silica for use in Australia as a cementitious material in concrete, mortar, and related applications, for example, needs to comply with AS 3582.3 (Standards Australia, 2016c). The SF content of a blended cement based concrete typically varies between 5–15%. For example, some Road Authorities in Australia stipulate in their structural concrete specifications a 10% SF proportion (VicRoads, 2020).

1.4.5 Triple blends

Triple blend cement based concretes are also used for increased durability. For example, in Australia VicRoads in their 'Structural Concrete Specification 610' (2020) stipulate that in a triple blend concrete mix, the Portland cement shall be a minimum of 60% and the individual contribution of slag, FA or Amorphous Silica shall be a maximum of 40%, 25% or 10% respectively.

For marine durability, triple blend cement based concretes such as 52% SL cement, 25% FA and 23% slag have been used (Green et al., 2009) in Australia, for example.

1.5 AGGREGATES

1.5.1 General

Since aggregate forms perhaps three quarters of the volume of the cured concrete, its properties play a major role in determining the durability and structural performance of the concrete. Aggregate is normally arbitrarily divided into 'fine' and 'coarse' aggregate. Fine aggregates or sand pass a 5 mm mesh sieve, while coarse aggregates do not usually exceed a nominal size of 50 mm. Naturally occurring coarse aggregate may consist of crushed rocks such as basalt, granite, quartzite, diorite, limestone, or dolomitic limestones. Other aggregates include BFS, scoria, expanded shale, and foamed slag. Lightweight and ultra-lightweight aggregates are available. Manufactured aggregates also and crushed concrete is used as an aggregate. The requirements of aggregates for use in concrete in Australia, for example, are specified in AS 2758.1 (2014a).

The durability, strength, shrinkage, wear resistance, and other mechanical properties of concrete will be influenced by aggregate characteristics such as particle size, shape, and surface texture, hardness, strength, elastic modulus, porosity, contamination, and chemical reactivity with the cement paste. Aggregates do not normally influence the chemistry of cement unless they participate in alkali-aggregate reactions (AAR), as will be elaborated later. Therefore, the chemistry of concrete is usually the chemistry of cement. Use of chloride contaminated aggregates such as sea sand or other materials from saline origins can have dire consequences with respect to reinforcement corrosion, as will be elaborated later. Organic matter may interfere with the hydration of the cement and reduce the final strength of the concrete. Clay and other fine material should be avoided wherever possible as it may coat the aggregate, be chemically reactive, and/or form soft inclusions in the concrete. It may also increase the water demand of the concrete.

1.5.2 The design of a concrete mix

Important qualities of concrete are workability, density, strength, and durability. In a hardened concrete, the cement gel that holds the particles together fills the space between the aggregate particles. The total amount of this space and hence, the amount of cement gel required to fill the space, is determined by the shape and size, or more particularly, the distribution of sizes of the aggregate particles. Since the relative amounts of aggregate and cement gel in the hardened concrete will also control the amount of water in the original mix, the particle size distribution, or grading is also a major factor in determining the workability of the concrete and hence, is of particular importance in the design of a concrete mix.

'Workability' is not a rigorously defined term but is used to describe the ease with which the concrete can be placed in position. Of particular importance is the ability of the concrete to form a coherent mass in between reinforcing bars without leaving holes or cracks which may subsequently be a site for the initiation of failure. The workability is primarily controlled by the size of the aggregate and the water content. The maximum size of the aggregate will be controlled by the structure into which the concrete will be placed. Large well-graded aggregates have fewer and smaller voids than smaller sizes, but the aggregate must be small enough to fit easily between the reinforcing bars if gaps in the concrete are to be avoided. It must also be small enough to fit easily into the formwork. A commonly accepted measure of the ability of the concrete to adapt itself to the shape into which it is to be poured is given by the 'slump' test. In this test described in Australian Standard 1012.3.1 (2014b) for example, a cone is filled with concrete, compacted in a

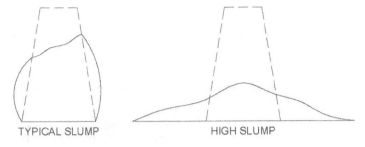

TYPICAL SLUMP   HIGH SLUMP

Figure 1.5 The slump test. (Courtesy of Standards Australia, 2014b)

standard manner and then the cone is inverted on to a flat surface. When the cone is removed the mass of concrete slumps as can be seen in Figure 1.5 and the vertical subsidence of the concrete is defined as the slump.

The slump is primarily controlled by the shape of the aggregate and the amount of water with which it is mixed. The slump that is required of a concrete will depend upon the conditions of the placement, but the American Concrete Institute (ACI) Manual of Concrete Practice – Section 211.1 (ACI 211.1-91, Reapproved 2009) gives the following general guidelines, Table 1.8.

An approximate relationship between the size of the aggregate and the water content is given in the ACI Manual and it can be seen, refer Table 1.9, that the larger the aggregate size, the less water is required to achieve the same slump.

The strength of the concrete is primarily controlled by the water/cement ratio (water/binder ratio). The ACI Manual gives the following table (Table 1.10) and so from the known water content and the water/cement ratio (water/binder ratio) the mass of cement (binder) required to achieve a given strength can be calculated.

Table 1.8 Example concrete slumps for various types of construction

TABLE A1.5.3.1 – RECOMMENDED SLUMPS FOR VARIOUS TYPES OF CONSTRUCTION (SI)

	Slump, mm	
Type of construction	*Maximum**	*Minimum*
Reinforced foundation walls and footings	75	25
Plain footing, caissons, and substructure walls	75	25
Beams and reinforced walls	100	25
Building columns	100	25
Pavements and slabs	75	25
Mass concrete	75	25

Source: (ACI 211.1-91, Reapproved 2009)
* May be increased 25 mm for methods of consolidation other than vibration

Table 1.9 Water contents based on aggregate size

APPROXIMATE MIXING WATER AND AIR CONTENT REQUIREMENTS FOR
DIFFERENT SLUMPS AND NOMINAL MAXIMUM SIZES OF AGGREGATES (SI)

Water, kg/m³ of concrete for indicated nominal maximum sizes of aggregate								
Slump, mm	9.5	12.5	19	25	37.5	50	75	150
Non-air-entrained concrete								
25 to 50	207	199	190	179	166	154	130	113
75 to 100	228	216	205	193	181	169	145	124
150 to 175	243	228	216	202	190	178	160	–
Approximate amount of entrapped air in non-air-entrained concrete, percent	3	2.5	2	1.5		0.5	0.3	0.2

Source: (ACI 211.1-91, Reapproved 2009)

Table 1.10 Strength and water/cement ratio

RELATIONSHIPS BETWEEN WATER/CEMENT RATIO AND COMPRESSIVE
STRENGTH OF CONCRETE (SI)

Compressive strength at 28 days, MPa	Water/cement ratio, by mass	
	Non-air-entrained concrete	*Air-entrained concrete*
40	0.42	
35	0.47	0.39
30	0.54	0.45
25	0.61	0.52
20	0.69	0.60
15	0.79	0.70

Values are estimated average strengths for concrete containing not more than 2% air for non-air-entrained concrete and 6% total air content for air-entrained concrete.
For a constant water/cement ratio, the strength of concrete is reduced as the air content is increased.
(ACI 211.1-91, Reapproved 2009)

The ability of the aggregate to fill the available space will depend upon its size and the coarseness (fineness ratio) of the fine aggregate. The grading of an aggregate is normally determined by sieving. That is, the aggregate is passed through a series of sieves with decreasing openings between the wires of the sieves. Recommended gradings for fine aggregate are given in AS 2758.1 (1998a) for example, see Table 1.11.

The volume of coarse aggregate that can be contained within unit volume of concrete is given in Table A1.5.3.6 of the ACI Manual of Concrete Practice (ACI 211.1-91, Reapproved 2009), refer Table 1.12 below. Given the density of the coarse aggregate the mass can be calculated.

Table 1.11 The grading of fine aggregates

FINE AGGREGATE-RECOMMENDED GRADINGS

| Sieve aperture mm | Mass of sample passing, per cent | |
	Natural fine aggregate	Manufactured fine aggregate
9.50	100	100
4.75	90 to 100	90 to 100
2.36	60 to 100	60 to 100
1.18	30 to 100	30 to 100
0.6	15 to 100	15 to 80
0.3	5 to 50	5 to 40
0.15	0 to 20	0 to 25
0.075*	0 to 5	0 to 20

Source: Standards Australia (2014a)
* Consideration may be given to the use of a manufactured fine aggregate with greater than 20% passing the 0.075 mm size, provided it is used in combination with another fine aggregate where the total percentage passing 0.075 mm of the fine aggregate blend does not exceed 15% and provided the fine aggregate components meet the deviation limits in all respects.

Table 1.12 Volume of concrete occupied by coarse aggregate

VOLUME OF COARSE AGGREGATE PER UNIT OF VOLUME OF CONCRETE (SI)

| Nominal maximum size of aggregate, mm | Volume of dry-rodded coarse aggregate* per unit volume of concrete for different fineness moduli[†] of fine aggregate | | | |
	2.40	2.60	2.80	3.00
9.5	0.50	0.48	0.46	0.44
12.5	0.59	0.57	0.55	0.53
19	0.66	0.64	0.62	0.60
25	0.71	0.69	0.67	0.65
37.5	0.75	0.73	0.71	0.69
50	0.78	0.76	0.74	0.72
75	0.82	0.80	0.78	0.76
150	0.87	0.85	0.83	0.81

Source: (ACI 211.1-91, Reapproved 2009)
* Volumes are based on aggregates in dry-rodded conditions as described in ASTM C 29.
These volumes are selected from empirical relationships to produce concrete with a degree of workability suitable for usual reinforced construction. For less workable concrete such as required for concrete pavement construction they may be increased about 10%. For more workable concrete. such as may sometimes be required when placement is to be by pumping, they may be reduced up to 10%.
[†] See ASTM Method 136 for calculation of fineness modulus.

1.5.3 Estimation of fine aggregate content and mechanical properties

Knowing the density of all the other components, the amount of fine aggregate required to make up the mix can be calculated and a trial mix prepared. From the results obtained with the trial mix the necessary adjustments can be made to ensure that the final mix meets the desired specifications.

1.6 MIXING AND CURING WATER

The fourth component of the recipe for concrete is the mixing water. The relative amounts of water and cement (binder) are in fact the major determinants of the strength, the durability, and the workability of the concrete and so will form the subject of this next section.

The use of unsuitable water for mixing and/or curing concrete may result in reinforcement corrosion as well as other problems with strength, setting, and staining. Sea water introduces chlorides into concrete which may cause corrosion of reinforcement and also reduce long term strength. Brackish water also contains chlorides in high concentration. Water contaminated with organic matter may result in retardation of setting, staining, and strength reduction. High concentrations of sodium and potassium in mixing water increases the risk of alkali-aggregate reaction and high iron concentrations in curing water result in brown stains on the concrete surface. Curing water should be free of high concentrations of other aggressive agents such as sulphates, magnesium salts, and carbonic acid.

AS 1379 (2007a), for example, notes that water is deemed acceptable if:

- Service records show it is not injurious to strength and durability of concrete or embedded items.
- If service records show that when using the proposed water, the strength is at least 90% of the control sample strength at the corresponding age.
- If service records show that the initial set is between –60 to +90 minutes of the control sample time.
- Impurities limits are in accordance with Table 2.2 of the standard i.e.: Sugar <100 mg/L; Oil and grease <50 mg/L; and, pH >5.0.
- Total dissolved solids, chloride content, sulphate content, and sodium equivalent is tested and recorded.

The AS 1379 (2007a) requirement for the acid-soluble chloride content of hardened concrete, for example, is <0.8 kg/m^3. Roads & Maritime Services (RMS) NSW (2019) in Australia, on the other hand set a requirement of <0.3 kg/m^3 in there 'B80 Concrete Work for Bridges Specification' for both reinforced and prestressed concrete for example. VicRoads (Victorian Road Authority) in Australia in there 'Specification 610 Structural Concrete' (2020) set a requirement for the maximum acid-soluble chloride ion content of concrete as placed, expressed as a % of the total mass of cementitious material in the concrete mix of 0.1% for prestressed concrete (equivalent to ~0.4 kg/m³) and 0.15% for reinforced concrete (equivalent to ~0.5 kg/m³).

The acid-soluble sulphate-ion content of the hardened concrete, reported as SO_3, requirement in AS 1379 (2007a) is <50 g/kg of cement. The requirement in the RMS 'B80 Concrete Work for Bridges Specification' (2019) is 3.0% as acid-soluble SO_3 to cement for heat accelerated cured concrete or 5.0% otherwise. VicRoads 'Specification 610 Structural Concrete' (2020) is 4% as acid-soluble SO_3 to the total cementitious material for steam and heat accelerated cured concrete or 5% otherwise.

1.7 ADMIXTURES

The properties, and in particular the curing properties of a concrete as well as being controllable by the selection of the cement and aggregate, can be modified by the addition of 'admixtures', that is chemicals that added in small quantities to the mix can change the characteristics of the cure.

Admixtures are materials incorporated in the concrete mix to alter the properties of fresh or hardened concrete. The most common admixtures are accelerators, set retardants, water reducing agents (plasticisers), superplasticisers, air entraining agents, and waterproofing agents. The extensive list of admixtures detailed in AS 1478.1 (2000), for example, is shown in Table 1.13.

Table 1.13 Admixtures for concrete

Admixture	Type symbol
Air entraining	AEA
Water reducing	WR
Set retarding	Re
Set accelerating	Ac
Water reducing and set retarding	WRRe
Water reducing and set accelerating	WRAc
High range water reducing	HWR
High range water reducing and set retarding	HWRRe
Medium-range WR, normal setting	MWR
Special purpose, normal setting	SN
Special purpose accelerating	SAc
Special purpose retarding	SRe

Source: Standards Australia (2000)

1.7.1 Air entraining admixtures

Air entraining agents are used to form an array of very small air bubbles in the concrete. This provides resistance to frost attack and improved workability. The strength is reduced. Air entraining additives usually consist of surface-active agents such as the soluble salts of sulphated or sulphonated petroleum hydrocarbons. Air entraining agents can also include animal and vegetable fats and oils, wood resins, and wetting agents such as alkali salts of sulphated and sulphonated hydrocarbons. The entrainment of air does not adversely affect the penetrability (permeability) of concrete because the entrained air has the form of discrete bubbles which are not interconnected.

1.7.2 Set retarding admixtures

The function of set retarding admixtures is to delay the onset of curing so that concrete placement is possible in hot weather when the speed of curing might otherwise cause problems. They are also used to delay curing when it is desired to obtain an architectural finish on the surface by working on the surface after the formwork has been struck. Typically set retarding admixtures consist of starches, soluble borates, and phosphates, and the like. In practice set retarding admixtures that are also water reducing agents are more commonly used.

1.7.3 Set accelerating admixtures

Accelerators are used to accelerate the development of early strength, with the aim of speeding up work and increasing productivity. They are predominantly used when concrete must be placed at low temperatures. Common set-accelerators are highly ionised inorganic salts such as sodium carbonate, potassium carbonate, sodium aluminate, and ferric salts. Organic compounds such as triethanolamine are also used as are hydroxycarboxylic acid salts.

The most common accelerator used to be calcium chloride which acts by increasing the rate of hydration of C_3S or C_2S by increasing the volume of the hydration products. It may retard hydration of C_3A. At a dosage rate of less than 1% it acts as a retarder, but at higher concentrations it accelerates the cure and at a dosage of 3% may even induce 'flash set'. Its use in reinforced concrete is now negligible because the chloride ion has a very deleterious effect on corrosion of any reinforcement and its use has been declared illegal in nearly all countries for this reason.

Other accelerators may be based on calcium formate though they are more expensive and less effective.

1.7.4 Water reducing and set retarding admixtures

These admixtures permit concrete to be made using a lower amount of water (and hence reducing the penetrability) while retaining the same workability as a mix containing more water. Thus, higher strength concretes can be produced by using a water reducing agent together with a reduced water/cement (water/binder) ratio. In unmodified form, such admixtures also retard the set and so in the interests of faster work on the construction site are more often used in conjunction with an accelerator. Set retarding admixtures or retarders are also used in high temperature situations to increase the setting time of the fresh concrete.

Calcium lignosulphonate (a by-product of the wood pulp industry) is widely used as a water reducing and set retarding admixture. Other examples of set retarders are lignosulphonic acids, carbohydrate derivatives, soluble zinc salts, and soluble borates.

The organic retarders appear to function by adsorbing on the active sites on the C_3S and inhibiting its hydration.

The inorganic retarders may function by precipitating a thin layer of compound on the grains of cement and similarly inhibit the hydration reaction.

1.7.5 Water reducing and set accelerating admixtures

These are set retarding admixtures (calcium lignosulphonates) to which has been added a set accelerator such as the salt of a hydroxycarboxylic acid or triethanolamine. Such admixtures permit a reduction in the water content while maintaining the same workability without a sacrifice of setting time.

1.7.6 High range water reducing admixtures

Superplasticisers (high range water reducing admixtures) are used to produce flowing concrete, which is easy to place, particularly in inaccessible areas. In addition, low water/cement (water/binder) ratio, high-strength concrete at workable consistency can be achieved by the use of superplasticisers. Examples of superplasticisers are sulphonated melamine formaldehyde condensates and sulphonated naphthalene formaldehyde condensates. Superplasticisers are anionic and give a negative charge to cement particles, thus making them self-repelling.

Hyperplasticisers are now also available to be produce highly flowable, self-levelling concrete.

1.7.7 Waterproofing agents

Waterproofing agents are used to reduce the water absorption of concrete. They consist of pore filling materials or water repellents that may or may not be chemically active. Active chemical pore fillers include alkaline silicates, aluminium, and zinc sulphates, and aluminium and calcium chlorides. These materials may accelerate setting time either alone or in conjunction with other compounds. Inactive pore fillers include finely ground chalk, talc and Fuller's earth. They are sometimes used with calcium and aluminium soaps and improve workability. Active water repellents are soda and potash soaps which are sometimes used in conjunction with lime, alkaline silicates, or calcium chloride. Inert water repellents include vegetable oils, fats, waxes, calcium soaps, and bitumen.

1.7.8 Other

There is a wide range of other admixtures and additives that may be added to concrete to have possible effects such as inhibiting corrosion (see Inhibitors at Section 8.7), to decrease penetrability (permeability), to increase resistivity, and enhance durability.

The possible effects of any admixtures and additives should be carefully examined from a scientific and engineering point of view not just in terms of durability but also in terms of possible future maintenance actions for the reinforced concrete element. Some admixtures and additives may for example adversely affect the performance of future protection and repair methods.

1.8 STEEL REINFORCEMENT

1.8.1 Background

As was indicated earlier in this chapter, concrete is strong under compression but has weak tensile and shear strength. Reinforcing including steel reinforcement is incorporated into concrete to increase the tensile and shear strength of the reinforced concrete element. Steel and concrete also have similar coefficients of thermal expansion.

Rebar (short for reinforcing bar), collectively known as reinforcing steel, reinforcement steel, or steel reinforcement, is a steel bar, or mesh of steel wires.

There is also other steel reinforcement including: Galvanised steel reinforcement, Stainless steel reinforcement and Metallic-clad steel reinforcement for use in reinforced concrete construction and each of these is discussed at Chapter 11.

Reinforced concrete construction using other alternate reinforcement such as fibre-reinforced polymer (FRP) reinforcement is also possible (and is discussed at Chapter 11).

1.8.2 Conventional steel reinforcement

The most common type of conventional rebar is carbon steel. Conventional steel reinforcement for use in concrete in Australia and New Zealand is subject to the requirements of AS/NZS 4671 (2019).

Table 1.14 Conventional steel reinforcement designations

Profile	Strength grade	Ductility class	Standard grades	Nominal diameters (Australia only)
R – plain Round	250 MPa	L – Low	250 N	10 mm
D – Deformed	300 MPa	N – Normal	300E	12 mm
ribbed	500 MPa	E – Earthquake	500 L	16 mm
I – deformed	600 MPa	(seismic)	500 N	20 mm
Indented	750 MPa		500E	24 mm
			600 N	28 mm
			750 N	32 mm
				36 mm
				40 mm
				50 mm

Source: Standards Australia (2019)

Designations of steel rebar to AS/NZS 4671 (2019) in terms of profile (shape), strength grade (by the numerical value of the specified lower characteristic yield stress expressed in MPa), ductility class, standard grades and nominal diameters (mm) are summarised in Table 1.14, for example.

Tables 1.15a and 1.15b summarise the chemical composition of conventional steel rebar from AS/NZS 4671 (2019) for steel grades $\leq$ 500 MPa and steel grades > 500 MPa respectively.

For ribbed bars, longitudinal ribs may or may not be present, but two or more rows of parallel transverse ribs or indentations equally distributed around the circumference and with a uniform spacing along the entire length (excepting markings) are present. Examples of rib geometry with two rows of transverse ribs is shown at Figure 1.6.

Reinforcing steels are typically identified by either an alphanumeric marking system on the surface of the bar that identifies strength grade and ductility class or by a series of surface features on the product. Examples of surface

Table 1.15a Chemical composition of reinforcing steel grades $\leq$ 500 MPa

	Chemical composition, % max.							
	All grades			Carbon equivalent value (CEV) for standard grades				
Type of analysis	C	P	S	250 N	500 L	500 N	300 E	500 E
Cast analysis	0.22	0.050	0.050	0.43	0.39	0.44	0.43	0.49
Product analysis	0.24	0.055	0.055	0.45	0.41	0.46	0.45	0.51

Source: Standards Australia (2019)

Table 1.15b Chemical composition of reinforcing steel grades > 500 MPa

| | Chemical composition, % max. | | | | |
| | All grades | | | Carbon equivalent value (CEV) | |
Type of analysis	C	P	S	600 N	730 N
Cast analysis	0.33	0.050	0.050	0.49	0.49
Product analysis	0.35	0.055	0.055	0.51	0.51

Source: Standards Australia (2019)

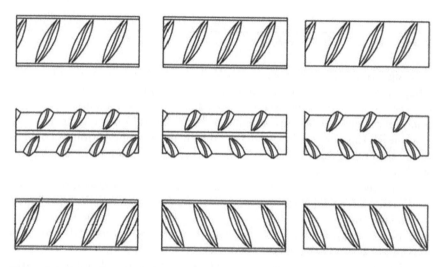

Figure 1.6 Examples of rib geometry for deformed bars. (Courtesy of Standards Australia, 2019)

features (grade identifiers) for deformed bars are presented at Figure 1.7 (Standards Australia, 2019), for example.

Surface features for plain bar in AS/NZS 4671 (2019) include the following:

a. *Plain grade 250 N and plain grade 500 L* – No particular identifying features.
b. *Plain grade 300 E* – Identified by a raised dot.
c. *Plain Grade 500 E* – Identified by a raised dot and dash.

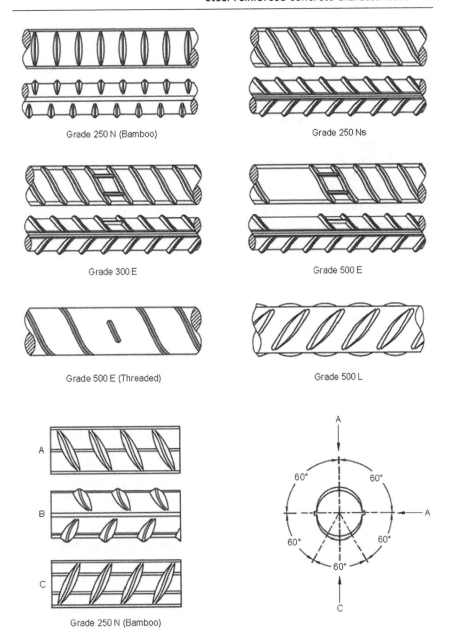

Figure 1.7 Examples of grade identifiers for ribbed deformed rebars. (Courtesy of Standards Australia, 2019)

d. *Plain Grade 500 N* – Identified by a minimum of two short transverse markings.

e. *Right-hand-threaded grade 500 E* – Identified by one short transverse rib on one side of the bar.

1.8.3 Prestressing steel reinforcement

Prestressed concrete is composed of high-strength concrete and high-strength steel. The concrete is 'prestressed' by being placed under compression by the tensioning of high-strength 'steel tendons' located within or adjacent to the concrete volume.

Prestressed concrete is used in a wide range of building and civil structures where its improved performance can allow longer spans, reduced structural thicknesses, and material savings compared to conventionally reinforced concrete. Applications can include high-rise buildings, residential buildings, foundation systems, bridge, and dam structures, silos and tanks, industrial pavements, and nuclear containment structures.

Prestressed concrete members can be divided into two basic types: pretensioned and post-tensioned.

Pretensioned concrete is where the high-strength steel tendons are tensioned *prior* to the surrounding concrete being cast. The concrete bonds to the tendons as it cures, following which the end-anchoring of the tendons is released, and the tendon tension forces are transferred to the concrete as compression by static friction.

Post-tensioned concrete is where the high-strength steel tendons are tensioned *after* the surrounding concrete has been cast. The tendons are not placed in direct contact with the concrete but are encapsulated within a protective duct typically constructed from plastic or galvanised steel materials. There are then two main types of tendon encapsulation systems: those where the tendons are subsequently bonded to the surrounding concrete by internal grouting of the duct after stressing (*bonded* post-tensioning); and those where the tendons are permanently debonded from the surrounding concrete, usually by means of a greased sheath over the tendon strands (*unbonded* post-tensioning).

The requirements for high tensile strength steel tendons to be used for prestressing concrete and for other similar purposes (e.g., masonry structures) are specified for example in AS/NZS 4672.1 (2007b). AS/NZS 4672.1 does not, however, cover requirements for anchorage devices and materials used in conjunction with the prestressing steel in structural components.

The specific properties for each type of prestressing steel are given, namely:

- As-drawn (mill coil) wire;
- Stress relived wire;

- Quenched and tempered wire;
- Strand; and
- Hot-rolled bars with or without subsequent processing.

The standard identifies that the chemical composition is related to the type of product and its size and tensile strength. In cast analyses, the content of both sulphur and phosphorus shall not exceed 0.04% in AS/NZS 4672.1 (2007b).

Geometrical properties in AS/NZS 4672.1 (2007b) are based upon:

- Nominal diameters; and
- Where the definition of geometrical properties by nominal diameters is insufficient or not appropriate, the geometrical properties of steel pre-stressing materials may be defined by nominal cross-sectional area with specified tolerances and appropriate details of the surface configuration of the wire, strand, or bar.

Mechanical properties identified in AS/NZS 4672.1 (2007b) include:

- Tensile strength;
- Proof force (or proof stress);
- Percentage total elongation at maximum force;
- Modulus of elasticity;
- Ductility; and
- Isothermal relaxation.

Wire strand construction for prestressing steel reinforcement can vary and examples of typical seven wire strand constructions and typical 19 wire strand construction from AS/NZS 4672.1 (2007b) are provided at Figures 1.8 and 1.9, respectively.

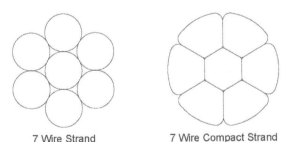

7 Wire Strand   7 Wire Compact Strand

Figure 1.8 Typical 7 wire strand constructions for prestressing steel. (Courtesy of Standards Australia, 2007b)

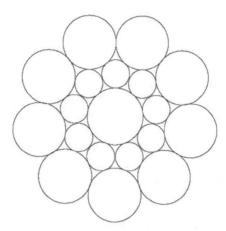

Figure 1.9 Typical 19 wire strand construction for prestressing steel. (Courtesy of Standards Australia, 2007b)

1.9 FIBRE-REINFORCED CONCRETE

Fibre-reinforced concrete is defined as concrete made with hydraulic cement, containing fine or fine and coarse aggregate, and discontinuous discrete fibres. It is usually used without massive reinforcement. The fibres can be made from natural material (e.g. sisal, cellulose) or are a manufactured product such as glass, steel, carbon, and polymer (e.g. polypropylene, kevlar). The purposes of reinforcing the cement-based matrix with fibres are to increase the tensile strength by delaying the growth of cracks, and to increase the toughness by transmitting stress across a cracked section so that much larger deformation is possible beyond the peak stress than without fibre reinforcement. Short fibre reinforcement results in enhanced strength and toughness in flexure and enhanced toughness in compression. Fibre reinforcement improves the impact strength and fatigue strength, and also reduces shrinkage. The quantity of fibres used is small, typically 1–5% by volume as higher volumes decrease the workability substantially.

REFERENCES

American Concrete Institute (2009), '*Standard Practice for Selecting Proportions for Normal, Heavyweight, and Mass Concrete*', ACI 211.1-91 (Reapproved 2009), Farmington Hills, MI, USA.

ASTM C150/C150M-16 (2016), '*Standard Specification for Portland Cement*', ASTM C150/C150M-19a, ASTM International, West Conshohocken, PA, USA.

Green, W, Riordan, G, Richardson, G and Atkinson, W (2009), '*Durability assessment, design and planning – Port Botany Expansion Project*', Proc 24th Biennial Conf Concrete Institute of Australia, Paper 65, Sydney, Australia.

Miller, D (2018), 'CEO's Report', *Concrete in Australia*, 34(3), September.

Neville, A M (1975), *'Properties of Concrete'*, 2nd Edition, Pitman International, London, UK.

Neville, A M (1981), *'Properties of Concrete'*, 3rd Edition, Pitman, Bath, UK.

Roads & Maritime Services (2019), *'Specification B80 Concrete Work for Bridges'*, Edition 7/Revision 2, November, Sydney, Australia.

Standards Australia (2000), *'AS 1478.1 Chemical admixtures for concrete, mortar and grout'*, Sydney, Australia.

Standards Australia (2007a), *'AS 1379 Specification and supply of concrete'*, Sydney, Australia.

Standards Australia (2007b), *'AS/NZS 4672.1 Steel prestressing materials Part 1: General requirements'*, Sydney, Australia.

Standards Australia (2010), *'AS 3972 Portland and blended cements'*, Sydney, Australia.

Standards Australia (2014a), *'AS 2758.1 Aggregates and rock for engineering purposes'*, Sydney, Australia.

Standards Australia (2014b), *'AS 1012.3.1 Determination of properties related to the consistency of concrete – Slump test'*, Sydney, Australia.

Standards Australia (2016a), *'AS 3582.1 Supplementary cementitious materials for use with Portland and blended cement – Fly ash'*, Sydney, Australia.

Standards Australia (2016b), *'AS 3582.2 Supplementary cementitious materials for use with Portland and blended cement – Slag – Ground granulated iron blast furnace slag'*, Sydney, Australia.

Standards Australia (2016c), *'AS 3582.3 Supplementary cementitious materials for use with Portland and blended cement – Amorphous silica'*, Sydney, Australia.

Standards Australia (2019), *'AS/NZS 4671 Steel for the reinforcement of concrete'*, Sydney, Australia.

Taylor, H (1967), *'Concrete Technology and Practice'*, 2nd edition. Angus and Robertson, Sydney, Australia.

Troxell, G E, and Davies, H E (1956), *'Composition and Properties of Concrete'*, McGraw Hill, New York, USA.

VicRoads (2020), *'Section 610 – Structural Concrete'*, February, Melbourne, Australia.

Chapter 2

Concrete deterioration mechanisms (A)

2.1 REINFORCED CONCRETE DETERIORATION

'Good steel in good concrete will last at least one person's lifetime' is a saying that gives credit to the ability of a properly formulated and constructed reinforced concrete structure to withstand the attack of most of the environments to which it might normally be exposed. But it is a matter of all too common experience that some reinforced concrete structures can show distress sometimes in as little as ten years. So, it is important to discover what makes the concrete, the steel, or the environment 'not good' from the point of view of the durability of a structure.

The ability of a reinforced concrete structure to maintain its function depends upon the integrity of the steel, the integrity of the concrete and the integrity of the bond between them. Not only can deterioration of any one of these components lead to a failure of the structure, but a deterioration of one component can lead to the failure of another. Thus, cracking in the concrete can provide a path through it by which aggressive agents from the environment can gain access to the steel and cause the corrosion of the steel that in turn leads to the development of bulky corrosion products that leads to further cracking and spalling of the concrete.

The most common cause of short-term failure of reinforced concrete is corrosion of the reinforcement. However, 'good steel in good concrete' is protected from corrosion by a so-called 'passive film' which is formed by the reaction of the iron with the highly alkaline environment of the pore water in the concrete:

$$2Fe + 6OH^- \rightarrow Fe_2O_3 + 3H_2O + 6e^-$$

This protective film may be attacked by aggressive agents from the surrounding environment. The most important of these aggressive agents are chloride ions which effectively produce holes in the film, and carbon dioxide gas which dissolves in the concrete pore water to produce an acid solution that neutralises the alkalis and so renders the passive film unstable.

There are three main routes by which aggressive agents can reach the metal surface, they can enter through cracks in the concrete or by penetrating through pores in the cement phase or they can permeate through the concrete under a pressure head. Furthermore, deterioration of concrete can be separated into two broad category types: (i) Degradation of concrete; (ii) Corrosion of steel reinforcement. Figure 2.1 summarises the causes of (i) and (ii) which can include one or more of mechanical, physical, structural, chemical, biological, and reinforcement corrosion mechanisms. The concrete degradation mechanisms need to be studied first before corrosion of reinforcement.

2.2 CRACKING

2.2.1 General

As the concrete cures, it loses water, and shrinks. If there is restraint to this shrinkage, then this can lead to the development of tensile stresses and cracking. Cracks are an intrinsic feature of concrete, even though much of

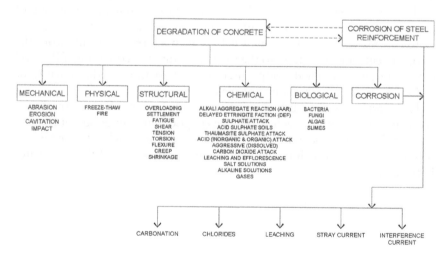

Figure 2.1 Summary of concrete deterioration types. (Adapted from Bertolini et al., 2004)

the cracking may be at a micro level and invisible to the naked eye. The causes of the shrinkage and hence of the cracking changes as the concrete cures. A number of cracking mechanisms can be differentiated at different stages of the cure, but it must be appreciated that there is a considerable overlap of the processes. Cracking can occur before the concrete has hardened and after the concrete has hardened, refer to a 'family tree' of crack types at Figure 2.2.

An understanding of the various causes of cracking is necessary before decisions can be made as to whether crack repairs are required, and what form they will take. Cracks may be classified as structural or non-structural and while structural cracking is typically due to overloading and may occur

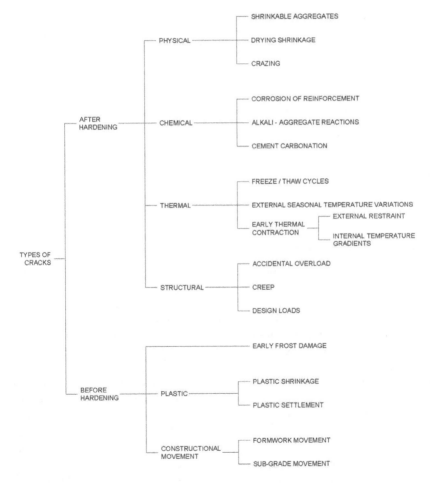

Figure 2.2 Type of cracks. (Courtesy of Concrete Society, 2010)

at any time after hardening of concrete, non-structural cracks are cracks which would have arisen irrespective of the applied loads and can usually be ascribed to deficiencies in design or construction practice.

The subject of cracking is covered in considerable detail by ACI Committee 224 in their report entitled 'Causes, Evaluation and Repair of Cracks in Concrete Structures' (American Concrete Institute, 1984) and also, in the paper 'Causes of Cracking' (Beeby, 1984) as part of the International Workshop 'Durability of Concrete Structures' from which Figure 2.3 is taken. This shows in diagrammatic form the appearance of the various forms of cracks which may appear in reinforced, prestressed, and post-tensioned concrete.

The types of cracks in Figure 2.3 may be characterised as follows:

Letter	Type of cracking	Primary cause	Time to appearance
A, B, C	Plastic settlement	Excess bleeding	10 minutes – 3 hours
D, E, F	Plastic shrinkage	Rapid early drying	30 minutes – 6 hours
G, H	Early thermal contraction	Excess heat generation	One day to three weeks
I	Long term drying shrinkage	Inefficient joints	Months
J, K	Crazing	Rich mixes	Days to months
L, M	Reinforcement corrosion	Poor concrete	Years
N	Alkali aggregate	Reactive aggregate	Years

2.2.2 Plastic settlement cracking

Soon after the concrete has been placed the heavy particles sink through the water and the water 'bleeds' to the surface. The initial shrinkage results from this compaction of the concrete as the excess water 'bleeds' to the surface. The earliest form of cracking is termed plastic settlement cracking and may be apparent within ten minutes to three hours after placement. Plastic settlement cracks are due to excessive bleeding accompanied by restraint. These cracks tend to run parallel to the restraining element such as reinforcement or formwork. The cracks may be quite wide at the surface of the concrete but rarely penetrate deeply into the mass of the concrete. If such cracks are noticed within about four hours of placing, the still plastic concrete can be re-vibrated to get rid of the cracks. Plastic settlement cracks can be avoided by using concrete mix proportions which give an acceptable rate of bleeding, use of air entraining agents and appropriate placement of restraints. Typical plastic settlement cracks are shown as A, B, and C in Figure 2.3.

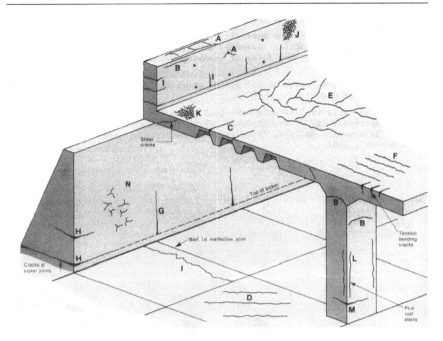

Figure 2.3 Typical cracks in reinforced concrete. (Courtesy of Beeby, 1984)

Although their effect is not always visible at the surface of the concrete as can be seen in Figure 2.4, they can still provide a channel for the transport of liquid through the concrete.

2.2.3 Plastic shrinkage cracking

Plastic shrinkage cracks also appear when the concrete is plastic, usually a half to six hours after placement. These cracks are caused by the shrinkage of the surface regions of the concrete when the water that bleeds to the surface is allowed to evaporate too quickly after placement. As opposed to plastic settlement cracks, plastic shrinkage cracks can be prevented by adequate curing particularly in the very early life of concrete to avoid high rates of evaporation at the surface. Curing must commence before free water on the concrete surface has been lost. Plastic shrinkage cracking is particularly prevalent on hot and windy days and can often be prevented by keeping the surface of the concrete wet or by covering it with wet hessian and/or by the use of curing compounds which are applied to the surface to prevent evaporation. Typical plastic shrinkage cracks are shown as D, E, and F in Figure 2.3 and Figure 2.5.

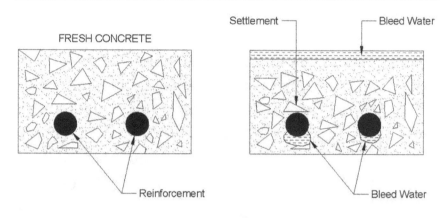

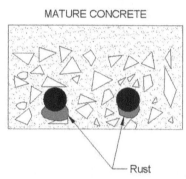

Figure 2.4 Void under reinforcing bar due to plastic settlement. (Courtesy of Gani, 1997, p.110)

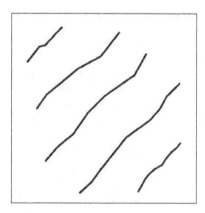

Figure 2.5 Plastic shrinkage cracks. (Courtesy of Bungey and Millard, 1996, p.9)

Restraint in the form of reinforcing bars allows the shrinking concrete to develop strains well in excess of that corresponding to the maximum tensile stress that un-reinforced concrete could sustain without causing wide cracks to form. It is for this reason that various national standards specify a minimum ratio of the area of the reinforcement to gross area of the concrete to prevent shrinkage cracking. For example, Australian Standard 3600 'Concrete structures' (2018) specifies the amount of steel that is necessary in order to prevent shrinkage cracking. This varies with the structural function that the reinforced concrete is required to perform and with the degree of exposure of the concrete but is comprehensively considered in the standard. It should be noted, however, that this is only a minimum which is specified and that steel in excess of the minimum is usually beneficial to the durability of the structure.

2.2.4 Early thermal contraction cracking

Early thermal contraction cracks appear at one day to two or three weeks after placement. Thermal cracks are a result of stresses brought about by temperature differentials. The usual cause of temperature differentials is the exothermic nature of cement hydration. The heat of hydration of ordinary Portland cement (OPC) is about 250 kJ/kg and as the specific heat of concrete is about 1 kJ/kg.°C there is approximately a 10°C rise in temperature for each 100 kg of cement in 1 m³ of concrete (Lees, 1992). When cooling follows after the hydration process the concrete contracts and if there is restraint to contraction then tensile stresses are developed and cracking may follow. A very common cause of thermal cracking is the casting of walls on already cured ground slabs, refer crack G at Figure 2.3.

Thermal contraction cracks can be prevented by reducing the amount of heat development by use of low heat cements, low cement content, blended cements, etc. Chilled aggregates and water may be necessary, insulation (of different types) is practised, construction at night, etc. Models are also used during design to predict concrete peak temperature, maximum temperature differentials, concrete strain, and concrete crack risk. Typical early thermal contraction cracks are shown as G and H in Figure 2.3.

2.2.5 Drying shrinkage cracking

As the water in the concrete dries out, the capillary forces set up in the pores increase and cause the concrete to shrink. Such 'drying shrinkage cracks' appear later than the previously mentioned cracks, usually weeks, or months after placement. Loss of moisture from cement paste results in shrinkage.

Figure 2.6 Drying shrinkage cracking to a potable water tank and associated lime leach deposition together with localised corrosion risk to reinforcement intersecting cracks. (Courtesy of Chris Weale)

If the shrinkage is restrained cracks may form because tensile stresses will be brought about by differential shrinkage. The risk of drying shrinkage cracks can be reduced by using the maximum possible amount of aggregate since it is only the cement gel that shrinks on curing. Reduced water content and adequate curing assist the prevention of drying shrinkage cracks. Typical drying shrinkage cracks are shown as I in Figure 2.3.

Figure 2.6 shows an example of drying shrinkage cracks to a potable water tank where unsightly lime leach staining has occurred and localised corrosion is a risk where the cracks intersect reinforcement due to reduced pH (i.e. pH < 10 leading to dissolution of the passive film).

2.2.6 Crazing

Crazing occurs whenever a weak surface layer is formed on the surface and this weak surface layer is unable to withstand quite small stresses which result from the differential shrinkage between the surface and the bulk. This may arise as a result of the separation of aggregate and cement in the very early stages of curing which leaves a thin layer of 'laitance' on the surface of the concrete or it may result from the very rapid evaporation of water from the surface of the concrete when it is placed under hot conditions. Crazing is shown at J and K in Figure 2.3.

2.2.7 Alkali aggregate reaction cracking

AAR cracking occurs because certain types of siliceous or carbonate rock aggregates can react with the sodium and potassium hydroxides in cement to cause concrete damage, i.e. alkali-silica reaction (ASR) and alkali-carbonate reaction (ACR). Such alkali aggregate reactions are much more common in Australasia with reactive silica aggregates than carbonate rock aggregates which are mainly dolomitic in origin. Reactive siliceous aggregates include opaline and chalcedonic cherts, tridymite, cristobalite, rhyolites, and andesites, and their tuffs, certain zeolites, and certain phyllites. Reactive silicas have a random network of silicon-oxygen tetrahedra, while unreactive silicas such as quartz have orderly tetrahedra. The ASR produces an alkali silicate gel. The gel attracts water from hardened cement paste and causes swelling. Osmotic pressure and expansion result, which in turn produces cracking and spalling of the cement paste, and exudation of gel. The rate of ASR depends on the fineness of siliceous material, alkali content, and water content of the cement paste, aggregate porosity, and cement paste penetrability (permeability). Typical AAR cracks are shown at N in Figure 2.3 and below in Figure 2.7.

Figures 2.8 and 2.9 show examples of ASR cracking to the reinforced concrete substructure elements of a coastal, marine, jetty structure in Western Australia where the effect of restraint to cracking is evident. In Figure 2.8 map cracking (three pronged, similar to crazing) is the

Figure 2.7 Alkali aggregate reaction cracking. (Courtesy of Bungey and Millard, 1996, p.9)

Figure 2.8 ASR map cracking to a transverse beam of a coastal jetty structure

Figure 2.9 ASR cracking to a longitudinal beam of a coastal jetty structure

characteristic pattern for ASR in unrestrained concrete. In other cases, where ligature reinforcement stretch and the element is in compression, longitudinal cracks may form, refer Figure 2.9, where the photograph shows a prestressed beam, whereby the ligatures have yielded and the crack is at the location of maximum stress.

As the above jetty structure was in a coastal marine environment, the ASR-induced cracking permitted the entry of water and chloride ions such that localised reinforcement corrosion presented a serious risk to the structure. With moisture ingress and residual reactive silica within the coarse and fine aggregate ongoing ASR expansion also needed to be managed for this structure.

Restraint has also had an effect with the ASR-induced crack patterns of other structures and elements. Examples include:

- Longitudinal cracking of prestressed road bridge planks.
- Mid-face splitting of road bridge headstock elements and map cracking at ends.
- Vertical cracking of prestressed bridge piles (also leading to pitting corrosion of prestressing steel).

In terms of testing for AAR in new construction, it is advisable to test for an aggregate's (coarse and fine aggregate) reactivity if this is not already known prior to selection for use in concrete. Various different tests are available for assessing the reactivity of aggregates prior to selection for use in concrete. These include petrographic examination of the aggregate, chemical analysis, and mortar bar, or concrete prism expansion tests based on standard mix designs. For mortar bar and concrete prism tests, expansion limits have been set to indicate the reactivity of the aggregate under test. Different AAR test methods can give contradictory results. There is ongoing research to improve test methods so that they better predict field performance of aggregates. However, interpretation must be based on local knowledge about the alkali reactivity of similar aggregates in the specific test used and in concrete structures.

Petrographic examination of aggregates is undertaken in accordance with ASTM C295-12 (2012) or AS 1141.65 (2008). Where previous use or petrographic tests do not provide sufficient evidence that the aggregates will be acceptable the Concrete Institute of Australia 'Performance Tests to Assess Concrete Durability' (2015) recommended practice document proposes that accelerated mortar bar testing to AS 1141.60.1 (2014a) be undertaken to determine the degree to which the aggregate is reactive. Furthermore, that recommended practice advises that where petrographic or mortar bar expansion tests are inconclusive then accelerated concrete prism testing to AS 1141.60.2 (2014b) can be used to provide specific evidence as to whether a proposed concrete mix and cement system will have acceptable resistance to AAR.

Results of accelerated testing can occasionally be misleading, and more than one test method might be needed to resolve the potential AAR risk. For example, in Western Australia a widely-used aggregate that is classified 'potentially reactive' by petrography, gives 'highly reactive' expansions by mortar bar test, but is classified 'innocuous' by concrete prism and has a 40+ years ASR problem free track record in marine and water retaining structures. Thus, wherever possible, the specific tests used should be those upon which knowledge about reactivity of local aggregates is based, and local experience should be utilised in interpreting test results (Concrete Institute of Australia, 2015).

Whichever test methods are used, the art of the testing is in designing a test programme that best meets project needs, rather than simply applying a pass/fail criterion associated with a particular test method. For example, with use of appropriate control mixes, test results can be evaluated by comparison with aggregates or concrete with known site performance (Concrete Institute of Australia, 2015).

Because significant variations in mineralogy can exist within quarries, material is typically characterised by petrography annually, and AAR testing is repeated at a frequency that reflects changes in the source material. For some sources this may be as often as yearly (Concrete Institute of Australia, 2015).

2.3 PENETRABILITY

An aggressive species can also penetrate the concrete and reach the surface of the metal by absorption, sorptivity, permeation, and/or diffusion through the bulk concrete. Although cracks usually form the easiest route for the entry of aggressive species absorption, sorptivity, permeation and/or diffusion through bulk concrete plays a major role in all those cases where intact concrete is involved.

Water absorption and sorptivity are measures of concrete pore volume (porosity) and structure as indicated by volume and rate of uptake of water by capillary suction respectively. Permeability is also a measure of pore structure but measures the influence of pore volume and connectivity on the rate of transport of water applied under pressure. Diffusion occurs due to concentration gradients of ions within the pore water of concrete.

Permeability is an important factor determining the ability of a concrete to withstand attack by an aqueous environment to the water in which it is immersed. Since aggressive agents in the water and even the water itself can only have access to the interior of the concrete through the capillary pores in the cement gel, the ability of a concrete to withstand chemical attack depends upon the pore structure. The permeability of the concrete determines the ability of aggressive ions to gain access to the interior of the concrete and to reinforcing bars that may be immersed in it. The permeability is measured as the volume of liquid passing per second through a cube of unit dimensions under unit pressure, that is, by Darcy's equation:

$$\frac{dq}{dt} = KA\frac{\Delta h}{L}$$

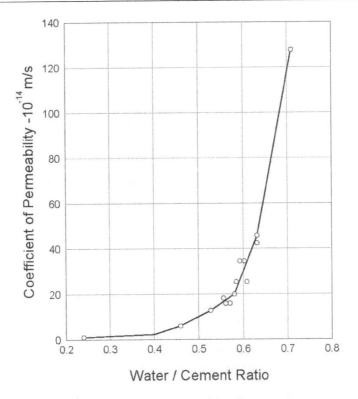

Figure 2.10 The relationship between water permeability and water/cement ratio for a cement paste. (Courtesy of Neville, 1975, p.386)

where dq/dt is measured in $m^3.sec^{-1}$, A in m^2, L in m and Δh (the hydraulic head) in m. K, the permeability coefficient, therefore has units of $m.sec^{-1}$.

Since the size and the degree of segmentation of the capillary pores is a function of the water/cement (water/binder) ratio and the age of the cementitious paste, the permeability is therefore a function of the water/cement (water/binder) ratio as can be seen in Figure 2.10.

A relationship between capillary porosity, concrete compressive strength and water permeability is shown in Figure 2.11.

2.4 CHEMICAL DETERIORATION

2.4.1 General

Concrete may be subject to chemical attack either from the natural environment or by industrial reagents stored in concrete tanks for example. The comparatively low cost of reinforced concrete compared with (say) stainless steel makes it the material of choice for 'bunds' in chemical plants. A bund

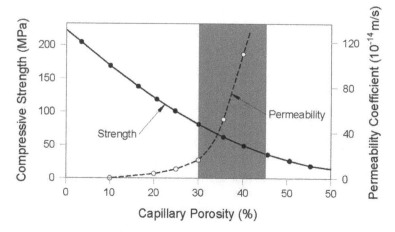

Figure 2.11 Compressive strength and water permeability. (Courtesy of Bertolini,
Elsener, Pedeferri, and Polder, 2004)

is a leakproof enclosure which must be constructed below hazardous chemical containing tanks to prevent the escape of such chemicals into the surrounding environment in the event of an accidental spill or other leakage.
This section will be primarily concerned with the unavoidable chemical
attack that arises from the exposure of reinforced concrete structures to the
natural environment.

Water passing through the concrete, either down cracks, through the capillary pores, or permeating through the bulk concrete may contain dissolved
aggressive agents that can attack the concrete. This attack may expose reinforcing bars to corrosive environments which in turn can cause further damage to the structure, but as this further damage must be regarded as a
consequence of the initial attack on the concrete, the nature of this attack is
considered first. Chemical deterioration mechanisms to be considered
include:

   a. AAR within concrete.
   b. Delayed ettringite reaction (DEF) within concrete.
   c. Sulphate ion attack on the concrete.
   d. Acid sulphate soils attack.
   e. Thaumasite sulphate attack (TSA).
   f. Acid attack.
   g. Aggressive carbon dioxide attack of concrete.
   h. Seawater attack.
   i. Leaching and efflorescence.
   j. Physical salt attack (PSA).
   k. Other chemical attack.

2.4.2 Alkali aggregate reaction

Refer to Section 2.2.7.

2.4.3 Delayed ettringite formation

The incorporation of calcium sulphate (gypsum) into ground cement to prevent 'flash set' was discussed earlier. Its function is to react with the tricalcium aluminate to produce ettringite $3CaO.Al_2O_3.3CaSO_4.31H_2O$ before the hydration of the C_3A can liberate enough heat to set off the cure of the rest of the concrete. Ettringite formed in the normal early way at ambient temperatures does not damage the concrete, however, if the temperature of the concrete exceeds about 70°C during the early stages of curing, for example in sizeable concrete elements, or in accelerated curing, a different hydration product forms. Once normal ambient temperatures are restored, it will convert back to ettringite if sufficient water is available. This process is known as DEF. Ettringite takes up more space than the original hydration products, and so the conversion generates internal stress, which, like AAR, can be enough to crack the concrete. Typical DEF cracks are shown at Figure 2.12.

Figure 2.12 Example of DEF cracking (unrestrained end of a bridge headstock).

DEF cracking may take 20 or more years to appear but can also be rapid and dramatic. Like other forms of cracking, it can increase the risk of secondary forms of deterioration such as reinforcement corrosion by allowing ingress of aggressive agents. Reported cases have included precast railway sleepers, and precast piles immersed in water. In these cases, accelerated heat curing is likely to be the cause.

DEF has also been reported in sizeable elements, where heat of hydration has elevated the early age concrete temperature. The risk of DEF is determined primarily by the early age curing temperature, although some cement compositions and aggregate types will further increase the risk (Kelham, 1996; Collepardi, 2003; FHWA, 2005; Brunetaud et al., 2007; fib, 2011). It is very sensitive to small changes in chemical and physical conditions so is often localised. This leads to inconsistent observations that make its cause difficult to identify (Thomas et al., 2008).

AAR is also induced by high early age curing temperatures. It is often associated with DEF, and there has been considerable debate about the relationship between the two reactions. Evidence suggests DEF is unlikely to cause damage unless preceded by AAR (or other mechanisms that produce microcracking in which the ettringite can crystallise), although it can occur by itself if all high-risk factors are present (Shayan & Xu, 2004).

DEF can almost always be prevented by limiting the maximum peak concrete hydration temperature to less than 70°C during initial early age. This can be achieved directly via the project specification. Concrete with specific proportions of ground granulated blast furnace slag (GGBFS) or fly ash (FA) can allow the initial early age peak concrete hydration temperature up to 80°C (Concrete Institute of Australia, 2015).

2.4.4 Sulphate attack

Hydrated tricalcium aluminate (C_3A) can also react with calcium sulphate to produce ettringite, $3CaO.Al_2O_3.3CaSO_4.31H_2O$. This renders the hardened concrete susceptible to attack by soluble sulphate ions which may be present in penetrating water. Initially a soluble sulphate, for example sodium sulphate, can react with the calcium hydroxide in the pore water to produce calcium sulphate:

$$Na_2SO_4 + Ca(OH)_2 \rightarrow CaSO_4 + 2NaOH$$

Then the calcium sulphate reacts with the hydrated tricalcium aluminate:

$$3CaO.Al_2O_3.12H_2O + 3(CaSO_4) + 19H_2O \rightarrow 3CaO.Al_2O_3.3CaSO_4.31H_2O$$

The solid ettringite that is formed has approximately three times the volume of the tricalcium aluminate hydrate from which it is formed.

Figure 2.13 Sulphate attack on concrete. (Courtesy of Skalny, 2002, p.6)

Consequently, the material swells and this leads to degradation of the cement (binder) paste phase (erodes with time), cracking, and surface spalling of the concrete, refer Figure 2.13.

The common sources of sulphates which may attack concrete are groundwater, soils, industrial chemicals, seawater, and products of sulphur oxidising bacteria. Groundwater and soils may contain sulphates of calcium, magnesium, sodium, potassium, and ammonium in various quantities depending on the location.

Calcium sulphate attacks only the hydrated tricalcium aluminate. Magnesium sulphate on the other hand attacks the other silicate hydrates. This is because magnesium hydroxide is highly insoluble and the lowering of the pH which results from the reaction of magnesium sulphate and calcium hydroxide serves to hydrolyse the hydrated calcium silicates so that the products of hydrolysis can react to produce ettringite. The reaction between magnesium sulphate solution and hydrated di calcium silicate (C_2S) or tri calcium silicate (C_3S) produces gypsum, magnesium hydroxide, and silica gel, e.g.:

$$3CaO.2SiO_2 + 3MgSO_4.7H_2O \rightarrow CaSO_4.2H_2O + Mg(OH)_2 + SiO_2$$

Magnesium hydroxide is an insoluble gelatinous material. It can react slowly with silica gel to form a hydrated magnesium silicate which does not have any binding properties. The calcium sulphoaluminate formed by the reaction of $CaSO_4.2H_2O$ and $3CaO.Al_2O_3$ due to the action of magnesium sulphate is further decomposed to gypsum, hydrated alumina, and magnesium hydroxide by magnesium sulphate for example. The SiO_2 is also made available for ASR.

Ammonium sulphate is also particularly aggressive towards concrete. This is thought to be connected with the increased solubility of gypsum in ammonium sulphate solutions (Lea, 1998). Another contributory mechanism is likely to be the formation of ammonia gas, e.g.:

$$Ca(OH)_2 + NH_4SO_4 \rightarrow CaSO_42H_2O + 2NH_3 \uparrow$$

Upon formation, ammonia gas will readily diffuse from the concrete enabling the above reaction to eventually proceed to completion rather than establish equilibrium. Diffusion of ammonia gas from the concrete will also render the concrete more porous and permeable and thus more susceptible to further attack from ammonium sulphate solution (Collins & Green, 1990).

Criteria which can be adopted to identify a site with a potential for the sulphate degradation of concrete can be based upon those of the U.S. Bureau of Reclamation and reported by Skalny (2002). These are detailed in Table 2.1. There is no exact Australian equivalent of American Type II cement and so, normally Australian Type SR cement would be recommended in all cases for which American Type II cement would be chosen on the basis of Table 2.1, for example. Type V American cement is equivalent to Australian Type SR cement.

British criteria, for example, is detailed in BS EN 206: Part 1 (2000) as provided at Table 2.2. Ammonium ion criteria is also provided as is criteria for magnesium ions as well as aggressive (dissolved) CO_2 (refer Section 2.4.8).

Table 2.1 Sulphate attack criteria

Exposure	Water soluble sulphate (SO₄) in soil %	Sulphate (SO₄) in water, mg/l	Cement	Water/cement ratio, maximum*
Mild	0.00–0.10	0–150	-	-
Moderate	0.10–0.20	150–1,500	Type II	0.50
Severe	0.20–2.00	1,500–10,000	Type V	0.45
Very Severe	Over 2.00	Over 10,000	Type V + Pozzolan†	0.45

Source: Skalny (2002, p235)
* A lower water/cement ratio may be necessary to prevent corrosion of embedded items.
† Use a pozzolan which has been determined by tests to improve sulphate resistance when used in concrete containing Type V cement.

Table 2.2 Sulphate attack criteria

Chemical species in groundwater	Weakly aggressive pH 5.5–6.5	Moderately aggressive pH 4.5–5.5	Strongly aggressive pH 4.0–4.5
Sulphate (mg/L)	200–600	600–3,000	3,000–6,000
Ammonium (mg/L)	15–30	30–60	60–100
Magnesium (mg/L)	300–1,000	1,000–3,000	3,000 up to saturation
Aggressive CO_2 (mg/L)	15–40	40–100	100 up to saturation

Source: (BS EN 206: Part 1, 2000)

Ground water attack on piles, for example, is dealt with in the 'Piling Code' AS 2159 (2009) and Table 2.3 from this code discusses attack by sulphates, chlorides, and acidic conditions.

Sulphate attack of concrete can be reduced by using low penetrability, low water/cement (water/binder) ratio concrete and a Sulphate Resisting (Type SR) cement.

In some cases, where sulphate concentrations are high, surface coatings, or waterproof linings can be used to prevent attack.

Fly ash (FA), blast furnace slag (BFS) and silica fume (SF) blended cement-based concretes exhibit increased sulphate resistance.

2.4.5 Acid sulphate soils

Acid sulphate soils is the common name given to naturally occurring soil containing pyritic sediments (iron sulphide, FeS). As long as pyritic sediment remains below the water table where it cannot be oxidised, it poses no problems to concrete. It is when pyritic sediment is exposed to air, that problems occur. Exposure to air results in oxidation of pyrite to sulphuric acid (H_2SO_4) which is highly aggressive to concrete because in addition to degrading the

Table 2.3 Piling criteria

EXPOSURE CLASSIFICATION FOR CONCRETE PILES-PILES IN SOIL					
Exposure conditions					*Exposure classification*
Sulphates (expressed as SO₄*)					
In soil (ppm)	*In groundwater (ppm)*	*pH*	*Chlorides in groundwater (ppm)*	*Soil conditions A†*	*Soil conditions B‡*
<5,000	<1,000	>5.5	<6,000	Mild	Non-aggressive
5,000–10,000	1,000–3,000	4.5–5.5	6,000–12,000	Moderate	Mild
10,000–20,000	3,000–10,000	4–4.5	12,000–30,000	Severe	Moderate
>20,000	>10,000	<4	>30,000	Very severe	Severe

Source: (AS 2159, 2009)
* Approximately 100 ppm SO₄ = 80 ppm SO₃.
† Soil Conditions A – High permeability soils (e.g. sands and gravels) which are in groundwater.
‡ Soil Condition B – Low permeability soils (e.g. silts and clays) or all soils above groundwater.

cement binder (acid attack), it also produces an expansive reaction within the concrete matrix (sulphate attack).

2.4.6 Thaumasite sulphate attack

At low temperatures below ~15°C prolonged effects of sulphate (in conjunction with carbonate) can produce the non-binding calcium carbonate silicate sulphate hydrate known as thaumasite ($CaSiO_3.CaCO_3.CaSO_4.15H_2O$). TSA occurs to buried concrete elements at low temperature (generally <15°C) that are exposed to mobile groundwater contaminated with sulphates in conjunction with carbonates, generally in aggregate but also within groundwater.

TSA has occurred to buried elements of road bridges in the UK.

2.4.7 Acid attack

Inorganic, or mineral, acids in general are considered to be problematical for concrete when the pH falls below 6.5 for prolonged periods. The key breakdown mechanism involves the reduction of pH due to consumption of calcium hydroxide which destabilises the calcium aluminate hydrates and the calcium silicate hydrates in the cement paste, resulting in breakdown of these cementing minerals. Ultimately the integrity of the concrete is diminished, and the surface concrete gradually erodes away with time. Some of the inorganic acids that are aggressive to concrete are summarised at Table 2.4. Criteria for assessing the environment aggressiveness in terms of pH are provided in Tables 2.2 and 2.3.

Table 2.4 A list of some acids that are aggressive to concrete

Acids	
Inorganic	*Organic*
Carbonic	Acetic
Hydrochloric	Citric
Hydrofluoric	Formic
Nitric	Humic
Phosphoric	Lactic
Sulphuric	Tannic
Other substances	
Aluminium chloride	Vegetable and animal fats
Ammonium salts	Vegetable oils
Hydrogen sulphide	Sulphates

Source: Neville (1995)

The effects of organic acids on concrete are much more difficult to predict. Simple guidelines based on pH values or concentrations of solutions, which may reasonably be applied to inorganic acids as a whole, are of little value where organic acids are concerned, and virtually each acid must be considered individually with regard properties such as its solubility in water and, most importantly, the solubility of its calcium salt (Lea, 1998). Some of the organic acids that are aggressive to concrete are also summarised at Table 2.4.

2.4.8 Aggressive (Dissolved) carbon dioxide attack

Attack of concrete by carbon dioxide (CO_2) is termed carbonation and is one of the commonest causes of reinforcement corrosion (refer later) because it destroys the alkaline environment that is responsible for the development of a protective passive film on the surface of the steel (refer later).

Carbonation may occur when carbon dioxide from the atmosphere dissolves in concrete pore water and penetrates inwards or when the concrete surface is exposed to water or soil containing dissolved carbon dioxide. As is the case with all aggressive agents the capacity of the environment to attack the concrete is dependent either on the existence of cracks or upon the penetrability of the concrete and therefore the initial water/cement (water/binder) ratio.

Carbon dioxide dissolves in the pore water to form carbonic acid by the reaction:

$$CO_2(g) + H_2O(l) \rightarrow H_2CO_3(aq)$$

Carbonic acid can dissociate into hydrogen and bicarbonate ions. The carbonic acid reacts with $Ca(OH)_2$ in the solution contained within the pores of the hardened cement paste to form neutral insoluble $CaCO_3$. The general reaction is as follows:

$$Ca(OH)_2(aq) + H_2CO_3(aq) \rightarrow CaCO_3(s) + 2H_2O(l)$$

And the nett effect is to reduce the alkalinity of the pore water which is essential to the maintenance of a passive film on any reinforcing steel (refer later) that may be present.

The attack of buried concrete by carbon dioxide dissolved in the groundwater is a two-stage process. The calcium hydroxide solution that fills the pores of the concrete, first reacts with dissolved carbon dioxide to form insoluble calcium carbonate. This strengthens and densifies the cement gel and reduces its penetrability. However, it then subsequently reacts with further dissolved carbon dioxide to form soluble calcium bicarbonate which is leached from the concrete. The extent to which each process takes place is a function of the calcium carbonate/calcium bicarbonate concentration of the ground water (which in turn is a function of the pH and the calcium content) and the amount of dissolved carbon dioxide.

A convenient criterion for the extent to which buried concrete may be subject to aggressive (dissolved) carbon dioxide/carbonic acid attack is the Langelier Saturation Index (LSI) for the ground water (Langelier, 1936). This measures the calcium carbonate dissolving properties of the groundwater and is defined in the following terms:

- The pH of a solution which contains the same concentration of calcium ions as the groundwater, but which is just saturated with dissolved carbon dioxide can be calculated. This is termed the saturation pH or pH_s.
- If the actual groundwater contains more dissolved carbon dioxide than the equilibrium concentration, it will have a lower pH and will attack the concrete. If the actual ground water has a lower concentration of dissolved carbon dioxide than the equilibrium saturation value, it will have a higher pH and will deposit calcium carbonate in the pores of the concrete.
- Consequently, the LSI is defined as the difference between the pH of the ground water and pH_s.
- If the difference is positive, that is the pH of the water under examination is greater than pH_s, then the water does not degrade concrete on the basis of its carbonic acid content.
- If the difference is negative, then the water is aggressive, and the degree of aggressiveness is measured by that difference.

In order to assess the potential for concrete corrosion (degradation) by dissolved carbon dioxide attack it is therefore necessary to determine calcium ion concentration and the alkalinity. The various forms of dissolved carbon dioxide can easily convert to each other or be lost from solution, consequently if on-site titration is not possible, samples taken from the boreholes should be delivered in sealed bottles with no air space above the liquid to analytical chemists for immediate analysis. The absence of an air space is necessary to prevent loss of carbon dioxide into that airspace.

AS 3735 (2001) 'Concrete structures for retaining liquids' gives a short method for determining the LSI. An approximate value of the LSI may be obtained from the equation:

$$L_1 = pH \text{ of water} - pH \text{ when in equilibrium with calcium carbonate}$$
$$= pH - 12.0 + \log_{10}\left[2.5 \times Ca^{++} \times \text{total alkalinity}\right]$$

(the concentration of Ca^{++} is expressed in mg/l and
that of the alkalinity as $mgCaCO_3/l$).

A negative value for L_1 means that the water has a demand for $CaCO_3$. A demand for $CaCO_3$ means that the water is aggressive to concrete.

2.4.9 Seawater attack

Concrete exposed to seawater can suffer deterioration by sulphate attack, particularly by the action of magnesium sulphate, and also by crystallisation of salts in the concrete pores. Seawater also contains dissolved carbon dioxide which may lead to carbonation. An extreme case of seawater attack of concrete because of a very poor choice of cement type (i.e. 13% C_3A content) is shown at Figure 2.14.

The rate of sulphate attack is, however, slower in seawater than in sulphate-bearing groundwaters or soils. The reactions that occur when concrete is exposed to sulphate ions or magnesium sulphate solutions are thought to be the same as in seawater. Calcium sulphate and calcium sulphoaluminate, however, have increased solubility in chloride solutions and tend to be leached out of concrete in seawater. Calcium hydroxide may also be leached out of seawater exposed concrete.

Salts may also crystallise in the concrete above the high-water mark when evaporation occurs. The resultant expansion may cause disruption of the concrete.

Resistance to seawater attack can be provided by low penetrability, low water/cement ratio, high cement content, concrete. Because of the sequestration of chloride ions by the tricalcium aluminate, AS 3735 (2001) for example discourages the use of Sulphate Resisting (Type SR) cements in concrete exposed to seawater.

Figure 2.14 Seawater attack on concrete. (Courtesy of Woods, 1968, p.139)

2.4.10 Leaching and efflorescence

Natural waters may be classified as 'hard' or 'soft' usually dependent upon the concentration of calcium bicarbonate that they contain. Hard waters may contain a calcium ion content in excess of 10 mg/l (ppm) whereas a soft water may contain less than 1 mg/l (ppm) calcium. The capillary pores in a hardened cement paste contain a saturated solution of calcium hydroxide which is in equilibrium with the calcium silicate hydrates that form the cement gel. If soft water can permeate through the concrete, then it can leach free calcium hydroxide out of the hardened cement gel so that the pore water is diluted and the pH falls. Since the stability of the calcium silicates, aluminates and ferrites that constitute the cement gel requires a certain concentration of calcium hydroxide in the pore water, leaching by soft water can result in decomposition of these hydration products. The removal of the free lime in the capillary solution leads to dissolution of the calcium silicates, aluminates, and ferrites, and this hydrolytic action can continue until a large proportion of the calcium hydroxide is leached out, leaving the concrete with negligible strength. This leads to a removal of the cement gel from the concrete and Biczok (1967) has suggested that when 20% of the available calcium hydroxide has been leached from the concrete, its strength will have been reduced virtually to zero. Prior to this stage, loss of alkalinity due to the leaching process will result in reduced corrosion protection to reinforcement.

Calcium hydroxide that is leached to the concrete surface reacts with atmospheric carbon dioxide to form deposits of white calcium carbonate on the surface of the concrete. Corrosion by soft water is usually only a serious problem if water can penetrate through the concrete and leach out the lime (calcium hydroxide) which is providing the alkaline environment which is keeping the steel in a passive state (refer later). This problem therefore tends to arise only in situations such as air-conditioned underground carparks and tunnels which are situated below the water table. In such circumstances a concrete of high penetrability (permeability) must eventually have the lime leached out of it and must permit the rebar to corrode. In tunnels these deposits of calcium carbonate often take the form of stalactites and have been called 'white death' as they indicate the decay of the concrete. The process of such deposition of salts on the concrete surface is termed efflorescence. As is the case with most forms of external chemical attack, resistance to leaching, and efflorescence can be achieved by using low penetrability, low water/cement (water/binder) ratio concrete.

2.4.11 Physical salt attack

Physical damage to concrete located above soil level can occur due to PSA. PSA is associated with salt crystallisation in the near surface pores of concrete or rock. PSA (also called 'salt weathering' or 'salt hydration distress') has the same appearance as surface scaling by cycles of freezing and thawing (Haynes & Bassuoni, 2011).

Salts responsible for PSA on concrete include sodium sulphate ($NaSO_4$), sodium carbonate (Na_2CO_3) and sodium chloride ($NaCl$) (in decreasing order of aggressiveness). Magnesium sulphate ($MgSO_4$), calcium sulphate ($CaSO_4$), calcium chloride ($CaCl_2$), and sodium nitrate ($NaNO_3$) are also known to cause PSA on stone. PSA is associated with wet soil containing these dissolved salts. The salt solution is absorbed and subsequently transported (sorptivity, capillary suction/rise, wicking) to evaporation surfaces (Haynes & Bassuoni, 2011).

At the evaporation front, a given salt solution becomes supersaturated, and the salts crystallise. Salt crystals that form on the surface (commonly called efflorescence) are generally not harmful, but salt crystals can also form below the surface when the rate of solution supplied to the surface is slower than the rate of evaporation from the surface. Crystals that form below the surface (subflorescence) can result in surface scaling (Haynes & Bassuoni, 2011).

An example of PSA is shown at Figures 2.15 and 2.16 for the reinforced concrete support plinths for an above-ground mild steel cement lined (MSCL) water pipeline, circa 1940s, at Port Pirie in the state of South Australia in Australia. The water pipeline along some of its length traverses low lying salt pans which are very wet during winter.

Figure 2.15 Water pipeline, circa 1940s, Port Pirie, South Australia, supported above-ground on reinforced concrete plinths. (Courtesy of Greg Moore)

Figure 2.16 Close-up of physical salt attack to Port Pirie, circa 1940s, water pipeline concrete support plinths. (Courtesy of Greg Moore)

Hime et al (2001) have described the mechanism of deterioration when concrete located above soil level is exposed to Na_2SO_4 solutions, and deterioration is evidenced as white efflorescence, as 'salt hydration distress' (SHD). They describe SHD as the process which essentially relates to the repeated reconversions of Na_2SO_4 between its anhydrous and hydrated forms. Conversion of thenardite (anhydrous Na_2SO_4) to the phase known as mirabilite ($Na_2SO_4.10H_2O$) involves an expansion of 317% in volume.

Na_2SO_4 and Na_2CO_3 can undergo hydration phase transformations by changes in either ambient temperature or humidity of both. The conversion of a lower hydrate phase to a higher hydrate phase is accompanied by a considerable increase in volume (Haynes & Bassuoni, 2011).

The solubility of NaCl is not affected much by temperature changes, so a highly supersaturated solution does not develop. Therefore, damage by NaCl occurs by deliquescence and reprecipitation as the ambient relative humidity goes above and below the equilibrium relative humidity – a different mechanism than that of Na_2SO_4 and Na_2CO_3 – although in every case, the damage is due to crystallisation pressure (Haynes & Bassuoni, 2011).

2.4.12 Other chemical attack

Concrete may also be attacked in a number of industrial situations, particularly when concrete tanks are used in the chemical and food processing industries, for example. The effect of some commonly used chemicals on concrete is shown in Table 2.5 (Neville 1995). The effects of some gases on concrete are also included.

Alkaline solutions such as sodium hydroxide (NaOH), caustic, in alumina refineries for example, are also aggressive to concrete (Green, 1994).

In industrial situations or where aggressive chemical attack is expected, protection to the concrete may be afforded by coating the surfaces with

Table 2.5 Effect of some commonly used chemicals on concrete

Rate of attack at ambient temperature	Inorganic acids	Organic acids	Salt solutions	Miscellaneous
Rapid	Hydrochloric Hydrochloric Nitric Sulphuric	Acetic Formic Lactic	Aluminium chloride	
Moderate	Phosphoric	Tannic	Ammonium nitrate Ammonium sulphate Sodium sulphate Magnesium sulphate Calcium sulphate	Bromine (gas) Sulphite liquor
Slow	Carbonic		Ammonium chloride Magnesium chloride Sodium cyanide	Chlorine (gas) Seawater Soft water
Negligible		Oxalic Tartaric	Calcium chloride Sodium chloride Zinc nitrate Sodium chromate	Ammonia (liquid)

Source: Neville (1995)

resins or with glass fibre reinforced plastics. A newer method of tanking is by the attachment of welded polyethylene sheets to the surface of the concrete.

REFERENCES

American Concrete Institute (1984), '*ACI Committee 224 Causes, Evaluation and Repair of Cracks in Concrete Structures*', May–June, Farmington, MI, USA.

American Society for Testing and Materials (2012), '*ASTM C 295/C295M-12 Standard Guide for Petrographic Examination of Aggregates for Concrete*', West Conshohocken, USA.

Bertolini, L, Elsener, B, Pedeferri, P and Polder, R P (2004), '*Corrosion of Steel in Concrete*', WILEY VCH Verlag GMbH & Co KGaA, Weinheim.

Biczok, I (1967), '*Concrete Corrosion and Concrete Protection*', Hungarian Academy of Sciences, Budapest.

Beeby, A W (1984), '*Causes of Cracking*' in '*Durability of Concrete Structures*', RILEM, Budapest.

Brunetaud, X, Linder, R, Divet, L, Duragrin, D and Damidot, D (2007), 'Effect of curing conditions and concrete mix design on the expansion generated by delayed ettringite formation', *Materials and Structures*, 40, 567–578.

British Standards Institution (2000), '*BS EN 206: Part 1*', BSI, London, UK.

Bungey, J H and Millard, S G (1996), '*Testing of Concrete in Structures*', Chapman and Hall, London, UK.

Collepardi, M (2003), 'A state-of-the-art review on delayed ettringite attack on concrete', *Cement & Concrete Composites*, 25, 401–407.

Collins, F G and Green, W K (1990), '*Deterioration of concrete due to exposure to ammonium sulphate*', Use of Fly Ash, Slag & Silica Fume & Other Siliceous Materials in Concrete, Concrete Institute of Australia & CSIRO, Sydney, Australia, 3 September.

Concrete Institute of Australia (2015), '*Performance Tests to Assess Concrete Durability*', Recommended Practice Concrete Durability Series Z7/07, Sydney, Australia.

Concrete Society (2010), '*Non-Structural Cracks in Concrete*', Technical Report No. 22 – Fourth Edition, Camberley, England.

FHWA/TX-60/0-4085-5 (2005), 'Preventing ASR/DEF in New Concrete: Final Report', Center for Transport Research, The University of Texas, USA, November.

fib Bulletin (Draft by TG5.10) (2011), 'Birth Certificate and Through-Life Management Documentation'. fib C5 Meeting, Prague.

Gani, M S J (1997), '*Cement and Concrete*', Chapman and Hall, London, UK.

Green, W K (1994), '*Design and Specification of Durable Concrete*', Concrete Institute of Australia/Institution of Engineers Australia, Bunbury, Western Australia, 27 May.

Haynes, H and Bassuoni, M T (2011), '*Physical Salt Attack on Concrete*', Concrete International, November, 38–42.

Hime, W G, Martinek, R A, Bakus, L A and Marusin, S L (2001), '*Salt Hydration Distress*', Concrete International, October, 43–50.

Kelham, S (1996), 'The effect of cement composition and fineness on expansion associated with delayed ettringite formation', *Cement and Concrete Composites*, 18, 171.

Langelier, W F (1936), J.Amer Water Works Association, 28, 1500.

Lea, F M (1998), '*The Chemistry of Cement and Concrete*', ed Peter C Hewlitt, Edward Arnold, London, UK.

Lees, T P (1992), '*Durability in Concrete Structures*', ed Geoff Mayes, E & F Spon, London, UK.

Neville, A M (1975), '*Properties of Concrete*', Second Edition, Pitman International, London, UK.

Neville, A M (1995), '*Properties of Concrete*', Fourth Edition, Longman Group Limited, Harlow, England.

Shayan, A and Xu, A (2004), '*Effects of cement composition and temperature of curing on AAR and DEF expansion in steam cured concrete*', *Proc. 12th international conference on alkali aggregate reaction (ICAAR)*, Beijing, China, 773–788, 15–19 October.

Skalny, J (2002), '*Sulphate Attack on Concrete*', Spon Press, London, UK.

Standards Australia (2001), '*AS 3735 Concrete structures for retaining liquids*', Sydney, Australia.

Standards Australia (2008), '*AS 1141.65 Methods for sampling and testing aggregates – Alkali aggregate reactivity – Qualitative petrological screening for potential alkali-silica reaction*', Sydney, Australia.

Standards Australia (2009), '*AS 2159 Piling – Design and installation*', Sydney, Australia.

Standards Australia (2014a), '*AS 1141.60.1 Methods for sampling and testing aggregates. Method 60.1: Potential alkali-silica reactivity – Accelerated mortar bar method*', Sydney, Australia.

Standards Australia (2014b), '*AS 1141.60.2:2014 Methods for sampling and testing aggregates Method 60.2: Potential alkali-silica reactivity – Concrete prism method*', Sydney, Australia.

Standards Australia (2018), '*AS 3600 Concrete structures*', Sydney, Australia.

Thomas, M, Folliard, K, Drimalas, T and Ramlochan, T (2008), '*Diagnosing delayed ettringite formation in concrete structures*', *Cement and Concrete Research*, 38, 841–847.

Woods, H (1968), '*Durability of Concrete Construction*', ACI Concrete Monograph, No 4 Iowa State University Press, USA.

Chapter 3

Concrete deterioration
mechanisms (B)

3.1 BIOLOGICAL DETERIORATION

3.1.1 Bacteria

The so-called sulphur cycle is illustrated in Figure 3.1. It plays a significant role in the metabolism of most if not all, living creatures. The role of sulphate reducing bacteria (SRB) in the corrosion of metals is well recognised, but in connection with the corrosion of concrete, sulphur oxidising bacteria (SOB) are a major factor to be considered in aerated environments in which sulphides or sulphur are present.

They thus play a considerable role in the corrosion of sewers where they act to metabolise hydrogen sulphide and in polluted city atmospheres:

$$2H_2S + 2O_2 \rightarrow H_2S_2O_3 + H_2O$$

$$5Na_2S_2O_3 + 4O_2 + H_2O \rightarrow 5Na_2SO_4 + H_2SO_4 + 4S$$

$$4S + 6O_2 + 4H_2O \rightarrow 4H_2SO_4$$

With the result that they secrete thiosulphuric acid ($H_2S_2O_3$) and sulphuric acid (H_2SO_4) and the concrete is attacked.

SOB pose a major problem in sewers where there is a warm environment with plenty of available sulphide and plenty of organic nutrients.

Power station cooling towers present a similar potential for the growth of aerobic bacteria, a similar warm environment, nutrients from carbonated concrete, sulphide generated by the establishment of anaerobic regions where SRB can produce the sulphide.

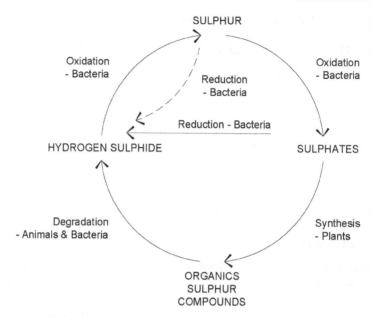

Figure 3.1 The sulphur cycle. (Courtesy of Videla, 1996)

The rate at which concrete is removed from the inner surface of the sewer is given by the equation due to Pomeroy & Parkhurst (1977):

$$C = 11.5k\Phi_{sw}\left(\frac{1}{A}\right)$$

Where C mm/yr is the rate of erosion of the surface. A is the calcium carbonate equivalent of the concrete (expressed as % by mass of concrete). It is a measure of the capacity of the concrete to neutralise the sulphuric acid produced as a result of bacteriological action and commonly taken as 0.2 for a granitic aggregate concrete. k is the acid efficiency factor, that is the efficiency with which the sulphuric acid can dissolve the cementitious material. It is a measure of the ease with which the generated acid can attack the cement paste. It is expected to lie between 0.3 and 1 and is often assumed for concrete to be 0.7. Φ_{sw} (g/m².hr) is the rate of approach of H_2S to the sewer wall and is given by:

$$\Phi_{sw} = 0.7\left(su\right)^{3/8} j[DS](b/P)$$

Where s is the energy gradient of the sewage and u the velocity, j is the fraction of dissolved sulphide present as H_2S, [DS] is dissolved sulphide, b/P is the ratio of surface width of sewage to exposed perimeter of pipe wall above surface. This factor will vary with the conditions of flow in the sewer.

3.1.2 Fungi

In sewer related studies, Mori et al. (1992) found an unidentified green fungus which grew at high pH levels and was capable of reducing the pH to levels suitable for colonisation and growth of bacteria.

Gu et al. (1998) go further in their explanation and identified the fungus as *Fusarium*. This fungus led to more rapid deterioration of concrete than SOB. It was also able to penetrate into the concrete. They also state that a wide range of organic acids are produced by fungi including acetic, oxalic, and glucuronic acids. Their study also suggested that interaction between fungal metabolites and calcium from cement results in the formation of soluble calcium organic complexes.

Cwalina (2008) reports that there are fungi other than *Fusarium* that are aggressive to concrete and lead to the production of organic acids and carbon dioxide gas.

3.1.3 Algae

Hughes (2013) advises that micro-organisms such as algae can lead to deterioration of concrete. Bacterially induced attack of concrete can also be induced by algae. Cwalina (2008) reports that many types of micro-organisms (bacteria, algae, lichens, yeasts, and fungi) can cause biodeterioration of concrete.

3.1.4 Slimes

It is not known whether slime(s) typically cause attack of concrete. However, it is noted that within the bottom of sewers slime layers form below the water level (e.g. Vincke et al., 1999) and that concrete within the slime layers remains uncorroded/unattacked.

3.1.5 Biofilms

Biofilm growth can deteriorate concrete, but they can also directly affect the operation of some reinforced concrete assets. As an example, ageing reinforced concrete canals (Figure 3.2) and flumes (Figure 3.3) make up a significant part of the water conveyance system supplying Hydro Tasmania's 29 hydroelectric power stations in Tasmania, Australia. These water conveyance structures are subject to harsh weather conditions, surface weathering, and biofilm growth, all of which contribute to a decreased life

Figure 3.2 Concrete lined trapezoidal canal section. (Courtesy of Andrewartha & Cribbin, 2009)

Figure 3.3 Concrete flume section. (Courtesy of Andrewartha & Cribbin, 2009)

expectancy of the reinforced concrete structures. Increased surface rough-ness is also of concern, as it reduces the hydraulic efficiency and hence the maximum power output of the particular power scheme (Andrewartha & Cribbin, 2009).

Biofouling is the undesirable growth of biological matter, including bacteria, algae, fungi, and invertebrate organisms at liquid-solid interfaces. The detrimental effect of biofilms on skin friction has been well established and the increases in frictional resistance and resultant energy losses due to biofilms are of major concern to structure owners including hydroelectric power generators and water supply industries. It is the presence of biofilms in a frictional sense rather than their effect on the concrete degradation that is the problem (Andrewartha & Cribbin, 2009).

It has been established that the effective roughness caused by biofilm growth is considerably larger than the absolute thickness of the biofilm layer. The increased drag cannot be attributed solely to the decrease in cross section due to biofilm accumulation and is due to the interaction of the biofilms with flowing water (Andrewartha & Cribbin, 2009).

Cleaning water conveyances is a proven method of short-term biofouling control and does temporarily increase the capacity and hence revenue potential. However, it does not provide lasting protection against biofilm growth, nor remove all of the fouling material from the surface. Mechanical cleaning methods gradually damage the surface and thus increasing the surface roughness and hence decreasing the capacity of concrete canals and flumes (Andrewartha & Cribbin, 2009).

Hydro Tasmania developed a 'smooth lining technique' for both controlling the growth of biofilms and improving the surface roughness of existing concrete water conveyances. The technique involves applying a smooth cementitious render or overlay to the existing concrete surface and then coating the render with a protective surface coating (poly siloxane typically) (Andrewartha & Cribbin, 2009).

3.2 PHYSICAL DETERIORATION

3.2.1 Freeze-thaw

In the alpine regions of countries freeze-thaw may be a physical deterioration risk to some reinforced concrete structures. This type of damage is caused by the penetration of water into the surface layers of the concrete and then this moisture being converted into ice by sub-zero temperatures. Water expands on freezing, increasing in volume by approximately 10%, and this causes disintegration of the surface layers of the concrete (Perkins, 1997).

Andrewartha and Cribbin (2009) report of severe frost attack to elements of a reinforced concrete flume in Tasmania, Australia. Liawenee Flume is approximately 650 m long and transfers water from the Ouse River to the Great Lake via the Liawenee Canal. It was constructed in 1918–1921 and raised in 1946 and frequently experiences some of the most extreme cold weather in Tasmania.

Figure 3.4 Severe frost attack at Liawenee Flume, Tasmania, Australia (note loss of concrete to tie beams). (Courtesy of Andrewartha & Cribbin, 2009)

Figure 3.5 Liawenee Flume post upgrade. (Courtesy of Andrewartha & Cribbin, 2009)

Eighty years of severe frost attack, concrete degradation, reinforcement corrosion, and biofouling reduced the asset to an unacceptable state, refer Figure 3.4, where the structural integrity was seriously compromised, prompting an extensive upgrade, refer Figure 3.5.

Table 3.1 Mitigation techniques for concrete deterioration mechanisms at Liawenee Flume, Tasmania, Australia

Deterioration mechanisms	Mitigation techniques
Freeze-thaw (frost attack)	Application of a silane-siloxane to all exposed concrete surfaces Use of air-entrained repair concrete
Mechanical stresses (water load and thermal movement)	Installation of adjustable galvanised steel tie beams Creation of sealed contraction joints in known crack locations
Concrete shrinkage and expansion(drying and wetting cycles)	Externally the application of a silane-siloxane to reduce moisture ingress Restoration of all external cracks Installation of a drip stop groove in underside of parapet No horizontal surfaces to allow ponding of water Internally the application of a two-coat poly siloxane paint system. This has the added benefit of improving the hydraulic efficiency of the flume
Carbonation(reduction of alkalinity)	Use of 40 MPa concrete which was placed using steel forms and internal vibration to ensure correct consolidation
Pure water leaching	Internally the application of a two-coat poly siloxane paint system. This has the added benefit of improving the hydraulic efficiency of the flume

Source: Andrewartha & Cribbin (2009)

Following restoration and upgrade (Table 3.1 and Table 3.2) the capacity of the flume was increased from 18.5 m³/s to 24 m³/s, an improvement of 30%. After seven years since the initial upgrade, the extent of biofouling is still at very low levels and no deterioration of the coating has been observed (Andrewartha & Cribbin, 2009).

3.2.2 Fire

Concrete, although not a refractory material, is non-combustible, and has good fire-resistant properties. The term fire-resistant should not, by standard definition, be applied to a material, but only to the structural element of which it forms a part (Lea, 1998).

Fire affected concrete can sustain various degrees of damage depending on the severity of the fire and the temperature levels reached. Damage from the fire can range from the presence of some soot and smoke deposits, some minor spalling, or exposed aggregate surface with no exposure of steel reinforcement; to soot blackening, colour changes to pink, significant spalling particularly at the edges and corners of structural elements with exposed ligatures/stirrups and some delamination; to a buff/grey colour with substantial

Table 3.2 Upgrade works undertaken at Liawenee Flume, Tasmania, Australia

Location	Upgrade work undertaken
Parapets and tie beams	The frost attacked and damaged concrete was removed and replaced with new air-entrained 40 MPa concrete, including a cast-in drip stop and diamond sawn construction joints
	The existing concrete tie beams were removed and replaced with adjustable galvanised steel tie beams
Cracks in walls	The eroded, frost attacked, and otherwise damaged concrete was removed, and the original profile was then restored using cementitious based products
	All existing cracks and contraction joints were mapped through the cementitious render and reinstated with diamond saw joints to control reflective cracking
	The joints were then sealed with fast cure polyurethane
Invert	The eroded and frost attacked concrete was removed. Construction joints were then installed, followed by the forming and placing of a 50 mm thick concrete overlay
	The construction joints were then sealed
External surfaces	All exposed concrete surfaces were cleaned by high pressure water blasting
	Two coats of silane/siloxane were then applied to prevent water ingress
Internal side walls	The eroded, frost attacked, moss covered, and damaged concrete was removed by high pressure water blasting
	The cracks were restored, and the profile of the wall rebuilt using cementitious products
	The internal walls were finished with a poly siloxane coating to provide a smooth finish with excellent hydraulic properties, see Figure 3.5

Source: Andrewartha & Cribbin (2009)

extensive spalling of the edges, sides, and soffit of structural elements, considerable exposure of steel reinforcement including some main bars and to a lesser extent prestressing tendons. The more severe fire damage would also involve total exposure of main reinforcing bars, significant exposure of prestressing tendons, significant cracking, and spalling, buckling of steel reinforcement and even significant fracture and deflection of concrete components (Andrews-Phaedonos, 2007).

The effects of high temperature on concrete components can also include: various degrees of reduction in compressive strength due to significant changes including microcracking within the concrete microstructure; colour changes which are consistent with strength reductions; reduction in the modulus of elasticity; various degrees of spalling; loss of bond between concrete and steel; possible loss of residual strength of steel reinforcement and possible loss of tension in prestressing tendons (Andrews-Phaedonos, 2007).

Further discussion of fire damaged concrete is provided at Section 3.5.

3.3 MECHANICAL DETERIORATION

Under many circumstances, concrete surfaces are subjected to wear. This may be due to attrition by sliding, scraping or percussion (e.g. pavements, slabs, industrial floors). In the case of hydraulic structures (pipelines, spillways, retained waterways, seawalls, etc), the action of abrasive materials carried by water leads to erosion. Another cause of damage to concrete in flowing water is cavitation (Neville, 1995).

3.3.1 Abrasion

Abrasion is thought to primarily involve high intensity stress applied locally so that the strength and hardness of the surface zone of concrete strongly influences resistance to abrasion. In consequence, the compressive strength of concrete is the principal factor controlling the resistance to abrasion (Neville, 1995).

The properties of the concrete in the surface zone are strongly affected by the finishing operations, which may reduce the water/cement (water/binder) ratio and improve compaction. Particularly good curing is of importance (Neville, 1995). Bleeding characteristics of the concrete are important (CC&AA, 2007). High strength silica fume blended cement concretes, for example, show improved resistance to abrasion and erosion (Lea, 1998).

There are many abrasion tests used for a variety of purposes. In general, the results of abrasion tests cannot be correlated directly with insitu durability performance, therefore their application is limited to comparative testing of (for example) different mix designs and construction techniques (Concrete Institute of Australia, 2015).

3.3.2 Erosion

Erosion of concrete is an important type of wear which may occur in concrete in contact with flowing water. It is convenient to distinguish between erosion due to solid particles carried by water and damage due to pitting resulting from cavities forming and collapsing in water flowing at high velocities (Neville, 1995). The latter is cavitation and is considered in the next section.

Erosion rate depends on the quantity, shape, size, and hardness of the particles being transported, on the velocity of their movement, on the presence of eddies, and also the quality of the concrete. As in the case of abrasion in general, the quality appears to best be measured by the compressive

Figure 3.6 Deteriorated concrete due to erosion and cavitation in water conveyance channels. (Courtesy of Andrewartha & Cribbin, 2009)

strength of concrete, but the mix composition is also relevant. In particular, concrete with large aggregate erodes less than mortar of equal strength, and hard aggregate improves the erosion resistance. However, under some conditions smaller size aggregate leads to a more uniform erosion of the surface (Neville, 1995). Wear resistance of coarse and fine aggregate is also important as is the smoothness of the surface of the concrete (Perkins, 1997).

In Australia in Hydro Tasmania's water conveyance structures the main surface deteriorating mechanisms are pure water leaching, erosion, cycles of freezing, and thawing, and the formation of biofilms on the surface. In terms of erosion, concrete surfaces in hydraulic structures are subject to flowing water containing fine particles and solids, which over time degrades away the surface, refer Figure 3.6 and Figure 3.7 (Andrewartha & Cribbin, 2009).

The damaged concrete in the case of internal wall surfaces of some of these hydraulic structures was severe enough to warrant repair. Proprietary, specialist, cementitious based repair materials were adopted and a methacrylic resin applied to protect the concrete from water ingress and erosion, and to help prevent the formation of biofilms (Andrewartha & Cribbin, 2009).

Water erosion-induced concrete deterioration can also be extensive, for examples in plunge pool structures of dams, refer Figure 3.8 and Figure 3.9.

3.3.3 Cavitation

While good quality concrete can withstand steady, tangential high velocity flow of water, severe damage rapidly occurs in the presence of cavitation. By this is meant the formation of vapour bubbles when the local absolute pressure drops to the value of the ambient vapour pressure of water at the

Figure 3.7 Extensive water erosion deterioration to internal wall surfaces of water conveyancing structures. (Courtesy of Andrewartha & Cribbin, 2009)

Figure 3.8 Repulse Dam in Tasmania, Australia, on spill. (Courtesy of Norm Cribbin, Hydro Tasmania)

ambient temperature. The bubbles or cavities can be large, single voids, which later break up, or clouds of small bubbles. They flow downstream and, on entering an area of higher pressure, collapse with great impact. Because the collapse of the cavities means entry of high velocity water into

Figure 3.9 Water erosion-induced concrete deterioration to the plunge pool of Repulse Dam, Tasmania, Australia. (Courtesy of Norm Cribbin, Hydro Tasmania)

the previously vapour occupied space, extremely high pressure on a small area is generated during very short time intervals, and it is the repeated collapse over a given part of the concrete surface that causes pitting. Greatest damage is by clouds of minute cavities found in eddies. They usually coalesce momentarily into a large amorphous cavity which collapses extremely rapidly. Many of the cavities pulsate at high frequency, and this seems to aggravate damage over an extended area (Neville, 1995).

The effect of cavitation can be very destructive to concrete even of the highest quality (Perkins, 1997). Surfaces become irregular, jagged, and pitted, in contrast to the smoothly worn surface of concrete eroded by waterborne solids. Cavitation damage does not progress steadily, usually after an initial period of small damage, rapid deterioration occurs, followed by damage at a slower rate (Neville, 1995).

Generally, cavitation damage has been found to occur to spillways, penstocks, aprons, energy dissipating basins and to syphons and tunnels in hydroelectric schemes (Perkins, 1997).

Best resistance to cavitation damage is obtained by the use of high strength concrete, possibly formed by an absorptive lining (which reduces the surface water/cement ratio). Use of polymers, steel fibres, or resilient coatings can improve cavitation resistance. The solution of the cavitation damage lies, however, in the achievement of smooth and well aligned surfaces free from irregularities such as depressions, projections, joints and misalignments, and by the absence of abrupt changes in slope or curvature that tend to pull the flow from the surface (Neville, 1995).

3.3.4 Impact

Impact deterioration of concrete is possible in areas of heavy traffic or loads. Impact strength is of importance when concrete is subject to a repeated falling object, as in pile driving, or impacts of a mass at high velocity (Neville, 1995).

In general, the impact strength of concrete increases with an increase in compressive strength, but the higher the static compressive strength of the concrete the lower the energy absorbed per blow before cracking. Figure 3.10 gives some examples of the relation between the impact strength and the compressive strength of concrete with different aggregates (Neville, 1995).

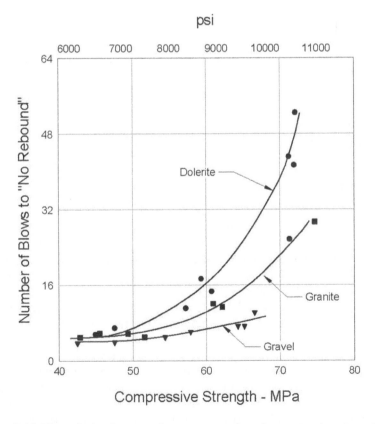

Figure 3.10 The relation between impact strength and compressive strength for concretes made with different aggregates. (Courtesy of Neville, 1995, p.343)

3.4 STRUCTURAL DETERIORATION

Structural deterioration of reinforced, prestressed, and post-tensioned concrete structure elements is the domain of the structural engineer and as such is beyond the scope of this book. Suffice to say that some brief background information is provided below in terms of some structural modes of deterioration.

3.4.1 Overloading

The operational requirements of structures may change during their design (service) life such that increased loading is presented to reinforced, prestressed, or post-tensioned elements. Overloading is therefore also possible.

Overloading of reinforced, prestressed, or post-tensioned structure elements would typically become evident as cracking and deformation. In some cases, e.g. localised corrosion of prestressed or post-tensioned elements, structural failure associated with overloading would be possible.

3.4.2 Settlement

Settlement is a possible mode of structural deterioration. For example, settlement of road bridges, due to changing conditions of lakes, can lead to significant cracking and deformation of structural reinforced and prestressed concrete elements.

3.4.3 Fatigue

In many structures, repeated loading is applied such as, for example, to bridges (road and rail, operational loading), mining structures (operational), wharves and jetties (operational, wind, wave), industrial buildings (operational), road, and airfield pavements, and railway sleepers. The number of cycles of loading applied during the life of the structure may be as high as 10 million, and occasionally even 50 million (Neville, 1995).

When a material fails under a number of repeated loads, each smaller than the static compressive strength, failure in fatigue is said to take place. Both concrete and steel possess the characteristics of fatigue failure and discussion of these is beyond the scope of this book. However, fatigue of reinforced, prestressed, or post-tensioned structure, and building elements would typically become evident as cracking. Cracking can, however, 'join up' to form 'blocks' as in the case of some road bridges in Australia subject to repetitive high loadings from very large trucks ('road trains'), refer Figure 3.11, and Figure 3.12. In some instances, such fatigued bridge decks require replacement.

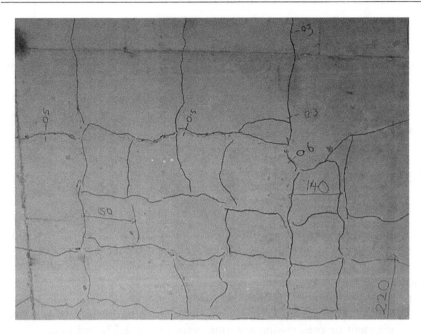

Figure 3.11 Fatigue cracking to a flat slab road bridge substructure. (Courtesy of Gavin Johnston, Main Roads Western Australia)

Figure 3.12 Fatigue cracking to a steel beam composite road bridge substructure. (Courtesy of Gavin Johnston, Main Roads Western Australia)

3.4.4 Other

The book's authors are scientists but are aware by having worked with structural engineers of other forms of structural deterioration due to shear, tension, torsion, and flexure.

Structural deterioration due to creep and shrinkage (deflection downwards of cantilevered sections of some road bridges for example) has also been known to the authors.

3.5 FIRE DAMAGED CONCRETE

3.5.1 General

Fire can damage reinforced concrete structures. Public safety may be compromised, disruption to operations may be marked, closing down of sections of structures may be needed, posting of load limits applied, etc. However, even after a severe fire, reinforced concrete structures are often capable of being repaired rather than demolished.

Fire may cause deleterious reactions to occur to both concrete and steel reinforcement or prestressing tendons, refer Figure 3.13, and Figure 3.14. Heating of concrete in a fire causes a series of mineralogical and strength changes. Heating of steel or prestressing tendons causes some changes to occur. Assessment by specialists is often required of concrete damaged structures. Procedures required include on-site inspection and testing techniques, laboratory testing, and structural fire analysis. The UK Concrete Society

Figure 3.13 Explosive spalling, exposure of steel ligatures, and main corner bars, partial exposure of prestressing tendons on the beam soffit, inner beam in the background, of a fire damaged road bridge, Victoria, Australia. (Courtesy of Andrews-Phaedonos, 2007)

Figure 3.14 View of the interior of a fire damaged reinforced concrete building showing a spalled slab soffit and columns. (Courtesy of Ingham, 2009)

Technical Report 68 (2008), for example, gives guidance on the assessment of fire damaged structures.

3.5.2 Effects on concrete

Heating of concrete in a fire causes a progressive series of mineralogical and strength changes that are summarised in Table 3.3.

The strength of concrete after cooling varies depending on temperature attained, the heating duration, mix proportions, aggregates present, and the applied loading during heating. For temperatures up to 300°C, the residual compressive strength of structural quality concrete is not significantly reduced, while for temperature greater than 500°C the residual strength may be reduced to only a small fraction of its original value. The temperature of 300°C is normally taken to be the critical temperature above which concrete is deemed to have been significantly damaged (Ingham, 2009).

Spalling of the surface layers is a common effect of fires and may be grouped into two or more types. Explosive spalling is erratic and generally occurs in the first 30 minutes of the fire. A slower spalling, referred to as 'sloughing off', occurs as cracks form parallel to the fire affected surfaces leading to a gradual separation of concrete layers and detachment of a section of concrete along some plane of weakness, such as a layer of reinforcement. Also, the thermal incompatibility of aggregates and cement paste causes stresses which frequently lead to cracks, particularly in the form of

Table 3.3 Summary of mineralogical and strength changes to concrete caused by heating

Heating temperature, °C	Changes caused by heating	
	Mineralogical	*Strength changes*
70–80	Dissociation of ettringite	Minor loss of strength possible (<10%)
105	Loss of physically bound water in aggregate and cement matrix commences, increasing capillary porosity	
120–165	Decomposition of gypsum	
250–350	Oxidation of iron compounds causing pink/red discolouration of aggregate Loss of bound water in cement matrix and associated degradation becomes more prominent	Significant loss of strength commences at 300°C
450–500	Dehydroxylation of portlandite. Aggregate calcines and will eventually change colour to white/grey	
575	5% increase in volume of quartz (-to-quartz transition) causing radial cracking around the quartz grains in the aggregate	Concrete not structurally useful after heating in temperatures in excess of 500–600°C
600–800	Release of carbon dioxide from carbonates may cause a considerable contraction of the concrete (with severe microcracking of the cement matrix)	
800–1,200	Dissociation of extreme thermal stress cause complete disintegration of calcareous constituents, resulting in whitish grey concrete colour and severe microcracking	
1,200	Concrete starts to melt	
1,300–1,400	Concrete melted	

Source: Ingham, (2009)

surface crazing. Thermal shock caused as rapid cooling from firefighting water may also cause cracking (Ingham, 2009).

Soot blackening and smoke deposits on concrete components are a direct by product of an intense fire, refer Figure 3.15. These can be deposited during the height of the fire although soot can also be deposited while the intensity of the fire is abating. Light grey areas are usually associated with exposure to high temperatures (Andrews-Phaedonos, 2007; Ingham, 2009).

A significant loss of strength of the order of 30–40% takes place once the temperature of the concrete has reached 300°C. This is the result of

Figure 3.15 Soot blackened road bridge in Victoria, Australia, also showing over-turned transport vehicle and fire damaged reinforced concrete sub-structure elements. (Courtesy of Andrews-Phaedonos, 2007)

significant internal cracking of the cementitious paste and aggregates due to thermal expansion, as well as the incompatibility between these and the steel reinforcement within the concrete. Above about 500–600°C more than 70–80% strength reduction takes place due to resultant friable and porous microstructure which lies in the grey to buff colour range. In the temperature range of 150–300°C the loss of strength ranges between 5 and 30% (Ingham, 2009).

In the temperature range of up to 300°C the loss in modulus of elasticity is similar to the loss in strength and in the order of 40%. At around 550°C the loss in modulus of elasticity is in the order of 50% (Ingham, 2009).

3.5.3 Visual concrete fire damage classification

The colour of concrete can change as a result of heating, which is apparent upon visual inspection. In many cases a pink/red discolouration occurs above 300°C, which is important since it coincides approximately with the onset of significant loss of strength due to heating. Any pink/red discoloured concrete should be regarded as suspect and potentially weakened (Ingham, 2009).

The colour change to pink or pink/red is a function of oxidisable iron content and it should be noted that as iron content varies, not all aggregates undergo colour changes on heating. In general, colour changes are most pronounced for siliceous aggregate and less so for limestone and granite (Ingham, 2009). At temperatures >600°C the colour of concrete changes to grey. At temperatures >900°C the colour of concrete changes to buff (Ingham, 2009).

An example of a damage classification scheme suitable for fire damaged concrete structures is shown in Table 3.4 (Concrete Society, 2008).

Table 3.4 Simplified visual concrete fire damage classification

Class of damage	Finish	Colour	Feature observed				
			Crazing	Spalling	Reinforcement	Cracks/deflection	
0 (Decoration required)	Unaffected	Normal	None	None	None exposed	None	
1 (Superficial repair required)	Some peeling	Normal	Slight	Minor	None exposed	None	
2 (General repair required)	Substantial loss	Pink/red*	Moderate	Localised	Up to 25% exposed	None	
3 (Principal repair required)	Local loss	Pink/red Whitish grey**	Extensive	Considerable	Up to 50% exposed	Minor/ None	
4 (Major repair required)	Destroyed	Whitish grey**	Surface lost	Almost total	Up to 50% exposed	Major/ Distorted	

Source: Concrete Society, (2008)
Notes:
* Pink/red discolouration is due to oxidation of ferric salts in aggregates and is not always present and seldom in calcareous aggregate.
** White/grey discolouration due to calcination of calcareous components of cement matrix and (where present) calcareous or flint aggregate.

3.5.4 Effect on reinforcement and prestressing steel

Steel reinforcement (depending on the type) can lose up to 50% of its yield strength at elevated temperatures of the order of 600°C. However, it can fully recover its yield strength on cooling at temperatures of up to 450°C for cold worked steel and up to 600°C for hot rolled steel. At temperatures higher than these the loss in yield strength is permanent. The modulus of elasticity of steel is also significantly reduced at elevated temperatures (Ingham, 2009).

Prestressing steel is more susceptible to fire damage and elevated temperatures, compared to normal steel reinforcement because loss in the strength of the order of 50%, occurs at the lower temperature of about 400°C. Loss of tension in the prestressing tendons can be a combination of the elevated temperature and loss in the modulus of elasticity of the concrete (Ingham, 2009).

The bond between steel and concrete can also be adversely affected at temperatures higher than 300°C because of the greater thermal conductivity of steel compared to the cover concrete (Ingham, 2009).

3.6 EXAMINATION OF SITES

In general, the most important consideration for the chemical and biological attack of concrete is when underground structures are subject to attack by aggressive ground waters. Structures which include hollow components such as underground car parks, road, or railway tunnels, culverts or pipes are often much more susceptible to attack by aggressive ground waters than are solid structures such as the piles supporting a building or concrete raft foundations. Water and the aggressive ions dissolved in it can be transported through the walls of hollow structures and attack more deeply into the concrete.

An examination of water/groundwater for aggressivity must include a determination of the pH, and the concentrations of chlorides or sulphates. In some circumstances an examination for the presence of sulphides and bacteria may be carried out to investigate the possibility of microbiologically influenced deterioration. For carbonic acid attack and leaching, the ground water has to be analysed for sulphates, calcium ions, magnesium ions, carbonates, bicarbonates, and the total ionic strength.

To assess the aggressivity of the ground water it is therefore necessary to determine:

- The pH.
- The concentration of chloride ions.
- The concentration of sulphate ions.
- The alkalinity.

- The concentration of calcium ions.
- The concentration of magnesium ions.
- The concentration of ammonium ions.

However, in general the analysis should also include a determination of the concentration of sodium and potassium so that an ionic balance can be struck. A discrepancy between the total cation and total anion concentration may reveal the presence of an ionic species which had not previously been suspected. If there is a discrepancy between the anion and cation concentrations of less than 2% then the analysis has probably revealed all the important ions in the water.

The techniques that are used to investigate the aggressivity of a site towards a structure which may be built upon it are based upon an analysis of all the factors which may affect the proposed structure. These include, the geology of the site, the chemical analysis of the soil and the groundwater and a consideration of the possible changes to the site that may be brought about by construction upon it. The planning of the site investigation should be carried out in conjunction with the geotechnical engineer who is investigating the subsoil conditions. This will ensure that important site characteristics such as the existence of underground streams and the delineation of the underlying strata are taken into account. It will also ensure that the work involved in the site investigations is optimised in that soil samples can be taken for analysis during the drilling of exploratory holes and that water samples can be taken from the piezometers that the geotechnical engineers may wish to install on the site.

The characteristics of the groundwater at any given point on the site may most readily be achieved by the installation of water sampling piezometers. These must be installed at various locations around the site, so situated that they sample the groundwater which will be in contact with each part of the structure. A simple water sampling piezometer is shown in Figure 3.16 (Cherry, 1998). Typically, a 55 or 60 mm diameter bore is either wash bored or diamond drilled to the depth at which it is proposed to sample the groundwater. The bottom of the bore is filled to a depth of about 200–300 mm with sand and then a 50 mm diameter PVC pipe is installed in the bore and the gap between the PVC pipe and the walls of the bore filled with a bentonite slurry. Water can then only enter the standpipe through the sand plug and the geotechnical engineer can be certain that the water being sampled has emanated solely from the stratum which it is intended to investigate. Water samples can be removed from the bottom of the bore using a closed cylinder which can be lowered to the bottom of the hole, opened at the bottom, and then closed again before the cylinder is taken to the surface. A simple model is shown in Figure 3.17 (Cherry, 1998).

It may take a considerable time for the water at the bottom of the standpipe to come to equilibrium with the ground water at the level being investigated. This will particularly be the case if the hole was drilled using wash

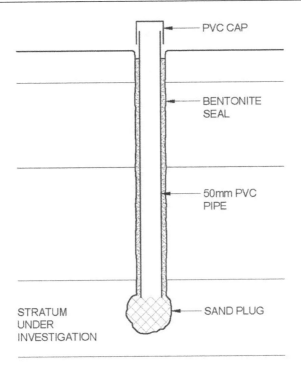

Figure 3.16 Water sampling piezometer. (Courtesy of Cherry, 1998)

boring and drilling water has permeated the area under investigation. A simple and quick technique to determine whether equilibrium has been established is to monitor the electrolytic conductivity of the groundwater and when this has stabilised, it may be assumed that the composition is constant. Samples are taken at intervals and the conductivity monitored until the conductivity of the water removed from a given bore hole has reached a more or less constant value, the water samples can then be analysed with some confidence that they represent the environment to which the structure will be exposed. Samples taken from the boreholes should be delivered in sealed bottles with no air space above the liquid to analytical chemists for immediate analysis for sodium, potassium, calcium, magnesium, ammonium, chloride, sulphate, free carbon dioxide, and pH. The absence of an air space is necessary to prevent loss of carbon dioxide into that airspace.

Soil samples can be removed during core drilling. As soon as they are removed from the core then the ends should be sealed to prevent the loss of any water from the sample or the ingress of any carbon dioxide and the samples can be analysed for chloride content, sulphate content, water content, organic matter content, and pH. Further samples may be used for a determination of the electrical resistivity of the soil and if necessary, a

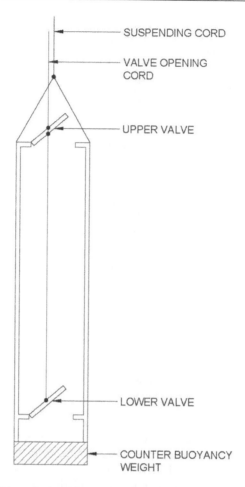

Figure 3.17 Sealable cylinder. (Courtesy of Cherry, 1998)

microbiological examination. Techniques for the determination of sulphate, moisture and organic matter content and pH as well as a vast range of other tests for the engineering properties of soils are detailed in, for example, the Australian Standard AS 1289 'Methods of testing soils for engineering purposes' series (various parts, 1993–2008).

Determination of bacteria types is the domain of the microbiologist. It is possible to determine the various types of bacteria and their likely numbers. For example, with SRB, qualitative determination (yes/no) is possible with portable site kits, serial dilution (factor of ten) determination can be undertaken and most probable number (MPN) of bacteria determined if necessary. The nutrients for bacteria in water, groundwater and soil should also be analysed such as: sulphate, sulphide, nitrate, nitrite, ammonia, orthophosphate, phosphorous, and organic carbon.

REFERENCES

Andrewartha, J and Cribbin, N (2009), '*When the Going Gets Rough, the Rough Stop Flowing*', *Proc. Concrete 09 Conf.*, *Concrete Institute of Australia*, Sydney, 17–19 September.

Andrews-Phaedonos, F (2007), '*Investigation, Assessment and Repair of Fire Damaged Pre-Stressed Concrete (PSC) Beams*', *23rd Biennial Conf of the Concrete Institute of Australia*, Perth, Australia.

CC&A Australia (2007), 'Effect of Manufactured Sand on Surface Properties of Concrete Pavements', Research Report, Cement Concrete & Aggregates Australia, 1 August.

Cherry, B W (1998), 'Cathodic protection of underground reinforced concrete structures', in '*Cathodic Protection Theory and Practice*' ed V Ashworth and C Googan, Ellis Horwood, London, UK, 326–350.

Concrete Institute of Australia (2015), '*Performance Tests to Assess Concrete Durability*', Recommended Practice Concrete Durability Series Z7/07, Sydney, Australia.

Concrete Society (2008), '*Assessment, Design and Repair of Fire-Damaged Concrete Structures*', Technical Report 68, Camberley, UK.

Cwalina, B (2008), '*Biodeterioration of Concrete*', Architecture Civil Engineering Environment, The Silesian University of Technology, No. 4/2008.

Gu, J-D, Ford, TE, Berke, N S, and Mitchell, R (1998), 'Biodeterioration of concrete by fungus Fusarium', *International Biodeterioration & Biodegradation*, 41, 101–109.

Hughes, P (2013), '*Bio-tenacious growth in subsea concrete*', World Tunnelling, April, 30–32.

Ingham, J (2009), 'Forensic engineering of fire-damaged structures', *Proc of the Institution of Civil Engineers – Civil Engineering*, 162, May, 12–17.

Lea, F M (1998), '*Lea's Chemistry of Cement and Concrete*', ed Peter C Hewlitt, Edward Arnold, London, UK.

Mori, T, Nonaka, T, Tazaki, T, Koga, M, Hirosaka, Y, Noda, S (1992), 'Interactions of nutrients, moisture and pH on microbial corrosion of concrete sewer pipes', *Water Research*, 26, 1, 29–37.

Neville, A M (1995), '*Properties of Concrete*', Fourth Edition, Longman Group Limited, Harlow, England.

Perkins, P H (1997), '*Repair, Protection and Waterproofing of Concrete Structures*', E & FN Spon, London, UK.

Pomeroy, R D, and Parkhurst, J D (1977), '*The forecasting of sulphide build-up rates in sewers*', *Prog. Water Techn.*, 9(3), 621–628.

Standards Australia, (1993–2008), 'AS 1289 Methods of testing soils for engineering purposes', Sydney, Australia.

Videla, H (1996), '*Manual of Biocorrosion*', CRC Press, Boca Raton, FL, USA.

Vincke, E, Verstichel, S, Monteny, J, and Verstraete, W (1999), 'A new test procedure for biogenic sulfuric acid corrosion of concrete', *Biodegradation*, 10, 421–428.

Chapter 4

Corrosion of reinforcement (A)

4.1 BACKGROUND

When suitably designed, constructed, and maintained, reinforced concrete provides service lives of numerous decades to structures and buildings. Concrete provides reinforcing steel with excellent corrosion protection. The highly alkaline environment in concrete results in the spontaneous formation of a stable, tightly adhering, thin protective iron oxide (passive) film on the steel reinforcement surface, which protects it from corrosion. In addition, well proportioned, compacted, and cured concrete has a low penetrability, thereby minimising the ingress of corrosion-inducing species via the aqueous phase. It also has a relatively high electrical resistivity (many tens of thousands ohm.cm), which reduces the corrosion current and hence the rate of corrosion if corrosion is initiated.

4.2 PORTLAND CEMENT AND BLENDED CEMENT BINDERS

As indicated in Chapter 1, the hydraulic binder of concrete commonly consists of Portland cement or of mixtures of Portland cement and one or more of fly ash (FA), ground granulated iron blast-furnace slag (BFS) or silica fume (SF) (Standards Australia, 2010). The latter then being referred to as blended cements.

AS 3972 (2010) defines blended cement as a hydraulic cement containing Portland cement and a quantity comprised of one or both of the following:

a. Greater than 7.5% (by weight) of FA or BFS, or both.
b. Up to 10% SF.

Furthermore, AS 3972 (2010) designates cements as general purpose and special purpose. By definition, general purpose cements may be Portland cements (Type GP) or blended cements (Type GB). Special purpose cements are also defined in AS 3972 (2010) and may be Portland or blended cements with restrictions being placed on their composition.

4.3 ALKALINE ENVIRONMENT IN CONCRETE

As mentioned above, the highly alkaline environment in concrete results in the spontaneous formation of a passive iron oxide film on reinforcing steel.

The reaction of the cement compounds of Portland cement or blended cement with water results in the setting and hardening of the cement paste so that it binds the aggregate (coarse and fine) of the concrete together. A product of the hydration of Portland and blended cements is $Ca(OH)_2$ which forms together with NaOH and KOH.

As a result, the pH of the pore solution of sound concrete is normally in the range of 12–14 (Page & Treadaway, 1982; Tuutti, 1982; Tinnea & Young, 2000; Tinnea, 2002; Broomfield, 2007; Abd El Haleem et al., 2010; Ghods, et al., 2011), and maintained due to the CaO-$Ca(OH)_2$ pH buffer until this is overcome by, say, carbonation (referred to later), reduced by, say, leaching (also referred to later), or neutralised by hydrolysis triggered by chloride ion-induced pitting corrosion reactions (also referred to later).

For iron (steel) in a pH 12–14 aqueous alkaline environment, the Potential-pH diagram, see Figure 4.1, shows that the metal should be stable as an oxide. The stability of the said oxide layer thereby rendering the steel passive.

4.4 PHYSICAL BARRIER PROVIDED BY CONCRETE

In addition to providing a high pH passivating environment for reinforcing steel, concrete also provides a physical barrier against the ingress of corrosion-inducing substances.

The quality of concrete as a physical barrier may be assessed from penetrability data (and chemical data, i.e. blended cements have better chloride ion binding/adsorption capacity as well as improving the porosity of the aggregate-paste transition zone, referred to later). Penetrability data is collected from various measurements of a concrete's water absorption, water permeability, chloride diffusion and gaseous diffusion/penetration (O_2 and CO_2) (Concrete Institute of Australia, 2015). Some of these penetrability tests are discussed in more detail at Chapter 7.

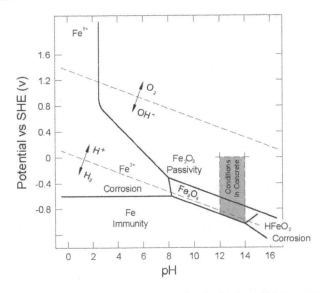

Figure 4.1 Potential-pH (Pourbaix) diagram for Fe-H$_2$O at 25°C and [Fe^{2+}], [Fe^{3+}] and [HFeO$_2^-$] = 10^{-6}M. (Courtesy of Green, 1991)

The factors affecting concrete penetrability include the type and quality of aggregates, type of cement (binder), cement (binder) content, water/cement (binder) ratio and production variables such as mixing uniformity, placement, degree of compaction, and adequacy of cure.

4.5 PASSIVITY AND THE PASSIVE FILM

4.5.1 Background

Corrosion is an electrochemical process. That is, the overall reaction consists of at least two sub-reactions (an oxidation reaction and a reduction reaction) that occur at different sites on the substrate/electrolyte interface. The oxidation or anodic reaction releases electrons and may be stylised as:

metal → oxidised state + electrons

The electrons released by this reaction are conducted through the metallic substrate and are consumed in the reduction or cathodic reaction which may be stylised as:

oxidant + electrons → reduced state

The two reaction sites are usually separated by distances that may vary between microns and many tens of millimetres, but because electronic currents have to flow between the sites through the substrate and ionic currents have to flow through the electrolyte the overall rate of reaction depends upon the ease with which these currents flow between the sites and the ease with which the oxidation and reduction reactions take place.

The corrosion of reinforcing steel in concrete is initially at least a very slow process because the highly alkaline (pH 12–14) concrete pore solution results in the formation of a passive iron oxide film.

4.5.2 Thermodynamics

Thermodynamics, the science of energy changes, is applied to corrosion studies to determine why a particular metal does or does not tend to corrode in a particular environment. The use of kinetics determines how rapidly the associated reactions occur.

Chemical and electrochemical thermodynamic data [chemical potential (μ_0) values, standard free energy change of reaction (ΔG_0) values, and standard electrode potential (E_0) values] provides the means for deciding which of a set of reactions is thermodynamically favoured and for predicting the most stable reaction products under specified conditions of electrode potential and solution composition.

Marcel Pourbaix recognised that the corrosion state of a metal could be represented by a point on a diagram the two axes of which were the potential and pH (Pourbaix, 1966). These diagrams are now often termed 'Pourbaix' diagrams.

The Pourbaix diagram plots electrochemical stability for different redox states of an element as a function of pH. The $Fe-H_2O$ system at 25°C, shown in Figure 4.1, is of most relevance to reinforcing steel in concrete. Here, the concentrations of dissolved ions, other than H^+ and OH^-, are taken to be $10^{-6}M$ and solid phases are assumed to be pure.

If the potential of iron immersed in neutral water (pH about 7) is measured using a standard hydrogen electrode (SHE), it will be seen in Figure 4.1 to have a value represented by, say −0.3 V. An examination of the Pourbaix diagram at Figure 4.1 suggests three possible means of achieving a reduction in the corrosion. First, the metal can have its potential so changed in the negative direction that it enters the domain of immunity, i.e. it can be cathodically protected. Secondly, the potential can be changed in the positive direction so that it enters the passivity domain. This is anodic protection. Thirdly, the pH of the electrolyte can be so adjusted that the metal enters the passivity domain.

To summarise, Figure 4.1 shows us that there are three basic zones of behaviour representing states of lowest energy for iron (steel):

1. Immunity – iron metal is thermodynamically stable and is immune to corrosion.
2. Corrosion – Fe^{2+}, Fe^{3+} and $HFeO_2^-$ ions are thermodynamically stable and corrosion will occur at a rate which cannot be predicted thermodynamically.
3. Passivity – iron oxides are thermodynamically stable. These oxides give rise to a condition termed passivity since significant corrosion may be stifled owing to the formation of a protective oxide layer on the metal surface (see shaded area on Figure 4.1 in the high pH region).

Having described the information that can be gleaned from Pourbaix diagrams with respect to deciding which of a set of reactions is thermodynamically favoured and which reaction products are most stable, it is necessary to recognise that certain limitations apply to the use of thermodynamics and Pourbaix diagrams. The kinetics (rates) of possible reactions are not considered, and therefore, it is impossible to predict whether a particular reaction, which is thermodynamically favoured, will occur at a significant rate in practice.

Further discussion on thermodynamics and how it applies to corrosion is provided at Chapter 5.

4.5.3 Kinetics

Discussion of kinetics and how it applies to corrosion is also provided in some detail at Chapter 5 suffice to say at this stage that Figure 4.2 shows the anodic polarisation curve for a passive metal, such as reinforcing steel in the alkaline environment of concrete.

At potentials more negative than the equilibrium potential or reversible potential for Fe/Fe^{2+} (i.e. E_o Fe/Fe^{2+}), iron is immune to corrosion or dissolution. Raising the potential to values more positive than E_o Fe/Fe^{2+} leads to active corrosion at a rate which initially increases with increasing overpotential (maximum corrosion current, I_3). At some potential more positive or noble than the equilibrium potential for the formation of a surface oxide film (i.e. E_o Fe/Fe_2O_3) active corrosion practically ceases (i.e. I_1) owing to surface oxide film formation which renders iron passive. Defects in that layer allows the passage of a small 'leakage current' (i.e. $I_1 \rightarrow I_2$). This 'leakage current' results from the necessity to reform the surface oxide passive film at

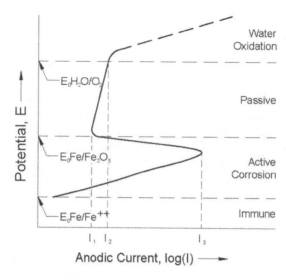

Figure 4.2 Anodic polarisation curve showing a transition from active dissolution to passivation. (Courtesy of Green, 1991)

points of local breakdown and to replace oxide lost by dissolution to the environment.

4.5.4 Film formation

As a result of the hydration reaction of the cement compounds as previously noted, the pore water surrounding the steel in concrete has a high pH (in the range 12–14) and so the steel is completely covered with a dense gamma ferric oxide (γ-Fe_2O_3, maghemite) protective passive film. This is formed by the anodic reaction:

$$4Fe(s) + 12OH^-(aq) \rightarrow 2\gamma Fe_2O_3(s) + 6H_2O(l) + 12e^-$$

Once the film has formed, the reaction continues at the metal/oxide inter-face as a result of hydroxyl ions diffusing through the film of iron oxide. The cathodic process is ascribed to the reaction:

$$3O_2(g) + 6H_2O(l) + 12e^- \rightarrow 12OH^-(aq)$$

This reaction takes place on the surface of the passive film. The cathodic process requires that oxygen molecules diffuse through the concrete to the passive film surface and are there reduced to hydroxyl ions. The electrical circuit is completed by the ionic transport through the passive film of the hydroxyl ions resulting from the cathodic reaction moving towards the metal/film interface and by electronic transport through the protective film to the film/electrolyte interface. Corrosion is proceeding in this state corresponding to the conversion of iron to iron oxide.

Passive (leakage) corrosion current density (i_{corr}) measured for steel reinforcement in undamaged concrete is typically of the order of 0.1 μA/cm^2 (Andrade & Gonzalez, 1978; Hansson, 1984). This current density therefore corresponds to an insignificantly slow corrosion (penetration) rate of 1 μm/year. This corrosion rate is of the order of 1 mm in 1,000 years and so it can be seen that the steel in a reinforced concrete structure or building element may be relatively unaffected by corrosion. The steel in this state is said to be 'passive' or protected by a 'passive film'.

4.5.5 Film composition

The precise nature of the passive film which is formed by the reaction of iron with the highly alkaline environment of the pore water in concrete varies. As above, it can be considered to be generally a dense gamma ferric oxide (γ-Fe$_2$O$_3$, maghemite).

The characteristics of oxides formed on steel under alkaline conditions and their electrochemical behaviour have been examined by, for example, Nasrazadani (1997), Cornell and Schwertmann (2000) and Freire et al. (2009). Iron-oxy-hydroxides like goethite (α-FeOOH), lepidocrocite (γ-FeOOH) and akagonite (β-FeOOH) below which more protective forms of iron oxides including magnetite (Fe$_3$O$_4$), maghemite (γ-Fe$_2$O$_3$) and haematite (α-Fe$_2$O$_3$) may be assumed to be the constituents of the passive film formed on steel electrodes under alkaline conditions.

Cohen (1978) indicated that the composition of the passive oxide layer on iron in alkaline aqueous solutions is a spinel α-Fe$_3$O$_4$-γ-Fe$_2$O$_3$ solid solution. Sagoe-Crentsil and Glasser (1989) attributed the passive action of pH>12 concrete pore solution to the formation of a surface layer of Fe$_2$O$_3$-Fe$_3$O$_4$. Ghods et al. (2011) also reported two-layer oxides in passive films, with an inner Fe^{2+} and Fe^{3+} layer and an outer pure Fe^{3+} layer. Al-Negheimish et al. (2014) have also proposed a two-layer oxide passive film but the top layer was composed of FeO and FeOOH with Fe$_2$O$_3$ close to the steel surface.

Ghods et al. (2013) report of a tri-layer model of oxides for the passive film on the steel when exposed to simulated concrete pore solution namely a layer of FeO at the metal surface, then a layer of Fe$_3$O$_4$ and then an outer layer of Fe$_2$O$_3$.

MnS inclusions are often found in steels and their effect is to change the Fe$_2$O$_3$ in the passive film from γ-Fe$_2$O$_3$ to α-Fe$_2$O$_3$ according to Al-Negheimish et al. (2014).

4.5.6 Film thickness

Singh and Singh (2012) proposed that the passive film formed in simulated concrete pore solution is of 'ultra-thin thickness' (i.e. <10 nm).

Ghods et al. (2013) reported that the thickness of the tri-layer (FeO/Fe_3O_4/Fe_2O_3) passive oxide film was uniform between 5 and 13 nm.

Al-Negheimish et al. (2014) have determined thicknesses of 6 to 7 nm for the passive film on steel in simulated concrete pore solution.

4.5.7 Models and theories

Veluchamy et al. (2017) indicate that the understanding of the passive state of metals started with Faraday (1844). Though several theories, models, and experimental works on passivity have been published in the literature, the mechanisms underlying the stability of the passive oxide over the metal still remains a mystery according to Veluchamy et al. (2017). They undertook a detailed review of theoretical and experimental results for the iron/electrolyte system invoking the high field model (ion-migration mechanism), modified high field model, point defect model, variants of the point defect model, diffusion Poisson coupled model, and the density functional theory based atomistic model. The experimental and model-predicted dependencies on applied voltage, pH, chloride and temperature have also been presented and discussed by Veluchamy et al. (2017).

4.6 REINFORCEMENT CORROSION

4.6.1 Loss of passivity and corrosion of steel in concrete

The passivity provided to steel reinforcement by the alkaline environment of concrete may be lost if the pH of the concrete pore solution falls because of carbonation or if aggressive ions such as chlorides penetrate in sufficient concentration to the steel reinforcement surface. Carbonation of concrete occurs as a result of atmospheric CO_2 gas (and atmospheric SO_x and NO_x gases) neutralising the concrete pore water (lowering its pH to 9) and thereby affecting the stability and continuity of the passive film. Leaching of $Ca(OH)_2$ (and NaOH and KOH) from concrete also lowers pH to allow corrosion of the steel reinforcement. Stray electrical currents, most commonly from electrified traction systems, and interference currents from cathodic protection systems, can also breakdown the passive film and cause of corrosion of steel reinforced and prestressed concrete elements.

Where the passive film is compromised, an anodic reaction is established of the form:

$$Fe \rightarrow Fe^{++} + 2e^-$$

Which involves dissolution of the iron. Although the passive film on the rest of the surface has a high ionic resistance it has a comparatively low electronic resistance and so the electrons liberated by this anodic reaction travel through the metal and the film to take part in the cathodic reaction:

$$\frac{1}{2}O_2 + H_2O + 2e^- \rightarrow 2OH^-.$$

This takes place on the unbroken passive film. In this stage, the cathodic, and anodic sites may be separated by millimetres and so this stage will be termed *minicell* corrosion. The circuit is completed by the passage of hydroxyl ions and ferrous ions through the water contained in the pores of the concrete. The hydroxyl ions have the greater mobility and so they meet the ferrous ions close to the anodic area and the rust forming reactions take place there:

$$Fe^{++} + 2(OH)^- \rightarrow Fe(OH)_2$$

$$3Fe(OH)_2 + H_2O \rightarrow Fe(OH)_3 + H_2O$$

$$2Fe(OH)_3 + Fe(OH)_2 \rightarrow Fe_3O_4 + 4H_2O$$

$$2Fe(OH)_3 \rightarrow Fe_2O_3 + 3H_2O$$

The corrosion mechanisms by chlorides, carbonation, leaching, stray, and interference electrical currents are provided in subsequent sections but discussion of the different forms of corrosion, composition of reinforcing steel corrosion products (rusts) and the consequence of the corrosion products is provided herewith.

4.6.2 Uniform (Microcell) corrosion and pitting (Macrocell) corrosion

According to the different spatial location of anodes and cathodes, corrosion of steel in concrete can occur in different forms (Elsener, 2002):

- As *microcells*, where anodic, and cathodic reactions are immediately adjacent, leading to uniform steel (iron) dissolution over the whole surface. This uniform (or general) corrosion is typically caused by carbonation of the concrete or by very high chloride content at the steel reinforcement.
- As *macrocells*, where a net distinction between corroding areas of the steel reinforcement (anodes) and non-corroding passive surfaces (cathodes) is found. Macrocells occur mainly in the case of chloride-induced corrosion (pitting) where the anodes are small with respect to the total (passive) steel reinforcement surface.

As previously, the description *minicell* corrosion is also proposed for uniform (or general) corrosion of steel reinforcement in concrete where the cathodic and anodic sites may be separated by millimetres (rather than microns).

4.6.3   Corrosion products composition – chloride-induced corrosion

The exact nature of reinforcing steel corrosion products (rusts) associated with chloride-induced corrosion varies depending on conditions. Corrosion products are of various layers and of various compositions. Generally speaking, they will be layers, and combinations of ferrous, and ferric hydroxides, hydroxyl/oxides and oxides each with possible varying degrees of hydration. $Fe(OH)_2$ is only chemically stable in acidic, deoxygenated, conditions that develop within propagating pits. $Fe(OH)_3$ is similarly only chemically stable in acidic and deoxygenated conditions. Fe_3O_4 or magnetite (black rust) could be expected to form under limited oxygen diffusion conditions associated with totally immersed or frequently wetted structures. Hydrated Fe_2O_3 or haematite (red rust) typically forms under conditions of good oxygen access.

Melchers and Li (2008) and Pape and Melchers (2013) have identified goethite (α-FeOOH), akagenite (β-FeOOH), lepidocrocite (γ-FeOOH) as corrosion products within chloride contaminated reinforcing steel samples from concrete structures. 'Green rusts' can also occur. Melchers and Li (2008) and Pape and Melchers (2013) have identified compounds such as iron oxide chloride (FeOCl), hibbingite (α-Fe$_2$(OH)$_3$Cl), iron chloride hydrate ($FeCl_2.4H_2O$), 'Green Rust I' (carbonate variety) and 'Green Rust II' (sulphate variety) within the corrosion products (rusts) of samples taken from steel reinforced and prestressed concrete structures in chloride-rich environments.

For corroding steel in mortar specimens subject to chloride ingress, Koleva et al. (2006) identified corrosion products mainly consisting of highly crystallised goethite (α-FeOOH), lepidocrocite (γ-FeOOH) and akagenite (β-FeOOH). Depending on the ratio of iron and chloride ions, the iron oxychlorides and iron oxyhydroxides present different morphologies and exert different influences on steel/mortar interface microstructure.

The corrosion products generally detected by Vera et al. (2009) from embedded steel in concrete cylinder samples exposed to simulated marine and industrial conditions and a natural marine atmospheric environment were: lepidocrocite (γ-FeOOH), goethite (α-FeOOH) and magnetite (Fe_3O_4); but in the chloride contaminated environments the presence of akagenite (β-FeOOH)was also detected; and, in the natural marine atmospheric environment the formation of siderite ($FeCO_3$) was also observed.

4.6.4 Corrosion products composition – carbonation-induced corrosion

In terms of carbonation-induced corrosion products, Kolio et al. (2015) on studies performed on existing reinforced concrete facades of 12 buildings of age 30–43 years, identified corrosion products that were mostly hydroxide type of rusts (i.e. goethite/α-FeOOH and lepidocrocite/γ-FeOOH).

Huet et al. (2005) identified corrosion products mainly composed of magnetite (Fe_3O_4) and lepidoocrocite (γ-FeOOH) in laboratory based studies of mild steel in carbonated concrete pore solution.

4.6.5 Corrosion products development – visible damage

The process of reinforcement corrosion often leads to corrosion products which will occupy a greater volume than the iron dissolved in its production, refer Figure 4.3 (Jaffer & Hansson, 2009). Furthermore, when the corrosion products become hydrated the volume increase is even greater (Broomfield, 2007), refer hydrated haematite (α-Fe_2O_3.$3H_2O$) (red rust) in Figure 4.3.

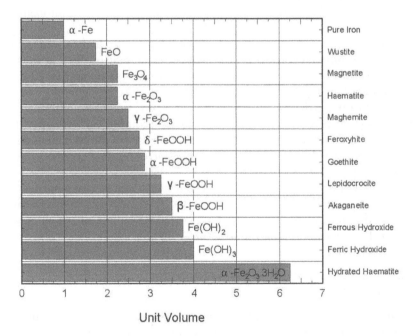

Figure 4.3 Relative volume of iron and some of its oxides. (Courtesy of Jaffer & Hansson, 2009)

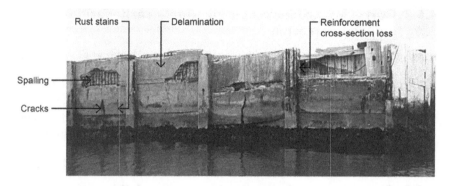

Figure 4.4 Reinforcement corrosion-induced deterioration.

The consequence of higher volume corrosion products is then to develop tensile stresses in the concrete covering the reinforcement. Concrete, being weak in tension, will crack as a consequence of the corrosion. Continuing formation of corrosion product(s) will enhance the expansion which ultimately leads to cracked pieces of concrete cover detaching, leading to delamination, and then spalling. Rust staining of the concrete may or may not occur together with the cracking or as a prelude to delamination and spalling. Section loss, bond loss, and anchorage loss of the reinforcement also occurs as a result of the corrosion.

Kolio et al. (2015) for example, describing work they had carried out on carbonation initiated corrosion on concrete façade panels, determined that the corrosion products (i.e. as previously discussed, goethite/α-FeOOH and lepidocrocite/γ-FeOOH) had a unit volume increase of roughly 3 times the volume of iron. Furthermore, they identified by taking into account the relative volume of the corrosion products that the required corrosion penetration (metal loss) to initiate visually observable cracks in the studied building façade panels was by average 54 µm with a corresponding corrosion (rust) products thickness of 112 µm.

Hydrated haematite (α-Fe$_2$O$_3$.3H$_2$O) (red rust) induced cracking, delamination and spalling of cover concrete can be frequently seen in deteriorated reinforced concrete structures and buildings, an example of which is provided at Figure 4.4 for a marine structure.

4.6.6 Corrosion products development – no visible damage

It is most important to note, however, that not all corrosion of reinforcement leads to rust staining, cracking, delamination, or spalling of cover concrete. Localised pitting, localised corrosion at cracks, localised corrosion at concrete defects, etc can result in marked section loss (loss of bond, loss of anchorage) and ultimately structural failure without the visible

Figure 4.5 Example of marked localised pitting (macrocell) corrosion and section loss (corrosion penetration) of steel reinforcement. Here pits would have started out narrow but with time have coalesced to result in section loss over a greater (anodic) area.

consequences of corrosion on the concrete surface, i.e. no rust staining, cracking, delamination, or spalling of cover concrete (Green et al., 2013). An example of marked localised section loss of reinforcement due to chloride-induced pitting (macrocell) corrosion is provided at Figure 4.5. The definition of macrocell corrosion being as at Section 4.6.2 where a net distinction between corroding areas of the steel reinforcement (anodes) and non-corroding passive surfaces (cathodes) is found.

Elsener (2002) pointed out that macrocell (pitting) corrosion is of concern because the local dissolution rate (reduction in cross-section of the carbon steel, prestressing, and post-tensioned steel reinforcement, loss of bond, loss of anchorage) may be accelerated due to the cathode/anode area ratio. Indeed, values of local corrosion (penetration) rates of up to 1 mm/year have been reported for bridge decks, sustaining walls or other chloride contaminated steel reinforced concrete structures according to Elsener (2002) because of cathode/anode area ratio effects.

Broomfield (2007) also notes that 'black rust' or 'green rust' (due to the colour of the liquid when first exposed to air after breakout) corrosion can also be found under damaged waterproof membranes and in some underwater or water saturated structures. He states that it is potentially dangerous as there is no indication of corrosion by cracking and spalling of the concrete and the reinforcing steel may be severely weakened before corrosion is detected. Reinforcement bars may also be 'hollowed out' in such deoxygenated conditions according to Broomfield (2007).

4.7 CHLORIDE-INDUCED CORROSION

4.7.1 General

It is known that chloride ions in sufficient concentration (Broomfield, 2007) can locally compromise the passivity of carbon steel, prestressed, and post-tensioned steel reinforcement in concrete leading to pitting corrosion. Pitting corrosion is localised accelerated dissolution of metal that occurs as a result of breakdown of the otherwise protective passive film on the metal surface (Frankel, 1998). Various mechanisms of chloride ion-induced corrosion of steel reinforcement in concrete are proposed in the literature.

Fundamental studies of pitting corrosion on engineering alloys have typically focused on the various stages of the pitting process, namely: passive film breakdown, the growth of metastable pits (which grow to about the micron scale and then repassivate) and the growth of larger, stable pits (Frankel, 1998).

Like Frankel (1998), Angst et al. (2011) propose from a fundamental viewpoint that chloride-induced pitting corrosion of steel reinforcement in concrete can be considered to occur as a sequence of distinct stages: *pit nucleation*, *metastable pitting* (including pit death) and *stable pit growth*.

4.7.2 Passive film breakdown/pit initiation

The American Concrete Institute (ACI) Committee 222 on 'Corrosion of Metals in Concrete' (1985) proposes three theories to explain the effects of chloride ions on steel reinforcement:

a. The Oxide Film Theory – postulates that chloride ions penetrate the passive film through pores or defects in the film, thereby 'colloidally dispersing' the film.
b. The Adsorption Theory – postulates that chloride ions are adsorbed preferentially on the metal surface in competition with dissolved oxygen or hydroxyl ions. The chloride ion promotes the hydration of the metal atoms, thus facilitating anodic dissolution.
c. The Transitory Complex Theory – postulates that chloride ions form a soluble complex with ferrous ions which then diffuses away from the anodic sites. Thereafter, the complex decomposes, iron hydroxide precipitates, and the chloride ions are released to transport more ferrous ions from anodic sites.

Frankel (1998), in a comprehensive review paper of pitting corrosion of metals, indicates that theories for passive film breakdown and pit initiation of pure metal systems (i.e. not alloys where pitting may also be associated with inclusions or second phase particles) have been categorised in three main mechanisms that focus on passive film penetration, film adsorption (and thinning) and film breaking, refer to Figure 4.6.

Soltis (2015) on the other hand, after a review of the above three main mechanisms that focus on passive film penetration, adsorption, and breaking, together with mechanisms such as the 'percolation model' and 'voids at the metal-oxide interface', proposes that despite a number of theories for passive film breakdown/pit initiation, this aspect of localised pitting

a) Penetration Mechanism

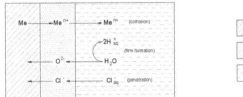

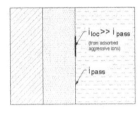 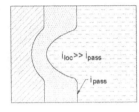 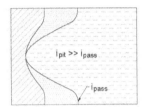

b) Adsorption Mechanism

c) Film Breaking Mechanism

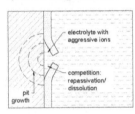

Figure 4.6 Schematic diagrams representing pit initiation by (a) penetration, (b) adsorption and thinning, and (c) film breaking. (Courtesy of Frankel, 1998)

corrosion remains still the least understood, with no generally accepted theory. Soltis (2015) then surmises that it is entirely possible that there is no single theory and the development of system-specific models are necessary because of differences in passive film properties.

Given the Soltis (2015) proposition that there may not be a single theory for passive film breakdown and pit initiation and that development of system-specific models may be necessary, in the remainder of this sub-section, literature has been reviewed with the aim of proposing a system-specific mechanism(s) for chloride ion-induced depassivation and pit initiation of reinforcement steel in concrete.

Bird et al. (1988) in laboratory based solution studies in the pH range 10–14 and containing NaCl in the range 10^{-3}–10^0 M, supports the idea that the first stages of localised passive film breakdown depend upon the competition for adsorption, at the oxide/liquid interface, between hydroxyl ions, and aggressive ions such as chlorides.

Leek and Poole (1990) for steel in concrete suggest that the breakdown of passivity by chloride ions is achieved by disruption of the passive film to substrate adhesive bond. An initial bond breakdown occurs beyond which internal pressure within the film due to surface tension effects, acts to disband chemically unaltered film, thereby expanding the size of the site of depassivation. Little or no chemical dissolution of the passive film occurs in this model.

Sagoe-Crentsil and Glasser (1990) for steel in chloride contaminated concrete show evidence for the 'Transitory Complex Theory'. X-ray diffraction analysis of steel from chloride contaminated cement paste samples showed the formation of an intermediate, soluble, $FeCl_2.4H_2O$ complex. Visually, they advise this was evident as a greenish-blue hue on the metal surface.

As noted previously, Pape and Melchers (2013) identified FeOCl, α-$Fe_2(OH)_3Cl$, $FeCl_2.4H_2O$, 'Green Rust I' (carbonate variety) and 'Green Rust II' (sulphate variety) compounds within the corrosion products of samples from reinforced and prestressed structures in chloride environments. It is thus proposed that the presence of 'green/greenish-blue rusts' may be evidence of the 'Transitory Complex Theory' above.

Ghods et al. (2013) state that the underlying mechanism for the depassivation process by chloride ions for steel in concrete is still not fully understood. They propose that one of the reasons behind this lack of understanding is the unavailability of data on the compositional and nanoscale morphological characteristics of the passive oxide film before and after exposure to chlorides. Furthermore, they advise since traditional electrochemical and most microscopic techniques do not provide such specific data, nanoscale surface characterisation techniques are required. Although such techniques have been widely used to characterise metal and metal alloy oxides, their application to study carbon steel and stainless steel in alkaline environments that are representative of concrete pore solution has been limited (Ghods et al., 2013).

In the study by Ghods et al. (2013), nanoscale transmission electron microscopy (TEM) samples prepared with a focused ion beam were used to investigate the passivity and depassivation of carbon steel reinforcement in simulated concrete pore solutions. It was found that the addition of chlorides at concentrations lower than the depassivation thresholds did not change the physical appearance of the passive oxide films. After exposure to chlorides in concentrations greater than depassivation thresholds, passive oxide films were no longer uniform, with some regions of the surface bare and some pit initiation sites formed on the surface of the steel.

Ogunsanya and Hansson (2020) suggest that the many mechanistic models of pitting can be summarised in two main groups. For the first group, it is proposed that there is a competition for anionic sites in the passive film in which chlorides are exchanged with hydroxide ions, causing local thinning, or dissolution. A second group proposed a mechanical breakdown of passive films by chloride ions, resulting from their penetration through defective sites or low energy areas (such as grain boundaries in the passive film). The occurrence of one model or the other will be determined predominantly by the nature (stoichiometry, defect density, composition, etc) of each passive film.

It is proposed that the study by Ghods et al. (2013) is evidence for both or either the 'Adsorption Theory' or 'Transitory Complex Theory' for chloride ion-induced passive film breakdown of steel reinforcement in concrete.

However, based on the majority of the above, it is hereby proposed that the 'Transitory Complex Theory' is the dominant mechanism for the chloride ion-induced depassivation/pit initiation of steel reinforcement in concrete and that the 'Adsorption Theory' may also be a contributing mechanism.

4.7.3 Metastable pitting

In fundamental terms, Frankel (1998) advises that metastable pits on an otherwise passive surface are pits that initiate and grow for a limited period before repassivating. For metals in solutions, pits can stop growing for a variety of reasons, but metastable pits are typically considered to be those from micron size at most with a lifetime of the order of seconds or less.

Angst et al. (2011) indicate for chloride-induced reinforcement corrosion that once a pit has formed, the anodic dissolution reaction has to be sustained, otherwise the pit might cease growing and repassivate within a short time. Nucleation/repassivation events occur during the phase of so called metastable pitting.

Furthermore, Angst et al. (2011) propose that it is generally accepted that many of the nucleated pits never achieve stability and that for a pit to survive the metastable state, it is necessary that the solution within the pit cavity (anolyte) maintains an aggressive chemical composition. This requirement is met by hydrolysis of the dissolving iron cations and the resulting increase in H^+ ion concentration as well as migration of chloride and other anions into the pit to maintain charge neutrality. As long as dilution of the aggressive pit chemistry is prevented, pit growth is self-sustaining and likely to achieve stability.

This mechanism of localised acidification due to metal ions hydrolysis is attributed to Galvele (1976). Newman (2010) then highlights that definition of the pit propagation process by the localised acidification model has since had an enormous influence on the development of corrosion science.

Lin et al. (2010) utilised a scanning-micro reference electrode (SMRE) technique to monitor localised corrosion processes of reinforcing steel in NaCl containing solution by in situ imaging of the corrosion potential and corrosion current. Metastable micro pits occurred randomly and instantaneously on the surface when the reinforcing steel was first immersed in the chloride containing solution. The number and size of active micro pits changed with immersion time; some micro pits lost their activity and some micro pits gathered into groups to form visible corrosion holes that were covered by yellow-brown corrosion products.

Furthermore, whether metastable pitting nuclei developed into macro corrosion pits or lost their corrosion activity, was closely associated with the local chloride concentration. Once stable pitting nuclei were formed, further hydrolysis reactions of Fe ions continued to lower the local pH, which accelerated dissolution of Fe at pits (Lin et al., 2010).

A porous cover of corrosion products and remnants of the passive film have also been reported for carbon steel and iron in chloride solutions (Alvarez & Galvele, 1984). Loss of this pit cover it is proposed will result in dilution of the anolyte and might stop anodic dissolution. The pit can survive rupture of the cover and continue to grow in a stable manner only if the pit geometry has already developed in such a way that the pit depth sufficiently limits transport of ionic species to and from the pit bottom, in other words: that the pit geometry (transport path) imposes a resistance against diffusion and migration (Angst et al., 2011).

4.7.4 Pit growth/pit propagation

4.7.4.1 General

Green (1991) has proposed that for chloride-induced pit growth to be sustained for reinforcing steel in concrete, the following conditions must be maintained:

a. A sufficient concentration of chloride ions at the pit.
b. The recycling of chloride ions during the corrosion process (i.e. from the hydrolysis of the intermediate iron chloride/oxy-chloride complexes).
c. Diffusion of chloride ions to the pit from the bulk pore solution.
d. Development of acidity within the pit.
e. Cathodic processes on the steel surface.
f. A continuous electrolytic path between cathodic sites and the pit.

Angst et al. (2011) advise that studying the transition from nucleation to stable pit growth for the case of reinforcement steel embedded in concrete is much more difficult owing to experimental reasons. For instance, the reference electrode cannot be placed as close to the pit as in the case of solution experiments, or the IR drop through the concrete might disturb electrochemical measurements. Also, visual examination of an electrode embedded in concrete is impossible during an experiment. In addition, concrete is an inhomogeneous material and thus local chemistry is not as well defined as in solutions. Corrosion kinetics are also clearly different for steel embedded in concrete compared with experiments in solution.

Angst et al. (2011) proposed, however, that in principle the localised pit growth/propagation mechanisms valid for stainless or carbon steel in solutions containing chlorides can be expected to apply also for reinforcement steel in concrete, namely: acidification of the anolyte and migration of chloride ions into the pit are required to maintain the aggressive local pit chemistry, which in turn is required to sustain anodic dissolution. In concrete, convection as a mechanism that promotes dilution of the anolyte can be neglected when compared with the situation of a metal surface directly exposed to bulk solution. However, considering the presence of soluble alkaline cement (binder) hydration products (i.e. NaOH, KOH, and $Ca(OH)_2$) providing a pH buffering capacity, local acidification can be considered a critical step (Angst et al., 2011). Particularly lime/portlandite $(Ca(OH)_2)$ segregated at the steel/concrete interface, and its ability to retain a fall in pH, as it has been suggested as a major reason for the inhibitive properties of concrete (Page, 1975). As long as OH- ions migrate preferentially into the pit cavity to maintain charge balance, they will inhibit the pH of the anolyte from falling too fast.

Alvarez and Galvele (1984) reported that for iron in alkaline solutions, acidification of the anolyte is the rate controlling step; once a low pH in the pit is reached, the behaviour becomes equal to that found in acid solutions where chloride accumulation in the pit is the rate limiting process.

4.7.4.2 Chemical conditions within propagating pits

Schematic diagrams of the reactions that occur within a propagating chloride-induced pit for steel reinforcement in chloride contaminated concrete are provided at Figures 4.7 and 4.8.

The chloride ions enter the passive film (via predominantly the 'Transitory Complex Theory') and give rise to soluble products such that a small bare

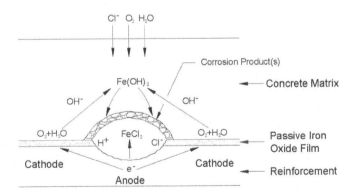

Figure 4.7 Schematic diagram of reactions associated with chloride-induced pit growth/pit propagation. (Courtesy of Treadaway, 1988)

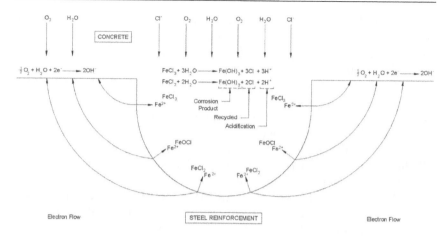

Figure 4.8 Schematic diagram of reactions within a propagating chloride-induced pit. (Courtesy of Treadaway, 1991)

area of metal surface is formed. Ferrous ions (Fe^{2+}) diffusing away from the anodic area will react with the high pH environment to form rust product which will shield the bare metal from oxygen. Further Fe^{2+} ions will now no longer be able to be oxidised by the dissolved oxygen and so may be hydrolysed, following the reaction:

$$3Fe^{2+} + 4H_2O \rightarrow Fe_3O_4 + 8H^+ + 2e^-$$

The production of H^+ ions in the incipient pit will reduce the pH. Tinnea (2002) has in fact measured reductions in pH to below 4 within pits for conventional steel reinforcement in chloride contaminated concrete.

A charge imbalance is also then generated by the H^+ ions production so that Cl^- ions will enter the pit which will now contain iron chloride (including the corrosive $FeCl_3$), iron oxy-chloride and iron chloro-hydroxyl compounds (Treadaway, 1991).

The rate of pitting corrosion is very large (locally) because the portion of the metal covered by the protective passive film is in general very much larger than the portion of the metal where the passive film has been damaged. Macrocell corrosion takes place where the area over which the reduction of oxygen takes place is much larger than the anodic area where metal is dissolved. Consequently, as the metal loss that has to balance the oxygen reduction must all take place in the localised area where the passive film has been damaged, the rate of penetration of the metal in that local area is very fast and rapid pitting is observed. As previously noted, values of pitting (penetration) rates of up to 1 mm/year have been reported for bridge decks, sustaining walls or other chloride contaminated steel reinforced concrete structures according to Elsener (2002).

4.7.5 Reinforcing steel quality

4.7.5.1 Metallurgy

Angst et al. (2011) indicate that the metallurgy (physical surface heterogeneities such as inclusions, lattice, or mill-scale defects, flaws, and grain boundaries) of the steel reinforcement also impacts the nucleation of chloride-induced pitting.

Bhandari et al. (2015) note that for engineering alloys, pits almost always initiate due to chemical or physical heterogeneity at the surface such as inclusions, second phase particles, solute-segregated grain boundaries, flaws, mechanical damage, or dislocations.

4.7.5.2 Defects

Lin et al. (2010) have undertaken work to image the chloride concentration on the surface of reinforcing steel, and to further study the effects of interfacial chloride ions on pitting corrosion in reinforcing steel by combining electron microprobe analyser (EMPA) ex situ imaging of the morphology and chemical composition of reinforcing steel at the same locations. Both the electrochemical inhomogeneities in the reinforcing steel and the non-uniform distribution of chloride at the surface were considered the most important factors leading to various localised attacks. They also developed a scanning-micro reference electrode (SMRE) technique to monitor localised corrosion processes of reinforcing steel in NaCl containing solution by in situ imaging of the corrosion potential and corrosion current. Combining the data of the SMRE technique with that of the EMPA measurements they confirmed that among the electrochemical inhomogeneities (defects) in reinforcing steel, MnS inclusions play a leading role in the initial corrosion because chloride prefers to adsorb and accumulate at the MnS inclusions, resulting in pitting corrosion.

Accordingly, dissolution of the MnS inclusion is generally accompanied by a local drop in pH around the MnS inclusion, as follows (Lin et al., 2010):

$$MnS(s) + 3H_2O(l) \rightarrow Mn^{2+}(aq) + HSO_3^-(aq) + 5H^+(aq) + 6e^-$$

Once a local reduction in pH and increase in chloride concentration is reached, nucleating pits can initiate, and form occluded pits, such that H^+, Cl^-, and HSO_3^- can accumulate inside pits. As previous, the metastable pitting nuclei either developed into corrosion pits or lost their corrosion activity. Furthermore, if stable pitting nuclei were formed, further hydrolysis reactions of Fe ions continued to lower local pH, which accelerated dissolution of Fe at pits (Lin et al., 2010).

4.7.6 Chloride threshold concentrations

It is known that chloride ions in sufficient concentration can destroy the passivity of steel reinforcement in concrete, thereby leading to corrosion initiation and propagation. The concentration of chloride ions that needs to arrive at the steel surface to destroy passivity is often termed the 'threshold' level/concentration, 'critical chloride level/concentration' or 'critical chloride content' (C_{crit}).

However, there is no universally accepted threshold concentration (and nor can there be) due to many factors (some of which are interrelated) including (Angst et al., 2009):

- Steel-concrete interface.
- Concentration of hydroxide ions in the pore solution (pH).
- Electrode potential of the steel.
- Binder type.
- Surface condition of the steel.
- Moisture content of the concrete.
- Oxygen availability at the steel surface.
- Water/binder ratio.
- Electrical resistivity of the concrete.
- Degree of hydration.
- Chemical composition of the steel.
- Temperature.
- Chloride source (mixed-in initially or penetrated into hardened concrete).
- Type of cation accompanying the chloride ion.
- Presence of other species, e.g. inhibiting substances.

Chess and Green (2020) point out that with all these variables it is perhaps not surprising that there would be a very significant range in the chloride threshold level at which corrosion is initiated. For laboratory studies these ranged from 0.04–8.4% chloride by weight of cement which is a 21,000% difference (Angst et al., 2009). This huge difference was also found in real structures by Angst et al. (2009) where the range was 0.1% to 1.95% chloride by weight of cement which is a 1,950% difference.

Angst et al. (2017) comment that it is well known that the onset of corrosion of unpolarised steel embedded in concrete might take place over a long period of time rather than a well defined instant. Steel may start corroding at relatively low chloride concentrations but if these are not able to sustain the corrosion process, repassivation will occur, which becomes apparent by a potential increase back to the initial passive level. Such depassivation-repassivation events have been observed in some studies. The chloride concentration measured at a time of stable corrosion is therefore more relevant in practice than the time at which the very first signs of potential deviations from the passive level become apparent and that C_{crit} should represent the chloride concentration at which corrosion initiates and also stably propagates.

Cao et al. (2019) have undertaken an updated review on C_{crit} which included the extensive body of literature published in Chinese since the 1960s. Some of the conclusions from this recent literature review include:

- There is a wide scatter in reported C_{crit} with values scattering by > 2 orders of magnitude. This is in agreement with previous literature reviews that mostly considered European and North American publications.
- After so many years and tremendous research efforts aimed at finding C_{crit}, the time may have come to critically question this concept. The wide variability in the literature suggest that a unique C_{crit} does not exist and that C_{crit} cannot be predicted as a function of parameters such as w/b ratio, binder type, steel surface condition, etc. Instead, Cao et al. (2019) suggest to determine C_{crit} experimentally for the actual systems of interest, considering both the concrete, and the rebars with their actual properties.
- A practice-related method is proposed for determining C_{crit} for actual structures (Angst et al., 2017).

Angst et al. (2017) note that while it is well known that the ability of a structure to withstand corrosion depends strongly on factors such as the materials used or the age, it is common practice to rely on threshold values stipulated in standards or textbooks. These threshold values for corrosion initiation (C_{crit}) are independent of the actual properties of a certain structure, which clearly limits the accuracy of condition assessments and service life predictions. The practice of using tabulated values can be traced to the lack of reliable methods to determine C_{crit} on-site and in the laboratory.

Angst et al. (2017) then propose an experimental protocol to determine C_{crit} for individual engineering structures or structural members. A number of reinforced concrete samples are taken from structures and laboratory corrosion testing is performed. The main advantage of this method is that it ensures real conditions concerning parameters that are well known to greatly influence C_{crit}, such as the steel-concrete interface, which cannot be representatively mimicked in laboratory-produced samples. At the same time, the accelerated corrosion test in the laboratory permits the reliable determination of C_{crit} prior to corrosion initiation on the tested structure; this is a major advantage over all common condition assessment methods that only permit estimating the conditions for corrosion after initiation, i.e. when the structure is already damaged.

The Angst et al. (2017) protocol yields the statistical distribution of C_{crit} for the tested structure. This serves as a basis for probabilistic prediction models for the remaining time to corrosion, which is needed for maintenance planning. Angst et al. (2017) propose that this method can potentially be used in material testing of civil infrastructures, similar to established methods used for mechanical testing.

Recently Green et al. (2019) report the most interesting findings of the condition survey of a 40+ year marine structure in Eastern Australia where the chloride levels at the depth of reinforcement within the reinforced concrete slab elements and reinforced concrete encased steel beam elements of some Dolphin structures are very high at ~0.32% to 0.69% (by weight of concrete); and that the electrode potential gradients and absolute values of potential, do not indicate marked corrosion associated with such very high chloride contents; and that the reinforcement condition at breakouts in sound concrete, also did not indicate marked corrosion and section loss that would be expected given such high chloride levels; and that the electrical resistivity of the concrete (at typically between 18,000 and 44,000 ohm.cm) is not abnormally high.

Historically, a threshold value as simplistic as 0.4% chloride (by weight cement) or 0.06% chloride (by weight concrete) has been utilised (Browne, 1982; Bamforth, 2004; Broomfield, 2007).

In terms of prestressed/post-tensioned steel in GP (Portland) concrete, a simplistic threshold chloride concentration has sometimes been assumed to be 0.04% chloride by weight of concrete (Stark, 1984).

For hot dipped galvanised steel reinforcement, the threshold level can be more than double that of the simplistic 0.4% by weight cement (0.06% by weight concrete) value for carbon steel, refer Chapter 11.

For stainless steel reinforcement, the threshold level is simplistically considered to be 5–10 times that of carbon steel depending on the type (alloy) of the stainless steel (Concrete Institute of Australia, 2015), refer also Chapter 11.

4.7.7 Chloride/hydroxyl ratio

It has already been mentioned that many factors (15 number at least) influence whether initiation of corrosion to reinforcement occurs. One of those factors is the ratio of the chloride ion concentration to the hydroxyl ion concentration in the immersing solution.

If the passivation reaction is written:

$$2Fe^{++} + 6OH^- \rightarrow Fe_2O_3 + 3H_2O + 2e^-$$

and the pitting reaction is written:

$$Fe_2O_3 + 6NaCl + 3H_2O \rightarrow 2Fe^{3+} + 6Cl^- + 6NaOH$$

it can be seen that the onset of pitting is governed by the ratio of the chloride ion concentration to the hydroxyl ion concentration in the immersing solution.

The inhibiting effect of hydroxide ions against chloride-induced corrosion as a factor influencing chloride threshold values was recognised in the early research of Venu et al. (1965), Hausmann (1967) and Gouda (1970).

Hausmann (1967) suggested the famous Cl^-/OH^- ratio of 0.6 based on probability considerations. Figure 4.9 indicates that this value is more a lower boundary than a clear threshold value (Angst et al., 2009).

Angst and Vennesland (2009) report that there is high overall scatter in reported Cl^-/OH^- ratios, however, ranging from 0.03 to 45 and thus over three orders of magnitude. They comment that remarkably higher chloride threshold values have been reported when mortar or concrete specimens are investigated in comparison with solution experiments. This then might be explained by the inhibiting effects of the interface of steel embedded in a cement matrix due to the formation of a portlandite $(Ca(OH)_2)$ layer at the steel surface.

Angst and Vennesland (2009) also discuss studies that show that the Cl^-/OH^- ratio increases with increasing pH of the pore solution, i.e. it is no constant value and the inhibiting effect of OH^- is stronger at higher pH values.

Cao et al. (2019) indicate that the reported threshold $[Cl^-]/[OH^-]$ ratios also scatter over a wide range of anywhere between 0.3 – 20 and so cannot be relied upon as an indicator of reinforcement corrosion initiation.

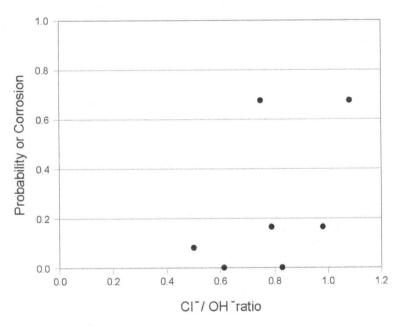

Figure 4.9 Results from Hausmann (1967) showing the probability of corrosion (number of corroding rods in a set of 12) vs. the Cl^-/OH^- ratio. (Courtesy of Angst et al., 2009)

Cao et al. (2019) also note that:

- Most of the [Cl-]/[OH-] ratios values in the literature were obtained from steel corrosion tests in simulated pore solutions, which presents conditions different from the case of steel embedded in concrete. This is mainly because of the ability of certain cement hydration phases to buffer the pH, and because of the relevance of ion transport processes, that are different in porous media from bulk solutions.
- Since it is difficult to test the pH value of concrete pore solution, especially at the concrete-steel interface and in a non-destructive manner, more investigations are needed to get a reliable critical [Cl-]/[OH-] ratio for steel bars long embedded in concrete.

4.8 CARBONATION-INDUCED CORROSION

The passivity provided to steel reinforcement by the alkaline environment of concrete can be lost due to a fall in the pH of the concrete pore solution because of carbonation and thereby damaging the passive film.

Carbonation may occur when carbon dioxide gas (and atmospheric SO_x and NO_x gases) from the atmosphere dissolves in concrete pore water and penetrates inwards or when the concrete surface is exposed to water or soil containing dissolved carbon dioxide, refer Section 2.4.8.

Carbon dioxide dissolves in the pore water to form carbonic acid by the reaction:

$$CO_2(g) + H_2O(l) \rightarrow H_2CO_3(aq)$$

Carbonic acid can dissociate into hydrogen and bicarbonate ions. The carbonic acid reacts with $Ca(OH)_2$ (portlandite) in the solution contained within the pores of the hardened cement paste to form neutral insoluble $CaCO_3$. The general reaction is as follows:

$$Ca(OH)_2(aq) + H_2CO_3(aq) \rightarrow CaCO_3(s) + 2H_2O(l)$$

And the net effect is to reduce the alkalinity of the pore water which is essential to the maintenance of a passive film on any reinforcing steel that may be present. While there is a pH buffer between CaO and $Ca(OH)_2$ that keeps the pH at approximately 12.5, the CO_2 keeps reacting until the buffer is consumed and then the pH will drop to levels no longer protective of steel.

As at Section 2.4.8, the attack of buried concrete by carbon dioxide dissolved in the groundwater is a two-stage process. The calcium hydroxide solution that fills the pores of the concrete, first reacts with dissolved carbon dioxide to form insoluble calcium carbonate. However, it then subsequently reacts with further carbon dioxide to form soluble calcium bicarbonate which is leached from the concrete. The extent to which each process takes place is a function of the calcium carbonate/calcium bicarbonate concentration of the ground water (which in turn is a function of the pH and the calcium content) and the amount of dissolved carbon dioxide.

The iron oxide passive film formed on steel in alkaline conditions is stable at pH levels greater than 10 (Broomfield, 2007) or greater than around 9.5 (Savija & Lukovic, 2016). Since the process of the carbonation of concrete lowers the pH of concrete pore water to lower than 9 (as low as 8.3) (Savija & Lukovic, 2016), the passive film is simply dissolved. The steel surface is then exposed to a pH~9 environment and corrosion thereby propagates.

Carbonation of concrete causes general corrosion where anodic and cathodic reactions are immediately adjacent (microcells or minicells), leading to uniform steel (iron) dissolution over the whole surface.

Localised carbonation-induced corrosion of steel reinforcement can, however, occur at cracks, concrete defects, low cover areas etc of reinforced concrete elements.

4.9 LEACHING-INDUCED CORROSION

The passivity provided to steel reinforcement by the alkaline environment of concrete may also be lost if the pH of the concrete pore solution falls because of leaching of $Ca(OH)_2$ (and NaOH and KOH). Leaching from concrete can lead to a lowering of the pH below 10 to cause corrosion of steel reinforcement.

As advised at Section 2.4.10, natural waters may be classified as 'hard' or 'soft' usually dependent upon the concentration of calcium bicarbonate that they contain. Hard waters may contain a calcium ion content in excess of 10 mg/l (ppm) whereas a soft water may contain less than 1 mg/l (ppm) calcium. The capillary pores in a hardened cement paste contain a saturated solution of calcium hydroxide which is in equilibrium with the calcium silicate hydrates that form the cement gel. If soft water can penetrate through the concrete (e.g. joint, crack, defect, etc) then it can leach free calcium hydroxide out of hardened cement gel so that the pore water is diluted and the pH falls.

Since the stability of the calcium silicates, aluminates and ferrites that constitute the hardened cement gel requires a certain concentration of calcium hydroxide in the pore water, leaching by soft water can also result in decomposition of these hydration products. The removal of the free lime in the capillary solution leads to dissolution of the calcium silicates, aluminates, and ferrites, and this hydrolytic action can continue until a large proportion of the calcium hydroxide is leached out, leaving the concrete with negligible strength.

Prior to this stage, loss of alkalinity due to the leaching process will result in reduced corrosion protection to reinforcement.

Calcium hydroxide that is leached to the concrete surface reacts with atmospheric carbon dioxide to form deposits of white calcium carbonate on

Figure 4.10 'Black rust' (magnetite) localised corrosion of steel reinforcement due to leaching at construction joints in the wall of a reinforced concrete tank containing potable water, age 72 years in rural near coast (~4 km) South Australia exposure. (Courtesy of Moore, 2017)

the surface of the concrete. These deposits of calcium carbonate may take the form of stalactites or severe efflorescence.

Leaching-induced corrosion of steel reinforcement can be localised and consequently more serious where it occurs at concrete joints, cracks, defects, etc. An example of localised corrosion of steel reinforcement in a potable water containing reinforced concrete tank is shown at Figure 4.10.

4.10 STRAY AND INTERFERENCE CURRENT-INDUCED CORROSION

4.10.1 General

Stray electrical currents, most commonly from electrified traction systems, can also breakdown the passive film and cause corrosion of steel reinforced, prestressed, and post-tensioned concrete elements.

Stray electrical currents are a potent source of corrosion problems in reinforced concrete. Currents flow through any environment in response to potential differences in that environment. If an extended reinforced concrete structure is in such an environment (i.e. in-ground or in-water), then these potential differences drive a current through the steel and this gives rise to corrosion. In the regions where the environment has the more positive potential, current will enter the metal by means of the reaction:

$$\tfrac{1}{2}O_2\,(g) + H_2O\,(l) + 2e^- \rightarrow 2OH^-\,(aq)$$

In the regions where the environment has the more negative potential, current will leave the metal by means of the reaction:

$$Fe\,(s) \rightarrow Fe^{++}\,(aq) + 2e^-$$

And as the metal is polarised in the positive direction that part of the structure is corroded.

4.10.2 Ground currents

Ground currents are particularly effective in causing corrosion in extended reinforced concrete structures such as concrete pipelines, tunnels and retaining walls because the potentials that can build up in the soil between the different portions of the structure are correspondingly larger. Although the driving force for the corrosion is greater, the resistance between the cathode and anode is not proportionately larger and so the corrosion currents are greater.

The most common cause of stray current corrosion is electrified traction systems. In general, the DC power for an electric train or tram is supplied by a sub-station and drawn from a positive overhead cable. The negative return to the sub-station is either through the rails on which the train/tram is running or through a separate conductor rail. Although most of the current driving the trains/trams returns to the sub-station from which it is drawn via the conductor rail of the traction system, it is inevitable that some of the current in the conductor rail leaks from that rail and goes back to the sub-station through the earth. This sets up potential gradients in the earth. If a low resistance path (such as metallic reinforcement in a concrete structure) is present, then the current may pass through the reinforcement. It can be seen from Figure 4.11 that where the current enters the structure, the reinforcement is cathodically protected. Where the current leaves the structure near the sub-station, corrosion is intensified.

The potential loss of reinforcing steel section due to stray current corrosion can be significant and design measures may need to be developed to mitigate such corrosion. As an example, suitable mitigation systems needed to be designed, implemented and maintained for the Sydney Harbour Tunnel in Australia as calculations carried out indicated a potential corrosion problem existed on the steel reinforcing of the immersed tunnel units and up to 150 kg of steel could be corroded per year by stray current (McCaffrey, 1991).

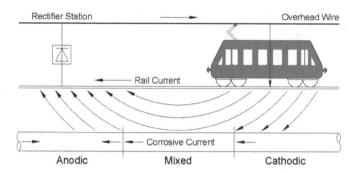

Figure 4.11 Ground currents set up by electric traction systems.

4.10.3 Interference currents

Interference corrosion currents are typically associated with cathodic protection schemes. The application of remote anode impressed current cathodic protection to (say) a buried pipeline involves the establishment of anodes (anode ground-bed) at some distance from the pipeline with the aim of passing current from the anodes (ground-bed) to the pipeline. If there is a reinforced concrete structure within this potential field, then potential gradients associated with the passage of current through the soil give rise to interference corrosion of steel reinforcement. The effect on the buried reinforced concrete structure is to set up cathodic areas near to the anode ground-bed and anodic areas closer to the structure which the cathodic protection system has been designed to protect.

4.10.4 Local corrosion due to stray or interference currents

Dependent on the type of reinforced or prestressed concrete structure and elements that are near an electrified traction system or cathodic protection schemes, stray current, or interference corrosion of steel reinforcement may be localised. As such, structurally significant section loss of reinforcement may occur within short periods of time, since current density can be converted to an equivalent mass loss or corrosion penetration rate by Faraday's Law.

For example, a section of reinforcing steel receiving, say 1 uA/cm² of stray current, equates to a mass loss or corrosion penetration rate of approximately 12 um/yr because of Faraday's Law (Fontana, 1987). If the stray current is 1 mA/cm², then the corrosion penetration rate will be approximately 12 mm/yr. Consequently, significant section loss of reinforcement may occur within short periods of time for reinforced and prestressed elements suffering stray current corrosion from ground currents or interference currents.

REFERENCES

Abd El Haleem, SM, Abd El Hal, EE, Abd El Wanees, S and Diab, A (2010), 'Environmental factors affecting the corrosion behaviour of reinforcing steel: I. The early stage of passive film formation in Ca(OH)$_2$ solutions', *Corrosion Science*, 52, 3875–3882.

Al-Negheimish, A, Alhozaimy, A, Rizwan Hussain, R, Al-Zaid, R and Singh, JK (2014), 'Role of manganese sulphide inclusions in steel rebar in the formation and breakdown of passive films in concrete pore solutions', *Corrosion*, 70, 1, January, 74–86.

Alvarez, MG and Galvele, JR (1984), 'The mechanism of pitting of high purity iron in NaCl solutions', *Corrosion Science*, 24, 1, 27–48.

American Concrete Institute (1985), 'Corrosion of metals in concrete', Committee 222, *ACI Journal*, 82, 1, 3–32.

Andrade, C and Gonzalez, JA (1978), 'Quantitative measurements of corrosion rate of reinforcing steels embedded in concrete using polarization resistance measurements', *Werkstoffe und Korrosion*, 29, 8, August, 515–519.

Angst, U and Vennesland, O (2009), 'Critical chloride content in reinforced concrete – State of the art', (In) *Concrete Repair, Rehabilitation and Retrofitting II*, (Eds) Alexander et al, Taylor & Francis Group, London, 149–150.

Angst, U, Elsener, B, Larsen, CK and Vennesland, O (2009), 'Critical chloride content in reinforced concrete – A review', *Cement and Concrete Research*, 39, 1122–1138.

Angst, U, Elsener, B, Larsen, CK and Vennesland, O (2011), 'Chloride induced reinforcement corrosion: Rate limiting step of early pitting corrosion', *Electrochimica Acta*, 56, 5877–5889.

Angst, UM, Boschmann, C, Wagner, M and Elsener, B (2017), 'Experimental protocol to determine the chloride threshold value for corrosion in samples taken from reinforced concrete structures', *Journal of Visualized Experiments* (126), August, 11.

Bamforth, PB (2004), '*Enhancing reinforced concrete durability – Guidance on selecting measures for minimising the risk of corrosion of reinforcement in concrete*', Concrete Society Technical Report No. 61, Surrey, UK.

Bhandari, J, Khan, F, Abbassi, R, Garaniya, V and Ojeda, R (2015), 'Modelling of pitting corrosion in marine and offshore steel structures – A technical review', *Journal of Loss of Prevention in the Process Industries*, 37, 39–62.

Bird, HEH, Pearson, BR and Brook, PA (1988), 'The breakdown of passive films on iron', *Corrosion Science*, 28, 1, 81–86.

Broomfield JP (2007), '*Corrosion of Steel in Concrete*', Taylor and Francis, 2nd Edition, Oxon, UK.

Browne, RD (1982), 'Design prediction of the life of reinforced concrete in marine and other chloride environments', *Durability of Building Materials*, Vol. 3, Elsevier Scientific, Amsterdam.

Cao, Y, Gehlen, C, Angst, U, Wang, L, Wang, Z and Yao, Y (2019), 'Critical chloride content in reinforced concrete – An updated review', *Cement and Concrete Research*, 117, 58–68.

Chess, P and Green, W (2020), '*Durability of Reinforced Concrete Structures*', CRC Press, Taylor & Francis Group, Boca Raton, FL, USA.

Cohen, M (1978), '*The Passivity and Breakdown of Passivity of Iron*', in: R P Frankeuthal, J Kurager, N J Princietor (Eds.), Passivity of Metals, The Electrochemical Society, 521.

Concrete Institute of Australia (2015), '*Performance tests to assess concrete durability*', Recommended Practice Z7/07, Concrete Durability Series, Sydney, Australia.

Cornell, RM, and Schwertmann, U (2000), '*The Iron Oxides*', Wiley-VCH, Weinhem.

Elsener, B (2002), 'Macrocell corrosion of steel in concrete – Implications for corrosion monitoring', *Cement & Concrete Composites*, 24, 65–72.

Faraday, M (1844), '*Experimental Researches in Electricity*', Vol II, Richard and Edward Taylor, London, UK.

Fontana, MG (1987), '*Corrosion Engineering*', McGraw-Hill, third Edition, Singapore.

Frankel, GS (1998), 'Pitting corrosion of metals a review of the critical factors', *Journal of the Electrochemical Society*, 145, 6, 2186–2198.

Freire, L, Novoa, XR, Montemor, M and Camezin, MJ (2009), 'Study of passive films formed on mild steel in alkaline media by the application of anodic potentials', *Materials Chemistry and Physics*, 114, 962–972.

Galvele, JR (1976), 'Transport processes and the mechanism of pitting of metals', *Journal of the Electrochemical Society*, 123, 464–474.

Ghods, P, Isgor, OB, Brown, JR, Bensebaa, F and Kingston, D (2011), 'XPS depth profiling study on the passive oxide film of carbon steel in saturated calcium hydroxide solution and the effect of chloride ion on the film properties', *Applied Surface Science*, 257, 10, March, 4669–4677.

Ghods, P, Isgor, OB, Carpenter, GJC, Li J, McRae GA and Gu, GP (2013), 'Nanoscale study of the passive films and chloride-induced depassivation of carbon steel rebar in simulated concrete pore solutions using FIB/TEM', *Cement and Concrete Research*, 47, 55–68.

Gouda, VK (1970), 'Corrosion and corrosion inhibition of reinforcing steel. I. Immersed in alkaline solutions', *British Corrosion Journal*, 5, 198–203.

Green, WK (1991), '*Electrochemical and chemical changes in chloride contaminated reinforced concrete following cathodic polarisation*', MSc Dissertation, University of Manchester Institute of Science and Technology, Manchester, UK.

Green, W, Dockrill, B and Eliasson, B (2013), '*Concrete repair and protection – overlooked issues*', Proc. *Corrosion and Prevention 2013 Conf.*, Australasian Corrosion Association Inc., Brisbane, Australia, November, Paper 020.

Green, W, Ehsman, J, Linton, S, Cambourn, N and Masia, S (2019), 'Chloride durability and future maintenance of a 40+ year marine structure', *Concrete in Australia*, Vol 45, No 4, 51–58.

Hansson, C M (1984), 'Comments on electrochemical measurements of the rate of corrosion of steel in concrete', *Cem Concr Res*, 14(4), 574–584.

Hausmann, DA (1967), 'Steel corrosion in concrete. How does it occur?', *Materials Protection*, 6, 19–23.

Huet, B, Hostis, VL, Miserque, F and Idrissi, H (2005), 'Electrochemical behaviour of mild steel in concrete: Influence of pH and carbonate content of concrete pore solution', *Electochimica Acta*, 51, 172–180.

Jaffer, SJ and Hansson, CM (2009), 'Chloride-induced corrosion products of steel in cracked-concrete subjected to different loading conditions', *Cement and Concrete Research*, 29, 116–125.

Koleva, DA, Hu, J, Fraaij, ALA, Stroeven, P, Boshkov, N and de Wit, JHW (2006), 'Quantitative characterisation of steel/cement paste interface microstructure and corrosion phenomena in mortars suffering from chloride attack', *Corrosion Science*, 48, 4001–4019.

Kolio, A, Honkanen, M, Lahdensivu, J, Vippola, M and Pentti, M (2015), 'Corrosion products of carbonation induced corrosion in existing reinforced concrete facades', *Cement and Concrete Research*, 78, 200–207.

Leek, DS and Poole, AS (1990), *'Corrosion of Steel Reinforcement in Concrete'*, ed. C L Page, K W J Treadaway and P B Bamforth, Society of Chemical Industry and Elsevier Applied Science, 65–73.

Lin, B, Ronggang, H, Chenqing, Y, Yan, L and Chagjian, L (2010), 'A study on the initiation of pitting corrosion in carbon steel in chloride-containing media using scanning electrochemical probes', *Electrochimica Acta*, 55, 6542–6545.

McCaffrey, WR (1991), *'Stray current mitigation systems in Sydney Harbour Tunnel'*, *Proc. Corrosion & Prevention 1991 Conf.*, Australasian Corrosion Association Inc., Sydney, Australia, November, Paper 27.

Melchers, RE and Li, CQ (2008), *'Long-term observations of reinforcement corrosion for concrete elements exposed to the north sea'*, *Proc. Corrosion & Prevention 2008 Conf.*, Australasian Corrosion Association Inc., Wellington, New Zealand, Paper 079.

Moore, G (2017), *Private communication*, February.

Nasrazadani, S (1997), 'The application of infra-red spectroscopy to the study of phosphoric acid and tannic acid interactions with magnetitie (Fe_3O_4), goethite (α-FeOOH) and lepidocrocite (γ-FeOOH)', *Corrosion Science*, 39, 1845–1859.

Newman, R (2010), 'Pitting corrosion of metals', *The Electrochemical Society Interface*, Spring, 33–38.

Ogunsanya, IG and Hansson, CM (2020), 'Reproducibility of the corrosion resistance of UNS S322205 and UNS S32304 stainless steel reinforcing bars', *Corrosion*, 76, 1, 114–130.

Page, C L (1975), 'Mechanism of corrosion protection in reinforced concrete marine structures', *Nature*, 258, 514–515.

Page, CL and Treadaway, KWJ (1982), 'Aspects of the electrochemistry of steel in concrete', *Nature*, 297, 109–115.

Pape, TM and Melchers, RE (2013), *'A study of reinforced concrete piles from the Hornibrook Highway bridge (1935–2011)'*, *Proc. Corrosion & Prevention 2013 Conf.*, Australasian Corrosion Association Inc., Brisbane, Australia, November, Paper 086.

Pourbaix, MA (1966), *'Atlas of Electrochemical Equilibria in Aqueous Solutions'*, Marcel Pourbaix, Pergamon Press, Oxford, UK.

Sagoe-Crentsil, KK and Glasser, FO (1989), 'Steel in concrete part II: Electron microscopy analysis', *Magazine of Concrete Research*, 213–220.

Sagoe-Crentsil, KK and Glasser, FP (1990), *Corrosion of Steel Reinforcement in Concrete*, ed. C L Page, K W J Treadaway and P B Bamforth, Society of Chemical Industry and Elsevier Applied Science, 74–85.

Savija, B and Lukovic, M (2016), 'Carbonation of cement paste: Understanding, challenges, and opportunities', *Construction and Building Materials*, 117, 285–301.

Singh, JK and Singh, DDN (2012), 'The nature of rusts and corrosion characteristics of low alloy and plain carbon steels in three kinds of concrete pore solution with salinity and different pH', *Corrosion Science*, 56, 129–142.

Soltis, J (2015), 'Passivity breakdown, pit initiation, and propagation of pits in metallic materials – Review', *Corrosion Science*, 90, 5–22.

Standards Australia (2010), 'AS 3972 Portland and blended cements', Sydney, Australia.

Stark, D (1984), '*Determination of permissible chloride levels in prestressed concrete*', *PCI Journal*, July–August, 106–116.

Tinnea, J and Young, JF (2000), '*The chemistry and microstructure of concrete: Its effect on corrosion testing*', *Proc. Corrosion & Prevention 2000 Conf.*, Australasian Corrosion Association Inc., Auckland, New Zealand, November, Paper 026.

Tinnea, J (2002), '*Localised corrosion failure of steel reinforcement in concrete: Field examples of the problem*', *Proc. 15th International Corrosion Congress*, Granada, Spain, Paper 706.

Treadaway, K (1988), 'Corrosion of steel in concrete', ed P Schiessl, *RILEM Report of Technical Committee 60-CSC*, Chapman and Hall, London, UK.

Treadaway, KWJ (1991), 'Corrosion propagation', COMETT Short Course on The Corrosion of Steel in Concrete, University of Oxford, UK.

Tuutti, K (1982), 'Corrosion of steel in concrete', Swedish Cement, and Concrete Association, Report Fo. 4.82.

Veluchamy, A, Sherwood, D, Emmanuel, B and Cole, IS (2017), 'Critical review on the passive film formation and breakdown on iron electrode and the models for the mechanisms underlying passivity', *Journal of Electroanalytical Chemistry*, 785, 196–215.

Venu, K, Balakrishnan, K and Rajagopalan, KS (1965), 'A potentiokinetic polarization study of the behaviour of steel in NaOH-NaCl system', *Corrosion Science*, 5, 59–69.

Vera, R, Villarroel, M, Carvajal, AM, Vera, E and Ortiz, C (2009), 'Corrosion products of reinforcement in concrete in marine and industrial environments', *Materials Chemistry and Physics*, 114, 467–474.

Chapter 5

Corrosion of reinforcement (B)

5.1 THERMODYNAMICS OF CORROSION

5.1.1 Background

As indicated in the previous chapter, thermodynamics, the science of energy changes, is applied to corrosion studies to determine why a particular metal does or does not tend to corrode in a particular environment.

5.1.2 The driving potential – the Nernst equation

The potential difference that drives the current around the electrochemical circuit has its origin in the energy that is released by the cathodic and anodic reactions. The total driving electromotive force (e.m.f.) is the sum of the Single Electrode Potentials set up at the cathodic and anodic sites. At the anodic site, free energy is released by the action of the oxidation reaction and so ions pass into solution. However, when an ion dissolves from the lattice and passes into solution then the solution acquires a positive charge and a negative charge is left on the electrode surface. This creates a so called 'double layer' of charge in which an electric field is set up with the positive end of the field in the solution and the negative end on the metal surface, refer Figure 5.1.

Because of the potential difference set up across the interface, energy is required to cause a positive ion to move against the potential gradient in order to go from lattice into solution. The system comes to equilibrium and there ceases to be a nett flow of ions into solution when the potential that is set up is such that the gain in electrical potential energy as the ion moves from lattice to solution is equal to the loss in chemical potential energy as it moves from lattice to solution. This equivalence may be written:

$$\Delta G = -zFE \tag{5a}$$

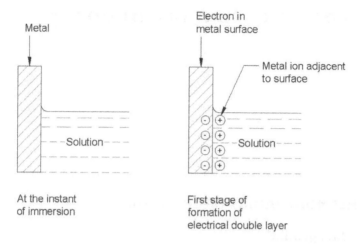

Figure 5.1 The electrical double layer. (Courtesy of Potter, 1961, p.73)

Where E is the potential developed across the interface and ΔG the loss in Chemical Free Energy when 1 gm.mole of the ion enters the solution. zF is the charge on 1 gm.mole of the ion. On this basis it is possible to derive an expression for the potential difference across the interface. The value of the potential difference in volts across an interface is given by an expression known as the Nernst equation. By convention the potential of the metal is measured with respect to the solution. (American usage is sometimes different and this can give rise to confusion, but throughout these sections this International convention will be used.) A positive (>0) potential would therefore indicate that as the metal is positive with respect to the solution and that there is a gain in electrical potential energy as a positive ion leaves the lattice and goes into the solution. The change in Gibbs Free Energy (chemical potential energy) in going from the solution to the metal must therefore because of Equation 5a, be negative, that is, the spontaneous direction of the reaction is:

$$M^{Z+} + ze^- \rightarrow M \tag{5.1}$$

And i.e. that the metal ions to plate out on to the lattice. Consequently, in all electrochemical work it is the reduction reaction rather than the oxidation that is considered.

By considering the balance between the chemical potential energy and the electrical potential energy it is possible to derive the expression:

$$E = E^0 + (RT/zF)\ln(a_r/a_p) \tag{5b}$$

In Equation 5b, E is the single electrode potential, E^0 is the standard electrode potential for the reaction that is taking place, a_r is the activity of the reactants (the oxidised species) and, a_p the activity of the products of the reaction (the reduced species). A number of standard electrode potentials is listed in Table 5.1.

Equation 5b is known as the Nernst equation and for reaction (5.1) may be written:

$$E = E^0 + (0.059/z) \log a_{M+}$$

Although the Nernst equation gives a theoretical value for the potential difference across a metal electrolyte interface this potential difference can only be measured by attaching the metal to another electrode which is immersed in the same electrolyte. Since this electrode will also have a potential difference across its interface with the electrolyte it is impossible to measure an electrode potential other than with reference to this second electrode. The standard electrode potentials reported in Table 5.1 below are reported with respect to a standard hydrogen electrode (SHE) (or normal hydrogen electrode, NHE) and consequently the standard electrode potential for that electrode is reported as 0.00 V. Potentials reported with respect to a standard saturated calomel electrode (SCE, $Hg/Hg_2CL_2/KCl$(saturated)) are all 0.24 V more negative (at standard temperature and pressure (STP), i.e. 25°C and 1 atm/101.3 kPa) than those reported with respect to the hydrogen electrode and potentials reported with respect to the copper sulphate electrode (CSE, $Cu/CuSO_4$(saturated)) are all 0.32 V more negative (at STP) than the same potentials reported on the hydrogen scale.

Table 5.1 Some standard electrode potentials

Reduction reaction	E^0 (vs SHE) (at STP)(V)
$MnO_2(s) + 4H^+ + 2e^- \rightarrow Mn^{2+} + 2H_2O$	1.23
$\frac{1}{2}O_2(g) + H_2O + 2e^- \rightarrow 2OH^-$	+0.40*
$Cu^{++} + 2e^- \rightarrow Cu(s)$	+0.34
$Hg_2Cl_2(s) + 2e^- \rightarrow 2Hg(l) + 2Cl^-$	+0.24
$H^+ + e^- \rightarrow \frac{1}{2}H_2(g)$	0.00
$Fe^{++} + 2e^- \rightarrow Fe(s)$	−0.44
$Zn^{++} + 2e^- \rightarrow Zn(s)_{,}$	−0.76
$Al^{+++} + 3e^- \rightarrow Al(s)$	−1.66

Source: Aylward & Findlay (1974)
* approximate value in concrete environment

Table 5.2 Potentials of various electrodes measured against Ag/AgCl/0.5 M KCl electrode

Type of electrode	Potential (E) at 25°C(mV)
Ag/AgCl/seawater	0
Ag/AgCl/KCl(saturated)	−54
Ag/AgCl/0.1 M KCl	+37
Ag/AgCl/1M KCl	−16
Zn/seawater	−1,033
Cu/CuSO₄(saturated)	+67
Zn⁺⁺ + 2e⁻ → Zn(s)	−0.76
Hg/Hg₂Cl₂/KCl (SCE)	−10

Source: Standards Australia (2008)

Some standard electrode potentials of various reference electrodes measured against a Ag/AgCl/0.5 M KCl electrode are summarised at Table 5.2.

In a similar fashion to the way in which a potential difference is set up across the interface between metal and electrolyte at the site of the anodic reaction a potential difference is set up at the site where the cathodic reaction takes place. By a similar reasoning a Nernst equation can be derived for the cathodic reaction. For the reaction:

$$O_2 + 2H_2O + 4e^- \rightarrow 4OH^- \tag{5.2a}$$

The equation may be written:

$$E = E^0 - (0.059/z)\log[a_{OH}/a_O]$$

Here, a_{OH} is the activity of the hydroxyl ions and a_O is the activity of oxygen in the portion of the solution which is immediately adjacent to the metal surface where this reaction is taking place.

It is the difference between the potentials for the cathodic and the anodic reactions that drives the corrosion current through the concrete. In the corrosion circuit the metal is effectively an equipotential surface and the main potential drop is at the interface where the cathodic and anodic reactions take place. This means that the electrolyte close to the anode is more positive with respect to the metal than the electrolyte close to the cathode and so the ionic current flows through the concrete with negative ions that move from cathodic area to anodic area through the concrete. This flow of current through the highly resistive concrete is associated with an iR drop in the concrete and the potential difference can be detected at the surface of the concrete by reference electrodes placed on the surface.

5.1.3 The potential – pH diagram

Let us now concern ourselves with the conditions under which the different mechanisms may operate. In general, the corrosion state of a metal in a specific environment will be determined by the pH of that environment and the potential of the metal with respect to the environment. As indicated previously, Marcel Pourbaix early on recognised that the corrosion state of a metal could be represented by a point on a diagram the two axes of which were potential and pH. These diagrams are now often termed 'Pourbaix' diagrams. The construction of such a diagram may be illustrated for the iron-water system.

It has already been noted that under some circumstances, the anodic reaction is:

$$Fe \rightarrow Fe^{++} + 2e^- \qquad (5.3)$$

For this the Nernst equation can be written:

$$E = E_0 + \frac{RT}{2F} \ln a_{Fe^{2+}}$$

And if the appropriate values are substituted:

$$E = -0.440 + 0.0295 \log a_{Fe2+}$$

If the concentration of ions in the rust product surrounding the anodic sites is assumed to be 10^{-6} gm.ions/litre (this is an appropriate value) then the potential at which the reaction takes place can be written as -0.62 V. If the iron is polarised to a potential that is more positive than this (for example by the action of a cathodic process on the same metal surface) then the reaction proceeds to the right and corrosion occurs. The metal is said to be in the 'active' state. If the potential of the electrode is made more negative than this value, then the reaction would proceed in the reverse direction and metal would be deposited rather than dissolved. The iron electrode would now be undergoing a cathodic reaction and would be said to be in the 'immune 'state. Once all the ferrous ions in the immediate environment had been discharged, (in a very short time), then the only way in which the negative potential could be maintained would be for an alternative cathodic process to take place. This process is the as previous oxygen reduction reaction as:

$$\tfrac{1}{2}O_2 + H_2O + 2e^- \rightarrow 2OH^- \qquad (5.2b)$$

On a potential-pH diagram the line represented by Equation (5.3) (*E = –0.62 V in our case with the assumptions so made*) would therefore mark the boundary between the region where the metal actively corrodes and the region in which it is immune from corrosion.

Another possible anodic reaction for iron in contact with a solution containing ferrous ions is:

$$2Fe^{++} + 3H_2O \rightarrow Fe_2O_3 + 6H^+ + 2e^- \tag{5.4}$$

As before writing the Nernst equation for this reaction:

$$E = E_0 + \frac{RT}{2F} \ln \frac{a_{Fe_2O_3} a_{H^+}^6}{a_{Fe^{++}}^2 a_{H_2O}^3}$$

And substituting the appropriate values:

$$E_o = 0.73 - 0.177pH - 0.59 \log\left[Fe^{++}\right]$$

At pH 7 and $[Fe^{++}] = 10^{-6}$, it can be seen that at potentials more positive than -0.16 V a passive film is formed. In this state the cathodic and anodic reactions are those described in Equations 5.2a and 5.3 respectively and the iron is in a 'passive' state with a very low corrosion rate. It can therefore be appreciated that at pH 7, the iron only actively corrodes between -0.62 V and -0.16 V and that the line on the potential-pH diagram which represents Equation (5.4) is the boundary between the active and passive states of iron.

Another anodic reaction is:

$$2Fe + 6OH^- \rightarrow Fe_2O_3 + 3H_2O + 6e^-$$

The Nernst equation for this passive film forming reaction yields:

$$E = E_0 + \frac{RT}{6F} \ln \frac{a_{Fe_2O_3} a_{H^+}^6}{a_{Fe} a_{H_2O}^3}$$

The activity of pure compounds may be put equal to unity and so substituting the appropriate figures this becomes:

$$E = 0.051 - 0.059pH \tag{5.5}$$

This is the potential at which the metal and the oxide are in equilibrium. If the metal is made more positive with respect to the solution the ratio of $a_{Fe_2O_3}/a_{Fe}$ must increase so that a protective oxide film is formed, and the metal is in the 'passive' state. However, since at low values of pH iron can only corrode if the potential is more positive than -0.62 V it can be seen by substituting this value for E in Equation (5.5) that at a pH value for the solution greater than 10 is derived and the iron is then either in the immune state or the passive state and high rates of corrosion cannot take place.

Other anodic reactions are possible. They can all be represented on a potential-pH diagram which indicates the range of potentials and pH over which a metal is 'active', 'passive' or 'immune' from corrosion. Such a diagram is shown in Figure 5.2 which is a simplified diagram which assumes that passivation can only be brought about by the formation of an Fe_2O_3 film. This potential-pH diagram also has two lines marked 'a' and 'b'. These are the lines marking the Nernst equation for the cathodic reactions. The 'a' line is the potential of the cathodic reaction:

$$2H^+(aq) + 2e^- \rightarrow H_2(g)$$

if it is assumed that the partial pressure of hydrogen is one atmosphere (atm/101.3 kPa). At potentials more negative than this water is decomposed to yield gaseous hydrogen and is stable at more positive potentials.

The 'b' line is the potential of the cathodic reaction:

$$\tfrac{1}{2}O_2(g) + H_2O(l) + 2e^- \rightarrow 2OH^-(aq)$$

if the partial pressure of oxygen is unity. At potentials more positive than this water is decomposed to yield gaseous oxygen. Water is therefore only stable between the 'a' and the 'b' lines.

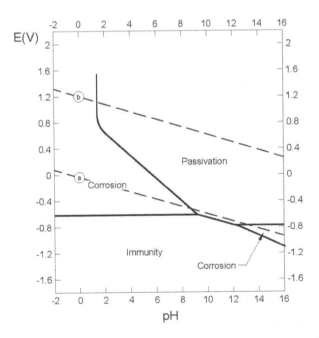

Figure 5.2 Potential-pH diagram for iron in pure water. (Courtesy of Pourbaix, 1966, p.314)

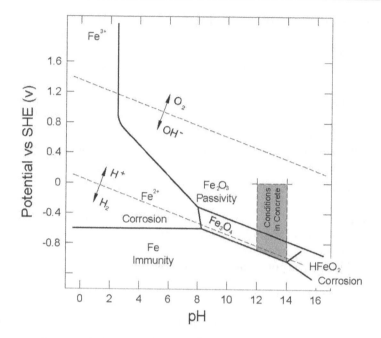

Figure 5.3 Potential-pH (Pourbaix) diagram for Fe-H_2O at 25°C and [Fe^{2+}], [Fe^{3+}] and [$HFeO_2^-$] = 10^{-6}M. (Courtesy of Green, 1991)

Figure 5.2 has been drawn for pure aqueous solutions.

A potential-pH or Pourbaix diagram that is of relevance to reinforcing steel in concrete is that for the system Fe-H_2O at 25°C, previously shown as Figure 4-1 and reproduced here as Figure 5.3, in which concentrations of dissolved ions, other than H^+ and OH^-, are taken to be 10^{-6}M and solid phases are assumed to be pure.

5.2 KINETICS OF CORROSION

5.2.1 Background

As indicated in the previous chapter, kinetics when applied to corrosion studies deals with the rates of possible reactions. Reactions might be thermodynamically favoured, but will the kinetics be such that they occur at a significant rate in practice?

5.2.2 Polarisation

The driving force for the flow of electrical current around the corrosion circuit is the difference between the single electrode potentials for the cathodic and anodic reactions. The magnitude of the corrosion current will be controlled by the ratio of this driving force (the so called Driving Potential) to the total resistance of the circuit. The resistance of the corrosion circuit has a number of components. Because these resistances are in general non-linear, that is the magnitude of the resistance varies with the magnitude of the current flowing across it, they are usually discussed in terms of the potential drop across them when a specified current is passing. These are termed the overpotentials and the sum of the separate overpotentials equals the Driving Potential.

Activation Overpotential. Often a dominant component of total resistance to corrosion current flow is the resistance associated with the activation overpotential. Table 5.1 of standard electrode potentials is the equilibrium potential when there is no nett flow of current across the interface between metal and concrete. If there is to be a nett flow of (say) positive ions away from the anodic areas then this current flow has to overcome the electrical resistance posed by the activation energy barrier at the metal electrolyte interface. To do this, the potential of the metal must be made more positive in order to repel the ions and maintain a flow of current. The relationship between the potential shift necessary to cause a certain current to flow and that current is given by:

$$\eta = a + b \log i$$

Where η is the shift in potential, (the activation overpotential), i is the current density that is being drawn from the system and a and b are constants. This equation is known as Tafel's equation and the process by which the potential of an anodic process is raised as a positive current flows from it is known as polarisation. In a similar fashion, if current is to be drawn from a cathodic process across the resistance posed by the activation energy barrier at that metal electrolyte interface, then the potential of that cathodic process must be made more negative to repel the negative ions. Tafel's law similarly applies to this reaction. Since in this case the ions that have to be repelled are negative, b has a negative coefficient.

When corrosion is occurring and positive current is flowing from the anode, then the anodic potential is polarised towards more positive values. The cathodic potential is similarly polarised towards more negative values. In the absence of a resistive component of the circuit, since the anodic current must equal the cathodic current, the system will settle down with the anodic reaction polarised positively and the cathodic reaction polarised negatively to the same potential. This potential is known as the rest potential or corrosion potential (E_{corr}) and is the potential which would be measured with a reference electrode immersed in the (theoretically zero resistivity) electrolyte. The current flowing from the anodic to the cathodic region through the electrolyte and from the cathodic region to the anodic

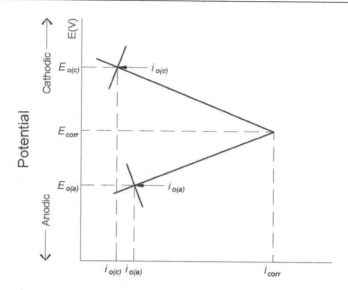

Figure 5.4 The Evans diagram. (Courtesy of Jastrzebski, 1987, p.579)

region in the metal is the corrosion current. If the polarisation curves are simplified so that they just consist of the Tafel lines, then the situation is as represented in Figure 5.4. This is usually termed an Evans diagram after Dr U.R. Evans of Cambridge.

Diffusion overpotential. A second component of the resistance in the corrosion circuit derives from the energy that has to be expended in bringing about the movement of ions to and from the substrate/electrolyte interface against a concentration gradient in the electrolyte. Typically, as a cathodic reaction proceeds, oxygen is consumed in the region close to the cathodic area. Because the diffusion of oxygen through concrete is slow, the oxygen which is consumed cannot be replaced immediately and so the oxygen concentration falls. From the application of the Nernst equation, it can be seen that the potential of the cathodic reaction becomes more negative. (Note that the term single electrode potential (SEP) is not used here because a current is flowing and the term SEP can only describe the equilibrium situation when there is no nett current flowing across the interface). The shift in Driving Potential brought about by the concentration changes at the interface is given by:

$$\eta_D^c = \frac{RT}{zF} \ln\left(\frac{c}{c_o}\right)$$

In this equation, c is the concentration of oxygen at the surface of the passive film and c_o is the concentration of oxygen in the bulk concrete adjacent to the reference electrode. If it is assumed that because the external surface of the concrete is exposed to the atmosphere, c_o is constant, then the diffusion overpotential is controlled by the value of c. This will change with time and its value at any moment will be controlled by the balance between the rate at which the oxygen is consumed by the cathodic reaction and the rate at which it is replaced by diffusion through the concrete. In the rest state an equilibrium will be established between these processes and the diffusion overpotential will assume a constant value.

A similar diffusion overpotential can be set up at the anodic area controlled by the speed at which (say) metal ions can diffuse away from the substrate/electrolyte interface.

Resistance Overpotential. In practice the concrete electrolyte between the cathodic and anodic areas has an electrical resistance which may be the only linear component of the total resistance of the circuit. It is given by:

$$\eta^r = i_c \rho$$

Where η^r is the resistance overpotential, i_c is the corrosion current density and ρ is the resistance of the concrete between cathode and anode. The electrical resistance of the metallic portion of the circuit between the cathode and the anode can usually be neglected.

5.2.3 Investigation of the corrosion state

The commonest experimental technique for the determination of the state of a metal surface, whether it is passive, active, or immune, is to polarise the metal to a more positive or negative potential and to measure the current required to bring about this polarisation. The polarisation may be carried out galvanostatically (in which the current is set to a constant value and the potential to which the metal comes at equilibrium is measured) or potentiostatically (in which the potential is set to a constant value and the current required to maintain this potential is measured), in the environment that is being studied. The experimental set up for potentiodynamic polarisation is shown in diagrammatic form in Figure 5.5. A current (cathodic or anodic) is impressed on the electrode and the current density that is required to polarise the electrode to a given potential is measured. The potential may be scanned at a constant rate (potentiodynamic polarisation) and the current density necessary to maintain any given potential gives an indication of whether the metal is in the passive or active range.

A potentiodynamic scan carried out on iron at pH 7 would have the form shown in Figure 5.6. At the rest potential (corrosion potential, E_{corr}) the cathodic current would equal the anodic current and the impressed current would be zero. If the potential is moved in the negative direction a small cathodic current would plate out any ferrous ions in solution and then as the iron moved into the immune region the whole current would stem from the oxygen reduction cathodic reaction:

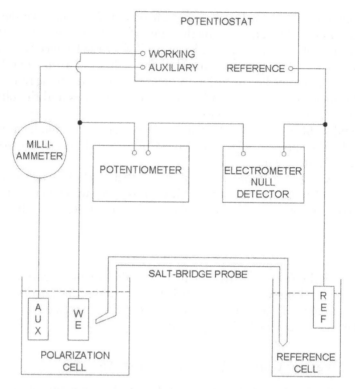

Figure 5.5 Experimental arrangement for the plotting of potentiodynamic scans.
(Courtesy of Ailor, 1971, p.181)

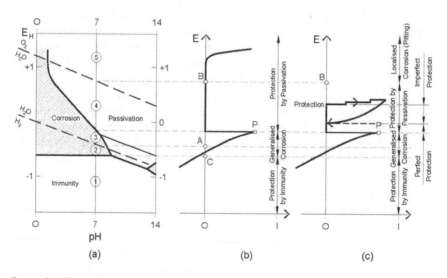

Figure 5.6 Potentiodynamic scan for a passive film forming metal. (Courtesy of
Pourbaix in Ailor, 1971, p.666)

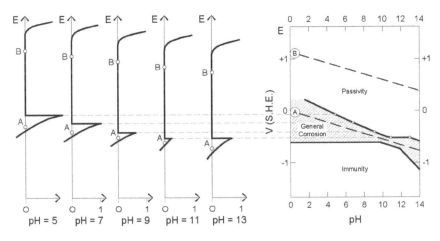

Figure 5.7 Experimental Potential-pH diagrams. (Courtesy of Pourbaix in Ailor, 1971, p.672)

$$\tfrac{1}{2}O_2(g) + H_2O(l) + 2e^- = 2OH^-(aq)$$

If the potential is made positive to the rest potential the anodic (metal dissolution) current increases until the 'primary passivation' potential (P) is reached. At this potential the current suddenly drops as the metal enters the passive region and remains at this low value until the 'rupture potential (r)' is reached. At potentials above the rupture potential, local dissolution of the passive layer takes place and pitting ensues. The potential range between the primary passivation potential and the rupture potential defines the passive range for that metal.

If a series of such scans is carried out over a range of pH, then a complete experimental potential-pH diagram can be plotted. When such scans are carried out, the transition from the active or immune state to the passive state is not immediate. If the rate at which the passive film forms is less than the rate at which it dissolves in the immersing solution, then corrosion may take place even though the metal is thermodynamically in the passive range. The practical potential-pH diagram can therefore look like Figure 5.7.

5.2.4 Pitting corrosion

Previously it has been shown that the ingress of chloride ions to the substrate/electrolyte interface can initiate pitting corrosion. This can be examined by the use of a potentiodynamic polarisation cycle which is continued by a return sweep in the negative direction after the largest positive potential has been attained. Typical results are shown in Figure 5.8.

If the scan is started at the corrosion (rest) potential, then as the potential is made more positive, corrosion increases, and the metal is said to be in the

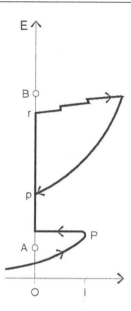

Figure 5.8 Potentiodynamic scan under pitting conditions. (Courtesy of Pourbaix in Ailor, 1971, p.670)

'active loop'. At P the 'primary passivation potential' the protective oxide film is formed, and the corrosion current diminishes to a small value. At r the rupture potential, the metal has been made so positive that the negatively charged chloride ion is able to enter the oxide film and produce a pit. Any positive movement of the potential will now produce a large increase in the corrosion current. If at this point, the direction of the potential movement is reversed, then in general, the current does not fall to its previous value at a given potential but is considerably higher. Because a pit has been formed, the metal can go on corroding at a high rate. As the potential is progressively moved in the negative direction it eventually reaches p the 'protection potential' at which the corrosion current has fallen to its passivation value. At this point the bare metal at the bottom of the pits has reformed its passive film and the pitting corrosion is suppressed. These potentials P, r, and p define three conditions for the metal surface. Positive to r the metal is in the pitting state. Between P and p the metal is in a perfectly passive (PP) state – even though pits may have formed, they will not grow. Between p and r, the metal is in the imperfectly passive (IP) state – if there are no pits, then the metal will remain passive, but if there are pits, then these pits will grow and contribute to corrosion.

If such cyclic polarisation scans are plotted over a range of values of pH, then a complete picture of the corrosion states of the metal can be drawn up as a modified potential-pH diagram as shown in Figure 5.9. The results can be seen in final form in Figure 5.10.

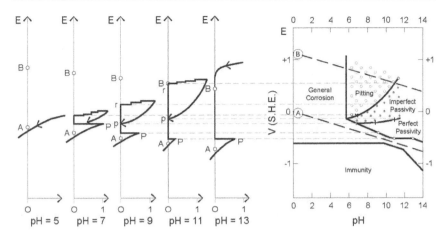

Figure 5.9 Production of an experimental potential-pH diagram. (Courtesy of Pourbaix in Ailor, p.673)

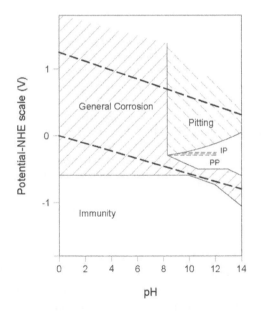

Figure 5.10 Modified potential-pH diagram for pitting conditions

The regions marked IP and PP represent the conditions of imperfect passivity and perfect passivity respectively.

5.2.5 Oxygen availability

The availability of oxygen is of considerable importance since the predominant cathodic reaction is the reduction of oxygen.

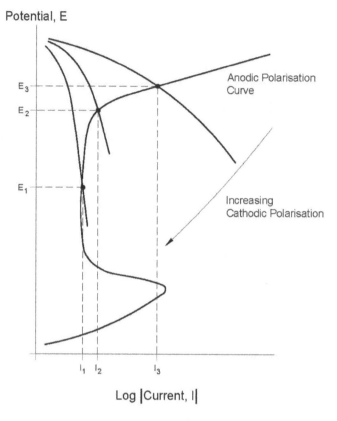

Figure 5.11 Evans diagram showing effect of increasing oxygen availability and increasing cathodic polarisation on a pitting anode. (Courtesy of Treadaway, 1991)

The polarisation of the oxygen reduction reaction will contribute substantially to the overall corrosion process, by influencing the corrosion current/rate (i.e. I_{corr}) and the corrosion potential (i.e. E_{corr}). Figure 5.11 conceptually shows the effect of increasing oxygen availability to the reinforcement surface and increasing cathodic polarisation. For the Evans diagram at Figure 5.11 the polarisation curves for the anodic and cathodic processes are plotted on the same axes (ignoring the difference in sign of the two currents) and for simplicity, it is assumed that both the exchange current (i.e. I_0) and equilibrium or reversible potential (i.e. E_0) for the oxygen reaction remains relatively constant.

Thus, for the oxygen reduction cathodic reaction operating with good oxygen supply the corrosion potential E_3 is developed and corresponding corrosion current (rate) I_3. Reduction in oxygen access to cathodic areas on the reinforcement surface causes cathodic polarisation of the corrosion process with a resultant reduction in both the corrosion potential to E_2, and corrosion current to I_2. Further reduction in oxygen access (cathodic polarisation) produces a more negative corrosion potential (i.e. E_1) and slower rate of corrosion (i.e. I_1). The effect discussed here whereby, an increase in oxygen access results in an increased corrosion rate and more positive corrosion potentials, is an illustration of a corrosion process being under cathodic control.

It should be noted that the corrosion current I_1 on Figure 5.11 is also the passive 'leakage current' (i.e. current required to maintain the passive film on reinforcement, refer Section 4.5.3). Furthermore, the corrosion potential E_2, is termed the pitting potential. As previous, at potentials more positive than this point, a small change in potential results in a significant increase in corrosion rate, bearing in mind the logarithmic relationship between corrosion current (rate) and corrosion potential.

Therefore, pitting corrosion of reinforcement can proceed (propagate) where oxygen supply is sufficient to ensure that the corrosion potential is always maintained above the pitting potential. However, by changing the environment, by making it more or less aggressive (e.g. chloride concentration, chloride/hydroxy ratio, etc), a change can occur in the pitting potential. Pitting corrosion of reinforcement can thus occur even under limited oxygen supply conditions. Discussion of this aspect has been provided previously at Section 4.6.6 and also refer Section 5.2.8.

From the practical point of view of most reinforced concrete structures, oxygen availability to cathodic sites is sufficient to ensure the corrosion potential is above the pitting potential. Where oxygen diffusion is limited, cathodic reactivity in relation to the resistance between anodic and cathodic sites (see Section 5.2.7) and the ratio of area of cathodic sites to area of anodic sites (cathode/anode ratio), (see Section 5.2.8), are important in encouraging the propagation of pitting attack (Treadaway, 1991).

5.2.6 Polarisation of the anodic process

The overall corrosion process is a mix of anodic and cathodic processes and therefore equal consideration needs to be given to anodic activity. The conceptually derived Evans diagram at Figure 5.12 represents the effect of increasing chloride concentration on reinforcement corrosion rate.

From the set of polarisation curves it can be seen that increasing chloride content at the reinforcement/concrete interface has a major effect in decreasing the range of potential over which passivity exists.

In the absence (below propagation levels) of chloride in concrete the passive current I_5 corresponds to the passive potential E_5. The range of potential over which passivity exists is that between b and E_5. An increase in the chloride content at the reinforcement/concrete interface causes the corrosion potential to decrease from E_{4p} to E_{1p} with a corresponding increase in corrosion current (rate) from I_{4p} to I_{1p}, and a decrease in the potential range over which passivity exists.

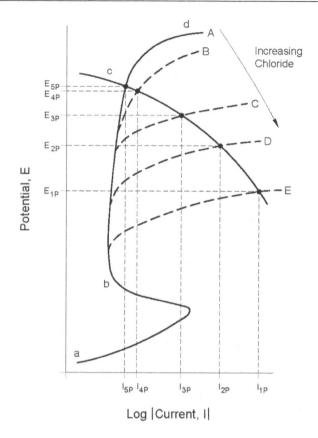

Figure 5.12 Evans diagram of the influence of increasing chloride content on corrosion rate of reinforcing steel in concrete. (Courtesy of Treadaway, 1991)

Corrosion current increasing as the corrosion potential becomes more negative is an illustration of a corrosion process being under anodic control.

5.2.7 Resistance between anodic and cathodic sites

In addition to the efficiency of the anodic and cathodic reactions the corrosion process depends on the interlinking and balancing of the two reactions, by equivalent electron flow through the metal and ionic current flow through the electrolyte. Electron flow through the metal essentially proceeds unhindered. Thus, any restriction in the electrolytic path between the anodic and cathodic sites will manifest itself as an increase in internal resistance (resistivity) of the corrosion cell, causing a reduction in current flow and hence rate of corrosion.

The conceptually derived Evans diagram at Figure 5.13 schematically shows the effect of increasing electrolytic resistance of concrete between

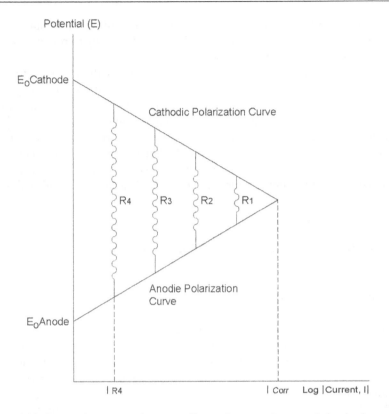

Figure 5.13 Evans diagram indicating effect of increasing anode/cathode resistance (concrete resistivity) on corrosion rate. (Courtesy of Treadaway, 1991)

anodic and cathodic areas on both corrosion current flow (rate) and anode and cathode potentials (Treadaway, 1991). An increase in the anode/cathode resistance leads to a reduction in corrosion current flow and hence corrosion rate [from I_{corr} (no resistance) to I_{R4} (greatest electrolytic resistance/resistivity)].

5.2.8 Potential difference between anodic and cathodic sites

A practical example in reinforced concrete structures of large potential differences between anodic and cathodic sites on reinforcement, as well as a case where strong cathodic polarisation (see Section 5.2.5) does not lead to limitation of pitting corrosion, occurs in severely chloride contaminated concrete, which has a high moisture content, and where, therefore oxygen diffusion is limited (Treadaway, 1991). Under these conditions ionic current flow can occur over large distances making a large cathode available such that while cathodic efficiency on a unit area basis of reinforcement surface is low, the combined effect over large areas of reinforcement surface can lead to high current flow at anodic sites, and additionally to potentials polarised to low negative values (Treadaway, 1991).

Pitting is highly likely in these circumstances and will be limited to a few anodic sites of high activity where corrosion will be intense and deep penetrating attack can occur. Highly acidic conditions will then be developed within the pits ensuring the stability of soluble iron (II) corrosion products. Furthermore, oxidation of corrosion products from iron (II) to iron (III) will be restricted due to limited oxygen supply. The importance of this corrosion situation is that little precipitation of corrosion product will occur and, therefore, the typical development of cracking, delamination, and spalling of cover concrete will be minimised.

5.3 REINFORCEMENT CORROSION PROGRESS

It has been suggested earlier that the two major causes of reinforcement corrosion are the general corrosion that results from carbonation and the pitting corrosion that results from attack by chloride ions. In each case it takes time for the aggressive agent to diffuse through the concrete to the surface of the steel and initiate corrosion. Tuutti (1982) recognised this and proposed the corrosion model in Figure 5.14 whereby there is negligible or no corrosion before the initiation time t_i immediately after which corrosion becomes active and propagates.

The Tuutti model has since been developed by others (e.g. Browne et al., 1983; Everett & Treadaway, 1985; Melchers & Li, 2006) with considerations of both the corrosion initiation and corrosion propagation periods. An example which gives rise to a variation of degree of deterioration with time is shown in Figure 5.15 (De Vries, 2002).

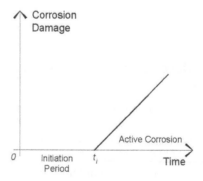

Figure 5.14 Classical Tuutti model showing no corrosion (damage) until t_i followed immediately by development of serious corrosion (damage). (Courtesy of Tuutti, 1982)

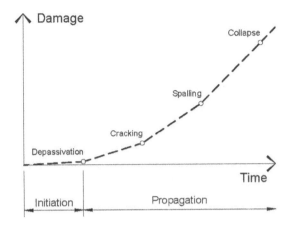

Figure 5.15 The progress of reinforcement corrosion. (Courtesy of De Vries, 2002)

In relation to the progress of chloride-induced reinforcement corrosion, Melchers (2017) has proposed on the basis of field studies (e.g. Melchers et al, 2009) and the examination of many reports (Melchers & Li, 2006) that the some limited amount of corrosion may commence at the initiation time t_i but that serious long-term corrosion generally does not commence until some later time t_{act}, refer Figure 5.16. In this model, the period $(0–t_i)$ has essentially zero or negligible corrosion. This is followed, commencing at the initiation time t_i by some corrosion for a relatively short period of time, after which there is no or only a very modest increase in corrosion, until the commencement, at the activation time t_{act}, of serious, damaging corrosion. Melchers and Li (2006) surmised that the corrosion process in the time period 0-t_i differs from that causing the serious long-term corrosion and that the active corrosion stage (t_{act} onwards) was the result of loss of concrete alkalis and in particular to the loss of $Ca(OH)_2$ (portlandite) as accelerated by the presence of chlorides.

'Metastable pitting and pit growth mechanistic considerations' have also been considered in chloride-induced reinforcement corrosion models, refer Figure 5.17 as an example (Green et al., 2019a, b).

Ehsman et al. (2018) have recently surmised for marine structures (up to 109 years of age) they investigated, that the 'active corrosion model' proposed by Melchers (2017) and the 'metastable pitting and pit growth mechanistic considerations' proposed by Green et al. (2017) are important explanations as to why little visible damage to concrete surfaces was observed despite very high chloride contents (up to 0.39% by weight concrete) at the depth of reinforcement.

Green et al. (2019a, b) report of an investigation and subsequent localised repair of a reinforced concrete wharf structure in the Port of Newcastle, New South Wales, Australia, constructed circa 1977. Determined chloride

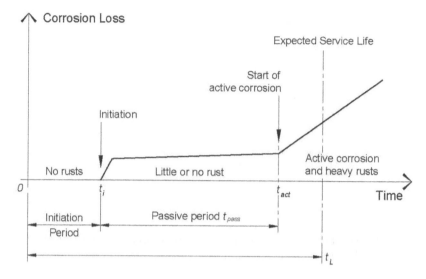

Figure 5.16 Modified corrosion model with separation between initiation at t_i and start of active corrosion at t_{act}. (Courtesy of Melchers, 2017)

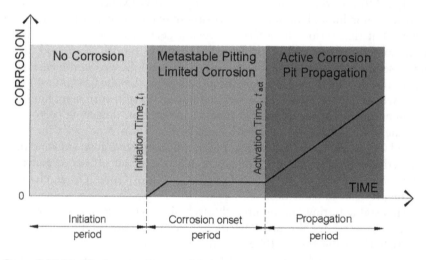

Figure 5.17 Modified corrosion model incorporating 'metastable pitting and pit growth mechanistic considerations'. (Courtesy of Green et al., 2019a, b)

levels at reinforcement depth in substructure beams and slabs are considered very high (i.e −0.32% to 0.69% by weight of concrete), however, the diagnosed and observed corrosion to reinforcement was less than expected and typically localised to areas of cracked cover concrete and poor concrete quality areas. It is surmised that the 'active corrosion model' and 'metastable pitting and pit growth mechanistic considerations' are important explanations as to why little visible damage to most concrete surfaces was observed despite very high chloride contents at reinforcement depth.

5.4 MODELLING CHLORIDE-INDUCED CORROSION INITIATION

Chloride ions can penetrate into hardened concrete under various transport mechanisms:

- Diffusion of chloride through the pore water due to a chloride concentration gradient.
- Permeation under a differential hydraulic pressure.
- Absorption (sorption or convection) into unsaturated concrete due to a differential moisture concentration.
- Wicking action (by sorption and by sorption and permeation where the latter may also be referred to as transpiration).
- Evaporation from the concrete surface can leave an elevated chloride concentration at the drying front when under wick action by sorption and permeation (transpiration). This chloride deposition is reduced at external and internal surfaces subject to frequent non-contaminated water washing the surface.

Diffusion of chloride ions into concrete is often considered a dominating mechanism in various conditions, while other transport mechanisms may only occur in some specific situations. Some models that have been developed for chloride penetration into concrete maintain the diffusion process is the key part of durability design for reinforced concrete structures exposed to chloride. In practice, however, many factors affect chloride ingress including (but not limited to): chloride binding/adsorption as determined by cement (binder) type; pore moisture content; carbonation; leaching; evaporation; etc. However, the chloride ingress models that are presented in *fib* (Fédération internationale du béton – International Federation for Structural Concrete) technical bulletins/codes (*fib* 34, 2006; *fib* 76, 2010) as well as concrete durability standards (ISO, 2012) are empirical models based on diffusion and use mathematical expressions developed to fit chloride profile data from many structures at different ages and using different materials. The expressions used resemble solutions to Fick's Second Law (Fick, 1855) for a constant diffusion coefficient (D) but this should be considered coincidental. The key parameter linked to chloride ingress in the models is 'D_{app}'. Because the expression looks like a diffusion expression, 'D_{app}' is sometimes mistaken for a diffusion coefficient. It is not. It is simply a term used to fit the profile data to the empirical expression (Papworth & Gehlen, 2016) and is more accurately described as a penetration coefficient. Understanding this helps resolve misunderstandings about the *fib*/ISO models.

The Concrete Institute of Australia have produced a comprehensive series of recommended practice (RP) documents on concrete durability and RP 5 (of 7) in that series is specifically on 'Durability Modelling of Reinforcement Corrosion in Concrete Structures' (Concrete Institute of Australia, 2020). This document covers thoroughly the various approaches that have been adopted to model chloride-induced corrosion initiation including deterministic analytical modelling, deterministic numerical modelling through to full probabilistic analysis. It is considered beyond the scope of this book to discuss these various approaches other than to advise of some of the contents of the Concrete Institute of Australia RP (Z7/05) as well as the recommendations made in terms of choice of models.

Discussion is provided in the Concrete Institute of Australia RP (Z7/05) on relevant matters such as:

- Durability design approaches (deemed-to-satisfy; avoidance of deterioration; full probabilistic; and, deterministic modelling (including partial factor)).
- General principles for modelling of reinforcement corrosion (corrosion models; limit states; design life; reliability assessment; conservative inputs, reliability, partial factors, and scaling factors).
- Modelling chloride ingress in concrete:
  - Basic models of chloride diffusion (Fick's First Law; Diffusion with free and bound chlorides; Fick's Second Law).
  - Initial chloride concentration (base chloride level).
  - Surface convection zone.
  - Time-dependent diffusion property (models using the time varying diffusion coefficient (Fick); 'Apparent' diffusion coefficient (D_{app}); Relationship between D_{app} and diffusion coefficient determined by Fick's Second Law).
  - Sources of input data related to time dependence of diffusion.
  - Measurement of the diffusion coefficient.
  - Effect of temperature.
  - Surface chloride concentration.
  - Threshold chloride concentration.
  - Example of full probabilistic analysis using aging factors.
  - Example of full probabilistic analysis using scaling factors.

It is then noted in the Concrete Institute of Australia RP Z7/05 (2020) that the solution to Fick's Second Law (Fick, 1855) for a diffusion coefficient which does not vary with time, refer Equation 5.6, has had wide exposure in the literature.

$$\frac{C}{C_s} = erfc\left(\frac{X}{2\sqrt{Dt}}\right) \tag{5.6}$$

Where
C is the concentration of chloride ions.
C_s is surface chloride concentration.
$erfc$ is the complement error function.
X is distance.
D is the diffusion coefficient.
t is time.

Equation 5.6 also appears in a modified form, refer Equation 5.7, in which the diffusion coefficient has been replaced by a time varying property differing from the Fick coefficient (as this would be applicable in Equation 5.6 only if it did not vary), described as the 'apparent coefficient of diffusion'. ISO 16204 (2012) uses a notation which conveniently suggests the name 'D_{app}'.

$$\frac{C-C_0}{C_s-C_0} = erfc\left(\frac{X-\Delta X}{2\sqrt{D_{app}t}}\right) = erfc\left(\frac{X-\Delta X}{2\sqrt{D_{ar}\left(\frac{t_r}{t}\right)^\alpha t}}\right) \tag{5.7}$$

Where
C_0 is the initial (background) chloride concentration in concrete.
ΔX is the depth of the convection zone.
D_{app} is the time varying apparent diffusion coefficient.
D_{ar} is the value of D_{app} at the reference exposure time t_r.
α = age factor.

This relationship (though in some instances with no allowance for a convection zone, i.e. ΔX effectively set to zero) may be found in Concrete Society Technical Report (TR) 61 (2004), *fib* Bulletin 34 (2006), and ISO 16204 (2012).

ISO 16204 (2012), *fib* Bulletin 34 (2006) and *fib* Bulletin 76 (2010) document how the model should be applied and within defined limits the Concrete Institute of Australia RP Z7/05 (2020) recommends it provides serviceable results whereby D_{app} can be determined directly from chloride profiles in exposed structures.

The Concrete Institute of Australia RP Z7/05 (2020) advises that the deterministic analytical solution to Fick's Second Law (Fick, 1855) for a time varying diffusion coefficient, refer Equation 5.8 (Tang & Gulikers, 2007), enables the concentration of chloride in concrete exposed to marine and other chloride environments to be calculated in accordance with diffusion theory, taking into account both the age of the concrete prior to exposure, and the period of exposure.

$$\frac{C-C_0}{C_s-C_0} = erfc\left(\frac{X-\Delta X}{2\sqrt{\dfrac{D_r}{1-n}\left[\left(1+\dfrac{t_{e0}}{t}\right)^{1-n}-\left(\dfrac{t_{e0}}{t}\right)^{1-n}\right]\left(\dfrac{t_r}{t}\right)^n t}}\right) \tag{5.8}$$

Where

t_{e0} is the age of concrete when exposure starts.

t_r is the reference concrete age at test or measurement.

t is the duration of exposure (and t_c equals $t_{e0} + t$, where t_c is the concrete age).

When designing for long service lives (say, 50 years or more), the expression approximates to a much simpler form, Equation 5.9, which also enables input values to be adopted from sources which have not used the exact Fick relationship. Concrete Society Technical Report TR61 (2004), referenced in ISO 16204 (2012), provides data which is readily translated for use in Equation 5.9.

$$\frac{C-C_0}{C_s-C_0} = erfc\left(\frac{X-\Delta X}{2\sqrt{Pt^{(1-n)}}}\right) \tag{5.9}$$

Where

P is a single time-independent parameter to represent the concrete's resistance to chloride ingress by diffusion.

Chloride ingress models are 'stochastic', i.e. they have no unique solution as the input variables are not fixed values but are distributed around mean values. Hence a full probability analysis is the only way to ascertain the reliability at which a specific design life will be achieved with particular concrete performance values.

Full probabilistic models for predicting the time to chloride-induced reinforcement corrosion initiation have been developed (e.g. *fib* 34, 2006; Kessler & Lasa, 2019; Concrete Institute of Australia, 2020). The corresponding limit state is defined as the state when a critical chloride concentration (threshold) reaches the reinforcement level. The outcome of the full probabilistic corrosion initiation prediction is a reliability index (β) or failure probability in dependence of time that can be compared with a specific target reliability (Kessler & Lasa, 2019).

Normally, reinforcement corrosion first impairs the serviceability of the structure. Serviceability limit states correspond to conditions beyond which specified service requirements for a structure or structural member are no longer met (EN 1990, 2002). Since reinforcement corrosion is an irreversible serviceability limit state, ISO 2394 (2015) proposes 1.5 as the target reliability (β). The *fib* Model Code for Service Life Design (*fib* 34, 2006) recommends 1.3 (β), which corresponds to a 10% probability of reinforcement depassivation.

5.5 MODELLING CARBONATION-INDUCED CORROSION INITIATION

Carbonation of concrete leads to the formation of a front, with carbonated concrete between the front and the concrete surface, which progresses to greater depth with the elapse of time.

The rate of progress of the carbonation front depends on the level of moisture in the concrete, the penetrability of the concrete, the cement (binder) composition, and the concentration of $CO_2(g)$ in the atmosphere. The rate can be accelerated by exposing a sample to an atmosphere with a greatly increased concentration of $CO_2(g)$. Standard tests (e.g. Duracrete, 1998) designed to accelerate the process in this way provide an accurate guide to the response of the concrete to conditions encountered in service (Concrete Institute of Australia, 2020).

fib Bulletin 34 (2006) provides a deterministic analytical model for the progress of carbonation in concrete, refer Equation 5.10, in which natural carbonation rate is related to the performance in an accelerated test.

$$X = \sqrt{2k_e k_c \left(k_t R_{ACC,0}^{-1} + \varepsilon_t\right) C_{ss}} \sqrt{t} \cdot W(t) \qquad (5.10)$$

Where
X is the carbonation depth (mm).
k_e is the environment factor to consider the influence of humidity in the exposure condition.
k_c is the execution transfer parameter to consider the curing condition before exposure.
$R_{ACC,0}^{-1}$ is the inverse carbonation resistance obtained in the DuraCrete (1998) standard accelerated test, $(mm^2/yr)/(kg/m^3)$.
k_t is the regression parameter to consider the carbonation difference between the accelerated test and that in the natural condition.
ε_t is the regression intercept to consider the carbonation difference between two test conditions, $(mm^2/yr)/(kg/m^3)$.
C_s is the carbon dioxide $(CO_{2(g)})$ concentration of the exposure environment, kg/m^3.
$W_{(t)}$ is the weather function to consider wetting events.

Carbonation models, like chloride models, are 'stochastic', i.e. they have no unique solution as the input variables are not fixed values but are distributed around mean values. Hence a full probability analysis is the only way to ascertain the reliability at which a specific design life will be achieved with particular concrete performance values.

Full probabilistic modelling is available through the *fib* Bulletin 34 (2006) procedure. The RP document Z7/05 (Concrete Institute of Australia, 2020) also provides full probabilistic modelling examples.

5.6 MODELLING CORROSION PROPAGATION

5.6.1 Background

The reinforcement corrosion propagation period can be short (e.g. GP cement-based concretes in chloride environs) but in some reinforced concrete elements it can be significant, for example, those with supplementary cementitious materials (SCMs) (blended cement-based) or relatively dry environments (i.e. insufficient water in the concrete for reinforcement corrosion). Durability modelling of the reinforcement corrosion propagation period is not well established (Concrete Institute of Australia, 2020).

To predict the duration of the corrosion propagation period, two major parameters need to be determined, namely (Concrete Institute of Australia, 2020):

i. The critical criterion for deterioration damage by reinforcement corrosion.
ii. Reinforcement corrosion rate in various conditions.

5.6.2 Corrosion damage criterion

The criterion for corrosion failure of a reinforced concrete structure depends on the permissible deterioration level for the limit states specified by the asset owner or any other stakeholders (Concrete Institute of Australia, 2020). Cracking of cover concrete may be the corrosion damage criterion (indicator) and consequently the serviceability limit state (SLS). Time to concrete spalling damage criterion/indicator may be the SLS (Li, 2004).

However, where the volume of corrosion products is small, such as in pitting corrosion, or in low oxygen/high chloride environments, significant section loss can occur without visible evidence such as rust staining, cracking, delamination, or spalling of cover concrete. If the deterioration continues undetected, the element may eventually fail suddenly and unexpectedly. Cracking, time to concrete spalling, etc as corrosion damage criterion and SLS would therefore be inapplicable in such circumstances.

Otieno et al. (2011) point out that prediction of corrosion propagation is a complex process mainly due to the difficulty in incorporating all the relevant factors affecting the process and the associated reinforcement corrosion-induced damage in a prediction model. The pre-defined corrosion damage criterion/indicator can then be said to denote the end of the corrosion propagation period, however, this criterion can range from loss of steel cross section, loss in stiffness, loss of steel-concrete interface bond, cracking of concrete cover, to local, or global failure of the structure or its members respectively. However, for repair purposes, Otieno et al. (2011) note that global failure (collapse of the structure) cannot be adopted as a limit state due to human safety reasons. They advise that a detailed coverage of these limit states can be found in the literature; but in summary, some of the basic requirements of a reinforcement corrosion-induced limit state criterion/indicator adopted should include:

i. It should be easy to assess and quantify.
ii. The level of damage should not compromise structural integrity such as its stability and hence safety of the users/occupants.
iii. The damage should be relatively easy to repair in terms of restoring both structural integrity and durability performance requirements.

5.6.3 Factors affecting corrosion rates

Various factors affect the corrosion rate as discussed earlier in this chapter as well as the previous chapter. The Concrete Institute of Australia durability modelling recommend practice document (2020) notes that environment conditions, concrete mix design, concrete quality, and the depth of cover concrete, can affect the availability of oxygen and moisture for example. The composition of the cementitious materials (including admixtures), influence the electrical conductivity of pore solutions and the corrosion rate also depends on the cause of depassivation (i.e. carbonation or chloride penetration). The corrosion rate therefore should be estimated using input data relevant to chloride-induced or carbonation-induced reinforcement corrosion as appropriate.

Furthermore, the Concrete Institute of Australia RP (2020) proposes that the frequency of wet/dry cycles is critical for the corrosion rate and must be well defined for specific structures and elements. For example, variable rainfall and ocean/inland water tidal range needs to be carefully considered for wet/dry cycle impact. Therefore, project specific assessment must be completed to determine the exposure type (e.g. a low frequency wet/dry may be durability assessed as moderate humidity) and corrosion rate impact for chloride contaminated and carbonated concrete (i.e. impact varies).

Otieno et al. (2011) mention also, for example, that the effects of both load- and corrosion-induced cracking on corrosion rate might need to be considered.

5.6.4 Corrosion rates – chloride contaminated concrete

There have been significant efforts in research to predict the corrosion rate in chloride contaminated concretes. However, there is no widely accepted

Table 5.3 Corrosion rates of reinforcement in chloride contaminated concrete

Exposure type	Corrosion rate (μm/y)	Standard deviation(μm/y)	Distribution
Wet-rarely dry	4	3	Lognormal
Cyclic wet-dry	30	20	Lognormal
Airborne sea water (saline aerosols)	30	20	Lognormal
Submerged	Not expected except bad concrete (i.e. poor quality) or lower cover	–	Lognormal
Tidal/splash zone	70	40	Lognormal

Source: Duracrete (1998)

model available for such a prediction due to the many factors that impact the kinetics of the reactions involved.

DuraCrete (1998) provided corrosion rate estimates at 20°C for chloride contaminated concrete in various exposure conditions as shown in Table 5.3. The corrosion rate is assumed to be zero in submerged conditions due to lack of oxygen to fuel the cathodic reaction. In wet conditions, the corrosion rate proposed is slightly higher, at 4 μm/yr. In cyclic wet-dry conditions or in airborne sea water (saline aerosols) conditions, a higher corrosion rate of 30 μm/y is proposed. In tidal/splash zone, the highest corrosion rate of 70 μm/y is estimated.

High standard deviation values are presented for the different exposure types because of the very high level of variation related to such factors as cover thickness, concrete w/b ratio, concrete temperature, chloride content and cementitious materials (binder) types. High end values of corrosion rate within the range are expected in conditions with low cover, high w/b ratio, high temperature, high chloride content near the reinforcement, and a GP cement-based concrete without SCMs. With SCMs, a significantly lower corrosion rate is expected due to for example, a higher electrical resistivity (Polder, 1996). Therefore, corrosion rates for specific conditions should be selected based on exposure, concrete cover, binder type and concrete quality etc (Concrete Institute of Australia, 2020).

5.6.5 Corrosion rates – carbonated concrete

Significant efforts have been made in research to predict the corrosion rate in carbonated concrete, but again no widely accepted model is available. Therefore, a simple method to determine the corrosion rate is considered to be more suitable (Concrete Institute of Australia, 2020).

DuraCrete (1998) provides corrosion rate estimations at 20°C for carbonated concrete under various moisture conditions as shown in Table 5.4.

Table 5.4 Corrosion rates of reinforcement in carbonated concrete

Exposure type	Corrosion rate (µm/y)	Standard deviation(µm/y)	Distribution
Dry (<60% humidity)	0	0	Lognormal
Wet-rarely dry (unsheltered)	4	3	Lognormal
Moderate humidity (sheltered)	2	1	Lognormal
Cyclic wet-dry (unsheltered)	5	3	Lognormal

Source: Duracrete (1998)

High standard deviation values are again presented as related to factors such as variation in cover thickness, concrete w/b ratio and concrete temperature. High end values of corrosion rate within each range are expected in conditions with low cover, high water/cement ratio and a high temperature. SCMs may increase or decrease the carbonation rate depending on their composition and addition rate. (Although not necessarily advantageous to reducing carbonation, SCMs may be used for other reasons e.g. mitigating potential damage caused by alkali silica reaction (ASR) or early age thermal cracking.) Therefore, the corrosion rate for specific conditions should be selected based on the specific exposure, concrete cover, binder type, concrete quality etc (Concrete Institute of Australia , 2020).

5.6.6 Length of corrosion propagation period

5.6.6.1 Background

Otieno et al. (2011) indicate that regardless of the corrosion damage criterion/indicator adopted, prediction models for the length of the corrosion propagation period can be grouped as either analytical, numerical, or empirical depending on the set of conditions used in their development.

Empirical models are based on assumed direct relationships between corrosion rate and basic concrete parameters, e.g. w/b ratio, binder type and environmental parameters (Otieno et al., 2011). They are usually developed using data from laboratory factorial experiments that, by design, isolate other corrosion-influencing parameters. Otieno et al. (2011) advise that empirical models are sub-divided into three types viz:

i. Expert Delphic oracle models.
ii. Fuzzy logic models.
iii. Models based on electrical resistivity and/or oxygen diffusion resistance of concrete.

Otieno et al. (2011) propose that three different approaches can be used to develop numerical models viz:

i. Finite element method (FEM).
ii. Boundary element method (BEM).
iii. Resistor networks and transmission line method.

Analytical models of the length of the corrosion propagation period are based on closed-form solutions to mathematical equations, i.e. the solutions to the (known) theoretical equations used to describe the system and/or the changes in the system can be expressed as a mathematical analytical function. This approach has been mostly used to model corrosion-induced cracking using a thick-walled cylinder approach (Otieno et al., 2011).

The RP Z7/05 document of the Concrete Institute of Australia (2020) has considered a 'General Model' and 'Andrade's (2014) Model' for predicting the length of the reinforcement corrosion propagation period.

5.6.6.2 General model

In some cases, the propagation period may be short because oxygen supply to cathodic areas will be readily available and the concrete resistivity readily supports ionic current flow (e.g. moist GP cement-based concrete). Where there is no information to suggest oxygen diffusion control kinetics or resistance control kinetics will limit the corrosion rate, the length of the reinforcement corrosion propagation period has been simply expressed as Equation 5.11 below (Concrete Institute of Australia, 2020):

$$T_1 = \frac{\delta_{CR}}{R_{corr}} \tag{5.11}$$

Where
T_1 is the length of corrosion propagation period, yrs.
R_{corr} is the rate of reinforcement corrosion, μm/year.
δ_{CR} is the reinforcement radial loss required for corrosion-induced cracking, μm, refer Equation 5.12, (Webster, 2000).

$$\delta_{CR} = 1.25 \cdot C \tag{5.12}$$

Where
δ_{CR} is the reinforcement radial loss required for corrosion-induced cracking, μm.
C is the concrete cover to reinforcement, mm.

For example then, if the concrete cover to reinforcement is 70 mm, the corrosion depth to initiate cracking in cover concrete is estimated to be

70 × 1.25 = 87.5 μm. DuraCrete (1998) recommends assuming a worst-case corrosion rate of 70 μm/year (see Table 5.3). Thus, for marine tidal/splash exposure conditions, the time from corrosion initiation to cracking for 70 mm cover depth is estimated to be 87.5 μm/70 μm/year = 1.25 years (Concrete Institute of Australia, 2020).

5.6.6.3 Andrade (2014) model

Andrade (2014) has proposed that the reinforcement corrosion propagation period can be predicted by Equation 5.13 as below (Concrete Institute of Australia, 2020):

$$T_1 = \frac{P_{corr} \rho_{ef} \left(\dfrac{t}{t_0}\right)^q \xi}{K_{corr} 0.00116} \tag{5.13}$$

Where

T_1 is the length of the corrosion propagation period, yrs.

P_{corr} is the maximum limit of steel cross section at the end of time T_1, cm (or mm).

t is the time during the corrosion, yrs.

t_0 is the initial age of the concrete (normally 28 days or 0.0767 years) at exposure, yrs.

q is the aging factor of the resistivity (normally taken as a value of 0.22 for Cement I, 0.37 for cement II/A-P and 0.57 for Cement II /A-V; these are European Cement Classifications).

ξ is the environmental factor of the corrosion rate (it can be of 10±2 for carbonation and 30±5 for chlorides).

K_{corr} is a constant with a value of 26 μA/cm²·kΩ·cm (= to 26 mV/cm) relating the resistivity and the corrosion current I_{corr}.

As the propagation period can be a significant part of a structure's design life in some conditions, this model may be useful in estimating propagation period. However, as this model does not have a significant amount of support data on real world structures, a high partial factor of at least 3 should be considered to the average propagation period estimated (Concrete Institute of Australia, 2020).

This model can also be used to calculate the required 28 day resistivity to achieve a certain propagation period and to develop mix design parameters to achieve the required resistivity (Concrete Institute of Australia, 2020).

5.6.6.4 Andrade (2017) model

More recently Andrade (2017) has considered that the accumulated corrosion (P_{corr}) is the damage accounted at a determined time while the corrosion

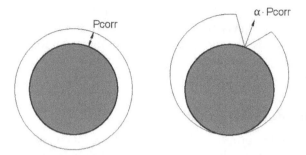

Figure 5.18 Homogeneous and localised corrosion. (Courtesy of Andrade, 2017)

rate (V_{corr}) is the velocity (corrosion by time unit). The values of accumulated corrosion should then be expressed in μm/year instead of μA/cm² in order to visualise that this parameter gives the loss in cross section calculated as equivalent homogeneous corrosion. Then, the simplest model of the propagation period, t_p, is given in Equation 5.14 for homogeneous corrosion (where ϕ_0= the initial diameter, ϕ_t= the diameter at time t (corroded)):

$$t_p = \frac{(\phi_0 - \phi_t)}{V_{corr}} = \frac{P_{corr}}{V_{corr}} \tag{5.14}$$

In the case of localised attack, a 'pitting factor' α can be introduced into that equation which takes into account the deepest corrosion spot, refer Equation 5.15 (Andrade, 2017). The remaining cross section is that having as its diameter that deepest pit (Figure 5.18) in order to be conservative. Andrade (2017) then advises that there may be several pits of different depths in the same bar, but that being the deepest is that which can induce a stress concentration leading to a premature failure. In the case of asymmetric corrosion, the remaining cross section should be the most conservative one due also to the uncertainty on the real shape of the area lost. It has been shown that a value of α=10 is a reasonable average value (Andrade, 2017).

$$P_{pit} = P_{corr} \cdot \alpha \tag{5.15}$$

5.7 DESIGN LIFE ACHIEVEMENT

Simply put, in terms of durability design for chloride- or carbonation-induced reinforcement corrosion, the philosophy can then be:

    a. To design for the time to corrosion initiation to equal the design life.
    or

b. To design for the time to corrosion initiation plus time to corrosion propagation to equal the design life (and to a particular limit state or corrosion damage criterion in the definition of design life).

Further discussion of design life achievement, service life design, limit states, durability design, and planning, etc is provided at Chapter 12.

REFERENCES

Ailor, WH (1971), '*Handbook on Corrosion Testing and Evaluation*', John Wiley and Sons, New York, USA.

Andrade, C (2014), '*Resistivity test criteria for durability design and quality control of concrete in chloride exposures*', Concrete in Australia, November.

Andrade, C (2017), 'Linear propagation models of deterioration processes of concrete', (In) *Reinforced Concrete Corrosion, Protection, Repair & Durability*, (Eds) W K Green, F G Collins, and M A Forsyth, ISBN: 978-0-646-97456-9, Australasian Corrosion Association Inc., Melbourne, Australia.

Aylward, GH and Findlay, TJV (1974), '*SI Chemical Data*', Jacaranda Wiley, West Perth, Western Australia.

Browne, RD, Geoghegan, MP and Baker, AF (1983), '*Corrosion of Reinforcement in Concrete Construction*', ed A.P. Crane, Ellis Horwood, London, UK, 193.

Concrete Institute of Australia (2020), '*Durability Modelling of Reinforcement Corrosion in Concrete Structures*', Concrete Durability Series Recommended Practice Z7/05, Sydney, Australia.

Concrete Society (2004), '*TR61 Enhancing Reinforced Concrete Durability*', Technical Report 61, The Concrete Society, Surrey, UK.

De Vries H (2002), 'Durability of concrete: A major concern to owners of reinforced concrete structures', (In) *Concrete for Extreme Conditions*, (Eds) R K Dhir, M J McCarthy & M D Newlands, Thomas Telford, London.

DuraCrete (1998), 'Modelling of Degradation - Probabilistic Performance Based Durability Design of Concrete Structures', EU - Brite EuRam III, Contract BRPR-CT95-0132, Project BE95-1347/R4-5, December, 174.

Ehsman, J, Melchers, R and Green, W (2018), 'Durability of Reinforced Concrete Marine Structures Up to 109 Years', *fib* Congress 2018, Melbourne, Australia, 7–11 October.

EN 1990 (2002), '*Eurocode – Basis of Structural Design*', Eurocode, London, UK.

Everett, LH and Treadaway, KWJ (1985), '*Deterioration Due to Corrosion in Reinforced Concrete*', BRE Information Paper 12/80, Building Research Establishment, Garston, UK.

fib Bulletin 34 (2006), '*Model Code for Service Life Design*', International Federation for Structural Concrete, Lausanne, Switzerland.

fib Bulletin 76 (2010), '*Benchmarking of Deemed to Satisfy Provisions in Standards: Durability of Reinforced Concrete Structures Exposed to Chlorides*', International Federation for Structural Concrete. Lausanne, Switzerland.

Fick, A (1855), 'On Liquid Diffusion', *Philosophy Magazine* (in English), S.4, Vol. 10, 30–39.

Green, W K (1991), '*Electrochemical and chemical changes in chloride contaminated reinforced concrete following cathodic polarisation*', MSc Dissertation, University of Manchester Institute of Science and Technology, Manchester, UK.

Green, WK, Collins, FG and Forsyth, MA (2017), 'Up-to-date overview of aspects of steel reinforcement corrosion in concrete', (In) *Reinforced Corrosion, Protection, Repair and Durability*, (Eds) W K Green, F G Collins, and M A Forsyth, ISBN 978-0-646-97456-9, Australasian Corrosion Association Inc., Melbourne, Australia.

Green, W, Linton, S, Katen, J and Ehsman, J (2019a), 'Do nothing and patch repair (without anodes): Engineered maintenance experiences of marine concrete structures', *Corrosion and Materials*, Vol 44, No 3, 60–67.

Green, W, Ehsman, J, Linton, S, Cambourn, N and Masia, S (2019b), 'Chloride Durability and Future Maintenance of a 40+ Year Marine Structure', *Concrete in Australia*, Vol 45, No 4, December, 51–58.

ISO 16204 (2012), '*Durability – Service Life Design of Concrete Structures*', International Organization for Standardization (ISO), Geneva, Switzerland.

ISO 2394 (2015), '*General Principles on Reliability of Structures*', International Organization for Standardization (ISO), Geneva, Switzerland.

Jastrzebski, Z D (1987), '*The Nature and Properties of Engineering Materials*', John Wiley, New York, USA.

Kessler, S and Lasa, I (2019), 'Study on probabilistic service life of florida bridges', *Materials Performance*, Vol 58, No 10, October, 46–51.

Li, CQ (2004), 'Reliability based service life prediction of corrosion affected concrete structures', *ASCE J Struct Eng*, 130 (10), 1570–1577.

Melchers, RE and Li, CQ (2006), 'Phenomenological modelling of corrosion loss of steel reinforcement in marine environments', *ACI Materials Journal*, 103(1), 25–32.

Melchers, RE, Li, CQ and Davison, MA (2009), 'Observations and analysis of a 63 year old reinforced concrete promenade railing exposed to the North Sea, *Magazine of Concrete Research*, 61(4), 233–243.

Melchers, RE (2017), 'Modelling durability of reinforced concrete structures', (In) *Reinforced corrosion, protection, repair and durability*, (Eds) W K Green, F G Collins, and M A Forsyth, ISBN 978-0-646-97456-9, Australasian Corrosion Association Inc., Melbourne, Australia.

Otieno, MB, Beushausen, HD and Alexander, MA (2011), 'Modelling of corrosion propagation in reinforced concrete structures – A critical review', *Cement & Concrete Composites*, 33, 240–245.

Papworth F and Gehlen C (2016), 'National and international code based deterministic and full probabilistic modelling to describe reliability of various Australasian marine structures', *fib* Symposium, Performance Based Approaches for Concrete Structures, Cape Town, November.

Polder, RB (1996), 'The influence of blast furnace slag, fly ash and silica fume on corrosion of reinforced concrete in marine environment', *HERON*, Vol 41, 287–300.

Pourbaix, M (1966), '*Atlas of Electrochemical Equilibria in Aqueous Solutions*', Pergamon Press, Oxford, UK.

Potter, EC (1961), '*Electrochemistry: Principles and Applications*', Newnes, London, UK.

Standards Australia (2008), '*AS 2832.5 Cathodic Protection of Metals Part 5: Steel in Concrete Structures*', Sydney, Australia.

Tang, L and Gulikers, J (2007), 'On the mathematics of time-dependent apparent chloride diffusion coefficient in concrete', *Cement and Concrete Research*, Vol 37, No 4, 589–595, April.

Tuutti, K (1982), 'Corrosion of Steel in Concrete', Report Fo. 4.82, Swedish Cement and Concrete Association.

Treadaway, KWJ (1991), 'Corrosion Propagation', COMETT Short Course on the Corrosion of Steel in Concrete, University of Oxford, UK.

Webster, MP (2000), 'The assessment of corrosion-damaged concrete structures', PhD Thesis to University of Birmingham, Birmingham, UK.

Chapter 6

Condition survey and diagnosis (A) – on-site measurements

6.1 PLANNING A CONDITION SURVEY

Knowledge of where corrosion is occurring, together with information regarding the causes of corrosion, is always necessary (essential) for effective repairs. Although the initial planning of a condition survey is likely to be based upon second-hand reports of signs of distress, no proper plan can be made until the investigator has made a personal examination of the site. The initial visual examination is primarily designed to determine which parts of the structure, if any, should be selected for a more detailed investigation. Electrical and electrochemical on-site measurements are designed to investigate the present corrosion state of the reinforcing metal without disrupting (much of) the surface of the concrete. The final stage of the investigation involves disruption of at least the existing concrete and the off-site physical and chemical analysis of the concrete and sometimes even the reinforcement in order to determine the mechanism(s) of the corrosion that has taken place. Only when this is understood can there be confidence in the remediation methods that are proposed. The on-site methods of visual examination, electrical and electrochemical methods and other non-destructive test (NDT) methods will be dealt with in this chapter and the chemical analysis and other laboratory-based methods in the subsequent chapter. The conclusion of any condition survey is the preparation of a report to the commissioning authority and this will be considered at the end of the next chapter.

The Concrete Institute of Australia (2014) defines a condition survey as a process whereby information is acquired relating to the current condition of the structure with regards to its appearance, functionality, and/or ability to meet specified performance requirements with the aim of recognising important limitations, defects, and deterioration. A wide range of parameters may be included within a condition survey with data being obtained by activities such as visual inspection and various forms of testing. A condition survey would also seek to gain an understanding of the (previous) circumstances which led to the development of that state, together with the associated mechanisms causing damage or deterioration (Concrete Institute of Australia, 2014).

A condition survey is also sometimes incorrectly referred to as a condition assessment and/or a condition evaluation. It is considered worthwhile here that the Concrete Institute of Australia (2014) terminology for condition assessment and condition evaluation be reproduced for information:

- **Condition Assessment:** A process of reviewing information gathered about the current condition of a structure or its components, its service environment, and general circumstances, whereby its adequacy for future service may be established against specified performance requirements for a defined set of loadings and/or environmental circumstances.
- **Condition Evaluation:** Similar to condition assessment but may be applied more specifically for comparing the present condition rating with a particular criterion, such as a specified loading. Condition evaluation generally considers the requirement for any later intervention which may be needed to meet the performance requirements specified.

Chess and Green (2020) then point out that a condition survey should only be undertaken by independent consultants and not consultants that are conflicted nor by contractors (who are obviously conflicted) or organisations that wear multiple hats (e.g. consult/supply/install).

6.2 VISUAL INSPECTION

A visual inspection will locate areas of advanced reinforcement corrosion, concrete defects, and deterioration. It will not give any information about reinforcement condition in apparently sound concrete. Visual inspections would typically focus on the following features:

1. Spalls.
2. Cracks.
3. Rust stains.
4. Weathering.
5. Insufficient compaction honeycombing.
6. Dampness.
7. Efflorescence.

The first three of these have been dealt with in previous chapters. Excessive weathering is usually an indication that far too lean a mix was used in the first place and honeycombing results from the capillary pores that are left in the concrete as a result of inadequate compaction during the construction process. Dampness in (say) a wall is usually an indication of severe cracking and/or poor quality concrete (e.g. inadequate compaction, mix segregation, etc) behind the visible surface. Similarly, efflorescence (white deposits on the surface of the concrete) arises when the water that results from excessive penetration through the concrete evaporates and leaves calcium carbonate deposited on the surface. This has been described previously.

Visual inspection may also involve examining surfaces for evidence of damage such as:

- **Surface scaling:** that may be related to traffic or natural weathering combined with freeze-thaw cycles, crystallisation of salts (wick action, capillarity, permeation and transpiration), or possibly construction-related defects.
- **Surface softening:** that may be related to attack by soft or aggressive water, acids, sulphates, and organics.
- **Surface erosion:** related to trafficking or water flow over an otherwise sound surface.
- **Surface deposits:** that may harbour bacteria and other microorganisms.

Visual inspection may also involve quantifying visible features such as:

- The location of defects and the area or length of concrete affected.
- The depth of surface scaling or softening.
- The orientation, length, and approximate width of cracks and their relationship to the position of reinforcement.

Visual inspection in terms of fire-damaged structures and buildings may include recording of such features as collapse, distortion, deflections, degree of damage to materials and smoke damage.

A significant issue for visual inspections is their subjectivity: they are undertaken by different people who even with training will give different ratings to the scale of deterioration. Hence, sound training, and provision of standard defect descriptions and ratings is required. Major asset owners may have their own inspection manuals and inspector accreditation schemes. General guides to assessing the visible condition of concrete structures are found in technical publications such as Concrete Institute of Australia Z15 (2011), Bamforth/CIRIA C660 (2007), UK Concrete Society Technical Report TR 22 (2010), and American Concrete Institute ACI 224.1R-07 (2012).

6.3 CRACKS

Cracks in concrete are inevitable really. Section 2.2 discussed the different forms of cracking to concrete when it is in the plastic or hardened state. During a condition survey or as part of quality assurance during construction, cracks in concrete will likely need to be recorded, widths measured, depths possibly, and also perhaps crack movement.

Recording of cracks is typically as a crack map where crack measurements (width, depth and orientation) are also undertaken. Crack width measurements are often initially carried out using a crack width gauge (or crack width meter). This can be a transparent gauge on which is marked a series of lines of increasing width and by positioning the gauge over a crack its width may be estimated as in Figure 6.1. A crack width gauge/meter is shown in Figure 6.2. In general, this method is accurate to about ±0.1 mm.

Measuring crack width at the concrete surface is not as straightforward as it seems because crack width varies along crack length. In addition, crack width changes with concrete temperature (i.e. cracks close as the temperature increases) and moisture content (cracks close as concrete moisture content increases). A representative number of measurements is also sometimes necessary including additional positions near maximum crack widths. The broken top corner arris of a crack must also be identified separately to the

Figure 6.1 Crack width gauge. (Courtesy of Shane Linton)

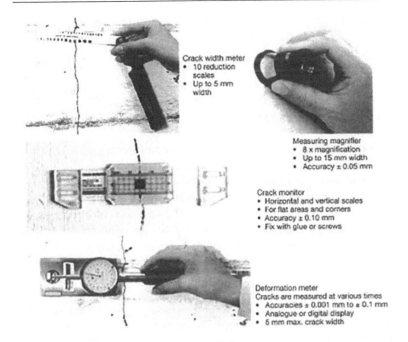

Crack width meter
• 10 reduction scales
• Up to 5 mm width

Measuring magnifier
• 8 x magnification
• Up to 15 mm width
• Accuracy ± 0.05 mm

Crack monitor
• Horizontal and vertical scales
• For flat areas and corners
• Accuracy ± 0.10 mm
• Fix with glue or screws

Deformation meter
Cracks are measured at various times
• Accuracies ± 0.001 mm to ± 0.1 mm
• Analogue or digital display
• 6 mm max. crack width

Figure 6.2 Different types of crack measuring equipment. (Courtesy of Concrete Institute of Australia, 2015, p.5:4)

crack width for example. Cracks can taper with depth and hence depth of crack may need to be established. A selection of crack measuring instruments is shown in Figure 6.2 (Concrete Institute of Australia, 2015).

Concrete Institute of Australia recommended practice document Z7/07 (2015) details the following when crack widths and crack depths are to be measured, for example, as part of quality assurance testing during construction:

a. The width values should be reported as maximum for each crack. A mean and distribution (with variance) can then be assessed for comparison with the design values. The mean and variance values require taking sufficient measurements to obtain a statistically representative sample (e.g. 30 measurements are likely to be sufficient). Planning for crack measurements is critical as the quantity needs to consider the number and length of cracks on specific structure elements or portions of large sized structures.

b. A measuring magnifier (refer Figure 6.2) be used to give an accuracy of ±0.05 mm. As previously, a crack width gauge (crack width meter) has an expected accuracy of ±0.1 mm. With an appropriate crack width distribution, the characteristic crack width can be assessed and used as the crack width.

c. Crack widths should be measured during the coolest part of the day when they have a maximum width. This will make decision making conservative. The impact of cooler weather crack widening needs to be considered when measurements are taken in hot weather conditions.

d. Concrete core samples of 50 mm diameter can be taken through representative crack positions to reinforcement depth to evaluate crack depth and width beneath the concrete surface. This approach is recommended where the concrete crack width measured at the surface is a concern or is non-compliant. When the core is removed from the structure, the loss of restraint from the surrounding concrete may result in widening of the crack, which requires evaluation for the specific crack and the measured crack width impact on durability, structural adequacy, and contract requirements.

Where changes in crack width are to be assessed for operational or structural purposes, demountable strain gauges (accuracy to ±0.01 mm) or surface installed vibrating wire strain gauges are often used. Where a lower accuracy is acceptable, the crack width can be monitored using a crack monitor ('tell-tale gauge'), refer Figure 6.2, or a deformation meter, refer Figure 6.2 (Concrete Institute of Australia, 2015).

Measuring microscopes are useful for more precise work such as in laboratory research, but their accuracy is often greater than the crack movement and the measuring process is too slow to be practical for use on-site (Concrete Institute of Australia, 2015).

6.4 DELAMINATION DETECTION

The swelling of the reinforcing bar that results from the corrosion of reinforcing steel may cause hollow areas or delamination of the surrounding concrete. Many devices may be used to detect delamination. The choice of device depends on the extent of delamination detection and geometry of the structure. Horizontal structures such as bridge decks can be examined using a chain-drag. This has been described by Van Daveer (1975) and consists of segments of link chain that are dragged along the concrete surface. A noticeably different sound is produced when the chain-drag encounters delaminated concrete. A hammer test can be used. As is the case with the chain-drag method, a noticeably different ('hollow/drummy') sound is produced when a hammer is struck against delaminated concrete.

Ultrasonic flaw detection, impulse radar, thermographic, and even X-ray techniques can also be used for the detection of delaminated areas, but in general the simpler methods such as hammer tapping (sounding) are adequate for the task in hand.

Hammer tapping is also utilised to detect hollow sounding delaminated material to fire-affected element surfaces. Simple invasive techniques such as hammer and chisel may also be used (Ingham, 2009).

6.5 CONCRETE COVER

Finding where the reinforcement is and what is the depth of concrete cover over it is often a part of a condition survey. (It would be wrong to assume that the drawings of the structure adequately represent what was actually built.) Both for detecting where the reinforcement is and its cover, a cover-meter is most commonly used. There are a number of portable, battery powered commercial instruments available with varying degrees of accuracy and varying abilities to allow for different diameters of reinforcing bar. The simplest of these generate a magnetic field by passing an alternating current through an induction coil and then detect the field generated by the core of the induction coil by means of a second detector coil, refer Figure 6.3. The presence of any steel within the magnetic field of the generator coil affects that magnetic field and these changes are detected by the search coil. The instrument can be calibrated so that from these changes in the magnetic field the location of the steel and its distance from the search head can be determined. Other instruments make use of the eddy currents generated within

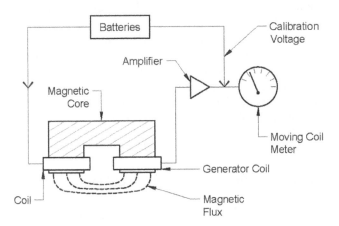

Figure 6.3 Typical simple covermeter circuitry. (Courtesy of Bungey & Millard, 1996, p.163)

Figure 6.4 Example of simple magnetic reluctance covermeter.

any metal that is in the magnetic field and measure the impedance changes. Basic calibration of the instruments is critical to their use, but once calibrated they are often accurate to ±2 mm.

This information can be used to assess whether inadequate cover was a factor contributing to corrosion problems and is particularly useful in estimations of when the carbonation front and/or critical chloride concentration may penetrate to the reinforcement level. Covermeter measurements are not easy and experience plays a large part in their interpretation. Instruments have to be calibrated for the diameter of the bar and particular problems arise when there is a mass of bars located near to one another. A typical simple magnetic reluctance covermeter is shown in Figure 6.4.

Reinforcement location/spacing/position and depth of cover over reinforcement may also be non-destructively determined using ground penetrating radar (GPR) and ultrasonic pulse echo (UPE) (Concrete Institute of Australia, 2015).

6.6 ELECTROCHEMICAL MEASUREMENTS

6.6.1 Electrode (half-cell) potential mapping

The corrosion state of a portion of reinforcing bar can be determined from its potential with respect to the concrete in which it is immersed. A reinforcing bar with an intact passive film will rest at a more positive (less negative) potential than reinforcing bar that is in an actively corroding state. In general it is not possible to place a reference electrode in the immediate environment of the reinforcing bar but it has been found that measuring the potential of the reinforcing bar with respect to a reference electrode located on the (accessible) surface of the concrete can give a valuable indication of what is going on several centimetres below the surface even when there are no signs of corrosion on the surface. The determination of the electrode (half-cell) potential between embedded reinforcing steel and the concrete surface has become the most widely used electrochemical technique for delineating where corrosion is active.

The technique is described in ASTM C876 (2015). Figure 6.5 shows the circuitry used to measure the electrode (half-cell) potential. The positive end of a high impedance ($>10^6$ ohms) voltmeter is electrically connected to the reinforcement. This involves locating the reinforcement with a covermeter and removal of the concrete by drilling, chipping, or coring to expose the steel. The condition of the rebar should be noted. A connection is made to the reinforcing steel. The negative end of the voltmeter is connected to a copper/copper sulphate reference electrode (half-cell) (CSE). (Although the standard refers to a copper/copper sulphate reference electrode, other standard reference electrodes such as silver/silver chloride or manganese/manganese dioxide can be used with appropriate corrections to the potential reading.) The reference electrode (half-cell) is moved to each measuring point on the concrete surface and the reading is recorded. The convention, described at Section 5.1.2, is that the potential of a metal is measured with respect to the environment so that the positive terminal on the voltmeter is

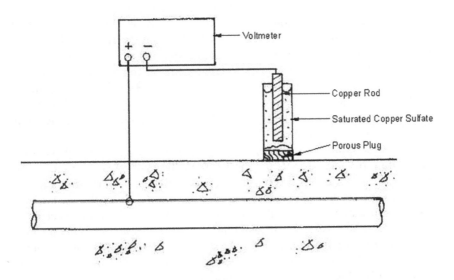

Figure 6.5 Measuring electrode (half-cell) potentials of reinforcing steel in concrete. (Courtesy of Bungey & Millard, 1996, p.166)

Table 6.1 The ASTM criteria for the probability of corrosion

Potential with respect to Cu/CuSO₄ electrode (CSE)	Corrosion condition
More positive than −200 mV	Less than 10% probability of corrosion
Between −200 mV and -350 mV	Between 10% and 90% probability of corrosion
More negative than −350 mV	Greater than 90% probability of corrosion

connected to the metal and the potential measured is usually negative. Positive readings may indicate that the concrete is too dry and corrosion currents cannot be detected. In most cases it is necessary to pre-wet the concrete surface with water before taking readings.

The interpretation of potentials suggested by ASTM C876 (2015) is given in Table 6.1. However, these criteria were originally developed for a specific situation. Factors such as the differing cover thickness and concrete resistivity can make a considerable difference to the actual potentials measured.

A major problem with the electrode (half-cell) potential survey technique therefore is the correct interpretation of results and the possibility of making a totally misleading evaluation. While the criteria may be true for American (Southern Californian) bridge decks that have received deicing salt treatment they are not always applicable to other types of structures. Practical experience has shown that an electrode potential survey alone cannot be used to judge the condition. Arup (1977) and others have suggested that different potential criteria are applicable to different structures and environments. Browne et al. (1983) found that carbonation of the concrete results in potentials less negative than expected, leading to the possibility of corrosion occurring at potentials more positive than −0.35 V CSE. The magnitude of the carbonation effect is around 100 mV. A variation in chloride concentration in the pore water between reinforcement and concrete surface may also give misleading results.

It has been noted by Stark (1984) and others that corrosion is also associated with differences in potential between locations of 100 mV or more. The pattern of the iso-potential contour map also provides information on type and location of corrosion, refer Figure 6.6. Anodic areas (because of the convention that the potential of the metal is measured with respect to its environment the most negative potential for the metal corresponds to the most positive for the concrete) are usually associated with steep potential gradients and characteristic whirlpools. Therefore, the shape of the contour map must be closely studied in addition to the values of potential.

Other difficulties with an electrode (half-cell) potential survey include access problems and the necessity to wet the concrete prior to measuring

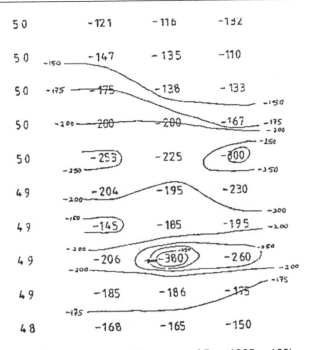

Figure 6.6 Potential contours map. (Courtesy of Foo, 1985, p.188)

potentials. However, despite the difficulties and shortcomings electrode (half-cell) potential mapping is probably the most valuable technique readily available to detect reinforcement corrosion activity. The results from potential measurements require careful analysis and should be used in conjunction with other tests such as delamination detection, cover measurements, concrete breakouts to reinforcement, and concrete resistivity measurements.

6.6.2 Polarisation resistance

A widely used method for the determination of the rate at which reinforcement may be corroding is the polarisation resistance technique. The polarisation resistance (R_p) technique (or sometimes called linear polarisation resistance, LPR) for the determination of the corrosion current was introduced by Stern and Geary (1957). If the specimen is polarised by a small amount to a new potential that is ΔE mV negative to the corrosion potential, then the ratio ΔE to i_i the impressed current, is termed the polarisation resistance, i.e:

$$\frac{di_i}{d\Delta E} = \frac{1}{R_p}$$

From the Tafel equation for the relationship between the current flowing and the applied potential for the individual (anodic and cathodic) corrosion reactions:

$$\Delta E = a + b \log i$$

and

$$i_p^a = i^a{}_{corr} \exp\left(\frac{2.3\Delta E}{b_a}\right)$$

and so, where i_p^a is the anodic current density at the shifted potential, i_{corr}^a the anodic current density at the corrosion potential and b_a the slope of the Tafel plot of E vs $\log i$.

As there is a similar expression for i_p^c and since $\Delta E_a = -\Delta E_c$ then if $i_i = i_p^a - i_p^c$:

$$i_i = i_{corr} \exp\left(\frac{2.3\Delta E}{b_a}\right) - i_{corr} \exp\left(\frac{-2.3\Delta E}{b_c}\right)$$

differentiating this equation with respect to ΔE yields at the corrosion potential when $\Delta E = 0$ and $i^a{}_{corr.} = i^c{}_{corr} = i_{corr}$:

$$\frac{di_i}{\Delta E} = 2.3 i_{corr}\left(\frac{1}{b_a} + \frac{1}{b_c}\right)$$

where i_i is the impressed current per unit area of electrode. If ω is the total area of corroding electrode involved, the ratio of the total applied current to the potential shift is given by the $\omega I/\eta$. Since:

$$\frac{1}{\omega R_p} = \left[\frac{dI_i}{\Delta E}\right]_{E=E_{corr}}$$

$$i_{corr} = \frac{1}{2.3\omega} \frac{b_a b_c}{b_a + b_c} \frac{1}{R}$$

if R is reported in ohms and ω in m², i_{corr} has the units of A/m². This equation has been simply written by a number of authors as:

$$i_{corr} = B/R$$

and has been used by a number of authors to calculate the corrosion current density.

Calculations of B have been made using b_a and b_c the constants of the Tafel equation plot of E against $\log i$. b_a is often assumed on the basis of experiments carried out in aqueous solution to be about 60 mV/decade. b_c

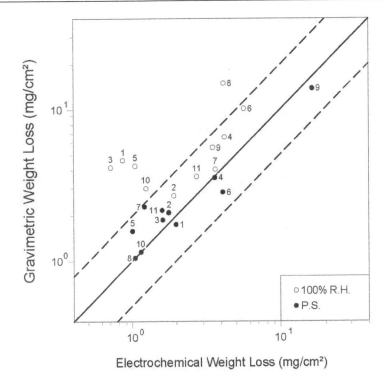

Figure 6.7 Relationship between R_p and rate of metal loss. (Courtesy of Andrade et al., 1983)

is often similarly assumed to be 120 mV/decade. B should then have a value of 40 mV and is often reported with values of between 14 and 52 mV. A typical set of results is shown in Figure 6.7 (Andrade et al., 1983). Professor Andrade is one of the foremost proponents of the use of polarisation resistance as a technique for assessing the corrosion rates of reinforcement in concrete. Her results indicate the accuracy with which the polarisation resistance can be related to a rate of metal loss.

If uniform metal loss is assumed the conversion factor relating corrosion current density to rate of metal loss (by Faraday's Law) is 1 μA.cm^{-2} = 11.6 μm.yr^{-1}. On this basis it is possible to relate the measured polarisation resistance to the rate of metal loss by means of the equation:

$$x = \left(11 \times 10^6 B\right) / \left(R_p . A\right)$$

Where x is the corrosion rate in μm.yr^{-1}, R_p is the polarisation resistance in ohms and A is the surface area of the steel under the influence of the polarising electrode in cm^2.

Broomfield (1997) has given the following 'broad criteria for corrosion':

Passive condition: $I_{corr} < 0.1$ μA.cm^{-2}.

Low to moderate corrosion: I_{corr}, 0. 1 to 0.5 μA.cm^{-2}.

Moderate to high corrosion: I_{corr}, 0.5 to 1 μA.cm^{-2}.

High corrosion rate: I_{corr}, > 1 μA.cm^{-2}.

The derivation of an expression for the polarisation resistance carried out by Stern and Geary was based upon simple activation polarisation in solutions whose electrical resistance could be neglected. Such conditions are not necessarily obtained for reinforced concrete and so, the value of B and ω may vary widely dependent upon the conditions of the survey. Thus, determinations of the R_p, while they do not necessarily yield accurate values for the rates of corrosion do yield comparative rates when the conditions of measurement are standardised.

6.7 CONCRETE RESISTIVITY

The concrete resistivity can also act as a measure of the ability of the concrete to act as an electrolyte and to carry corrosion (ionic) current. Resistivity can be measured by the Wenner four-electrode method as described by ASTM G57 (2012). The experimental set up is shown in Figure 6.8.

An AC or DC current is passed through the two outer electrodes of four equally spaced electrodes linearly arranged on the concrete surface. The potential drop across the two inner electrodes is measured. The resistance of the material between the inner electrodes is given by:

$R = V/I$

And the resistivity of the concrete is given by:

$\rho = 2\pi aR$

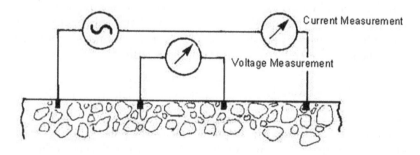

Figure 6.8 Measurement of concrete resistivity. (Courtesy of Pullar-Strecker, 1987, p.35)

Best position, minimal interaction with rebars

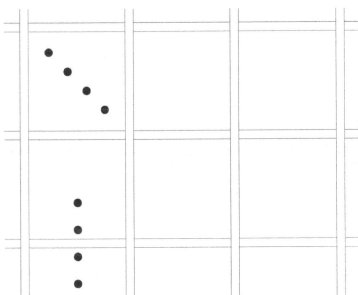

Least interaction if probe array is greater than bar spacing

Figure 6.9 Placing of Wenner electrodes to avoid rebar interference. (Courtesy of Broomfield, 1997, p.61)

Where a is the electrode spacing. Formulae for situations when there is an unequal spacing between the electrodes are given in ASTM G57 (2012). The measured resistivity is that of the concrete which lies with a distance about equal to the electrode spacing from the surface.

The electrodes must be placed so that there is little possibility of the current passing through the reinforcing bar and Broomfield (1997) has suggested the arrangements shown in Figure 6.9 as the best way of avoiding interference from the reinforcement.

An interpretation of resistivities is given in Table 6.2.

Table 6.2 Interpretation of concrete resistivities

Resistivity (Ω.cm)	Corrosion rate
<10,000	Cannot be Assessed (or very high if concrete not 'too wet')
10,000–50,000	Moderate to high
50,000–100,000	Low
>100,000	Generally non-corrosive

6.8 OTHER MEASUREMENTS

Other measurements are made to most commonly detect defects within the concrete. Defects may be detected to a limited extent by visual inspection and/or core sampling. A range of NDT methods is available for determining whether defects are present over a wider area within concrete structures and buildings.

6.8.1 Rebound hammer

The rebound hammer can be used to locate areas of poor quality concrete. A rebound hammer consists of a spring-loaded hammer which when released strikes a ball ended steel plunger which is placed in contact with the concrete surface, refer Figure 6.10. The rebound distance of the steel hammer from the plunger is measured and is an indication of the quality of the concrete (but more correctly the surface hardness or rebound of the concrete). The test is described in ASTM C805 'Standard Test Method for Rebound Number of Hardened Concrete' (2018).

Interpretation of rebound numbers as suggested by Browne et al. (1983) is given in Table 6.3.

The Rebound Number of the concrete depends not only on the compressive strength of the concrete but also the local hardness at the point of

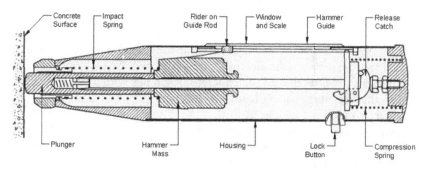

Figure 6.10 Rebound hammer. (Courtesy of Bungey and Millard, 1996, p.34)

Table 6.3 Interpretation of rebound hammer results

Average rebound number	Concrete quality
>40	Good
30–40	Fair
20–30	Poor
<20	Delaminated

Source: Browne et al. (1983)

impact. Thus, it is influenced by many factors including the presence of aggregate or paste at the point of impact; surface texture/finish, cleanliness, moisture condition, and carbonation; concrete age; and type of coarse aggregate. Although a measure of the quality and condition of the concrete surface, it is also affected by whether the element is supported or suspended, and on the thickness of suspended elements. Consequently, different concrete mixes of the same nominal strength, or the same concrete in different placements, will give different rebound numbers.

6.8.2 Ultrasonic pulse velocity

Ultrasonic Pulse Velocity (UPV) measures the average velocity of a compression wave over the path between two transducers – a source transducer and a receiver transducer. This velocity has a direct relationship to the compressive strength, elastic modulus, and density of concrete, thus it can be used to assess concrete quality. It is also affected by the presence of cracks, voids, and other discontinuities, therefore it may also be used to detect the extent of such defects.

The test can be carried out in several ways:

- Direct transmission, where the source transducer is opposite the receiver, refer Figure 6.11.
- Semi-direct transmission, where the source transducer and the receiver transducer do not face each other but are located on different faces of an element such as across a corner.
- Indirect transmission, where the source transducer, and receiver transducer are placed on the same surface, refer Figure 6.12.

Figure 6.11 UPV measurement – direct transmission.

Figure 6.12 UPV measurement – indirect transmission.

Standards providing guidance on both the use of and interpretation of UPV readings are BS EN 12504-4 (2004), ASTM C597–09 (2009), and ACI 228.1R (2003).

6.8.3 Ultrasonic pulse echo

Ultrasonic pulse echo (UPE) uses a device with numerous (48 or more) transmitter/receivers mounted in one head, refer Figure 6.13.

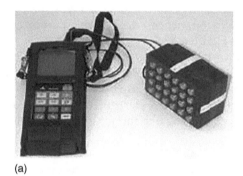

(a)

(b)

Figure 6.13 UPE measurement – numerous transmitter/receivers mounted in one head.

The small diameter shear wave transmitters are spaced at up to 30 mm centres and when triggered the array of signals produces a cross sectional image through the concrete. The frequency of the waves emitted can be increased or decreased dependent on the geometry of the scenario (i.e. thick element, large targets etc.).

In general use, the low frequency reflected waves are taken to represent flaws, voids or the concrete rear surface as these represent the largest change in impedance (which causes the waves to reflect).

In high frequency mode, the reinforcement is detected, and in some cases the device can be used to give a reasonable estimate of cover.

6.8.4 Impact echo

Elastic, low frequency transient stress waves are produced at a test point by tapping a small weight against the test element's concrete surface, refer Figure 6.14.

The waves travel through the concrete and are reflected by any discontinuities caused by changes in acoustic impedance (partially affected by concrete density). A transducer held on the surface adjacent to the impact point detects the reflected waves as they rebound between the concrete surface and discontinuities.

Where defects such as voids or delaminations provide a significant barrier to the wave, reflections of the wave energy cause an amplitude peak at the corresponding frequency and the distance to the defect can be calculated.

6.8.5 Ground penetrating radar

Ground Penetrating Radar (GPR) uses radio waves to pick up changes in the dielectric properties inside an element, refer Figure 6.15.

This technology makes use of a transmitter sending radio waves into the substrate, and a receiver recording the reflected waves. The waves are reflected where the dielectric constant changes.

Defects such as voids or delaminations can produce dielectric constant changes. The depth of such changes can be calculated using the time of flight and an estimation of wave velocity.

The location of steel embedments in concrete can be determined by GPR including concrete cover most particularly where reinforcing is particularly congested.

6.9 CARBONATION DEPTH

The attack of concrete by carbonation has been described in Section 4.8. As the pH of the concrete is lowered by the ingress of carbon dioxide gas so it loses its capacity to maintain a passive film on the surface of the steel. By

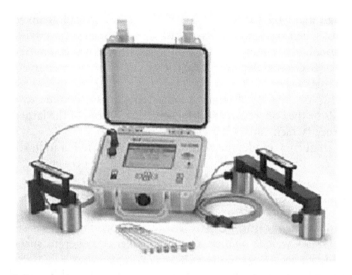

(a)

(b)

Figure 6.14 Impact echo (IE) measurement.

convention concrete is said to be carbonated when its pH has been lowered to such an extent that the acid-base indicator phenolphthalein (which is pink/purple in alkaline solution) becomes colourless. The depth of carbonation can therefore be useful in determining the cause of existing rebar corrosion and also may be useful to predict when corrosion may commence if carbonation continues at its present rate. The depth of carbonation is usually measured by spraying a phenolphthalein solution on a fresh surface of the concrete. The surface turns pink/purple where the pH is greater than around 9 whereas it remains uncoloured at lower pH values. Since the

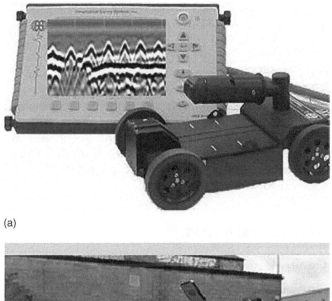

(a)

(b)

Figure 6.15 GPR measurement.

colour change for phenolphthalein is about 9 it does not indicate the regions where the alkalinity has been lowered to pH values between around 9 and 10 due to the early stages of the carbonation processes. Consequently, since the passive layer on steel breaks down (is dissolved) at pH 10 there may be a region perhaps 5 mm in depth ahead of the carbonation front where corrosion can take place, but this is rarely a disadvantage of the test.

A useful phenolphthalein solution may be prepared by dissolving 1 gram of phenolphthalein indicator in 100 ml of a 50:50 mixture of methylated spirits and water. Broomfield (1997) recommends a higher proportion of alcohol to water. This is then sprayed using a hand-held sprayer on to a freshly prepared surface. This should be done within a few minutes of exposing the surface as carbonation of the surface layer by carbon dioxide from the air takes place remarkably quickly. The carbonation depth can be measured on core samples, re-broken spalls, or in situ fresh fracture surfaces.

A common method of measuring the depth of carbonation is to drill a small (say 12 mm) hole into the surface of the concrete, spray the indicator solution into the hole and measure the depth at which the colour appears. If the test is carried out on the drillings from a single hole, the degree of alkalinity may be raised erroneously because the drill has ground and exposed uncarbonated cement particles. Another possible source of error can arise if the solution is sprayed into a drilled hole that is coated with dust. The dust should be removed before the test is carried out.

6.10 CONCRETE SAMPLING

6.10.1 General

Sampling from a concrete structure is often required to assess one or more properties. The sample type, size, locations, and frequency of collection will depend on the properties to be determined, the reliability required for the results and the practicalities and cost of testing.

Many test methods have specific requirements for the size and condition of samples. It is necessary to be familiar with such requirements before specifying a sampling programme. Unless sampling particular anomalies, a primary requirement is often to ensure that the extracted samples represent the bulk of the concrete in terms of properties such as aggregate distribution and voids.

The most common method for taking samples is by wet diamond coring. Where this is not convenient, drilled dust samples may be taken.

Before taking the sample, the reinforcement position should be located and marked so that the sample can be taken over the reinforcement to inspect the physical condition of the reinforcement and measure the cover depth for verification of covermeter readings, or to avoid the reinforcement as required.

Furthermore, before taking the sample the concrete surface should be inspected and any defects recorded.

6.10.2 Wet diamond coring

Diamond drilled cores, refer Figure 6.16, for chemical analysis of concrete commonly have diameters from 30 mm to 150 mm.

A 30 mm core is adequate for visual examination of deterioration induced by chemical attack. If the concrete maximum aggregate size is 20 mm or

Figure 6.16 Diamond drilled coring.

larger, a 30 mm core will not be suitable for chemical analysis (e.g. chloride or sulphate ion content) as the sample aggregate/cement (A/C) ratio might not be representative of the concrete. For example, a single piece of coarse aggregate could occupy the entire cross section of the core at one or more locations along the core length. If 30 mm diameter cores are the largest that can be taken for chemical analysis, three such cores should be extracted and combined to provide a representative sample.

Cores of 50 mm diameter provide sufficient area to give representative samples for chemical analysis provided that the depth increments selected are at least 10 mm thick.

Often it is preferable to cut a core in half longitudinally and measure chloride content on one half and cement content or carbonation on the other half. Cores of 75 mm diameter provide sufficient cross sectional surface area for both halves of the core to give representative samples.

Many physical test methods prescribe requirements for the geometry of the core sample. For example, for compressive strength testing AS 1012.14 (Standards Australia, 2018) states that 'The diameter of cores shall be not less than the greater of 75 mm or three times the nominal size of coarse aggregate in the concrete ...'. Other test methods specify the sample size by the face area of the core, or the volume of concrete comprising the core.

Cores of 100 mm diameter are often considered to be the standard size for diffusion, penetrability, and strength testing. However, 75 mm diameter cores can often be used as the minimum size for concrete with aggregate size up to 20 mm.

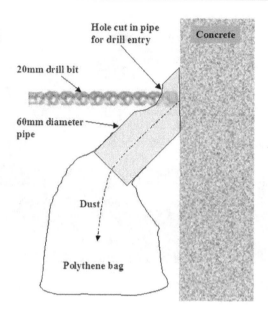

Figure 6.17 Method for taking drilled dust samples.

6.10.3 Drilled Dust Samples

A hand-held percussion drill with a 20–25 mm diameter tungsten carbide tipped drill bit is used for taking drilled dust samples. A 60 mm diameter pipe cut off at 45 degrees and with a drill bit entry hole is used for collecting dust, as shown in Figure 6.17.

Drilled dust samples extracted using a 20–25 mm bit will not provide a representative sample unless a number of adjacent holes are drilled, and the dust combined. The depth intervals will depend on the cover depth. Sampling beyond the cover depth will help determine the depth of contamination as well as background levels of the contaminant.

6.11 REPRESENTATIVENESS OF INVESTIGATIONS, TESTINGS, AND SAMPLING

It should be noted that in order to obtain representative test results, the locations of on-site testing and sampling must be representative of the population. In determining the locations to be tested, due consideration should be given to the configuration of the structure/element, the macro, meso and micro exposures and the method of construction (Concrete Institute of Australia, 2015). When developing the testing plan, the Concrete Institute of Australia (2015) proposes that the item to be assessed (structure or exposure) should be split into zones where the results can reasonably be expected to be similar. In that way they advise that the sample results expressed as a mean and variance will give a true statistical representation of the zone.

REFERENCES

American Concrete Institute (2003), '*In-place Methods to Estimate Concrete Strength*', ACI 228.1R:2003, Farmington Hills, Michigan, USA.

American Concrete Institute (2012), '*Causes, Evaluation, and Repair of Cracks in Concrete Structures*', ACI 224.1R-07, Farmington Hills, Michigan, USA.

American Society of Testing Materials (2009), '*Standard Test Method for Pulse Velocity through Concrete*', ASTM C597-09, West Conshohocken, Pennsylvania, USA.

American Society of Testing Materials (2012), '*Standard Test Method for Field Measurement of Soil Resistivity Using the Wenner Four-Electrode Method*', ASTM G57-06 (Reapproved 2012), West Conshohocken, Pennsylvania, USA.

American Society of Testing Materials (2015), '*Standard Test Method for Corrosion Potentials of Uncoated Reinforcing Steel in Concrete*', ASTM C876-15, West Conshohocken, Pennsylvania, USA.

American Society of Testing Materials (2018), '*Standard Test Method for Rebound Number of Hardened Concrete*', ASTM C805/C805M-18, West Conshohocken, Pennsylvania, USA.

Andrade, C, Molina, A, Huete, F and Gonzalez, JA (1983) 'Relation between alkali content of cements and corrosion rates of the galvanised reinforcements', in *Corrosion of Reinforcement in Concrete Construction* ed A.P. Crane, Ellis Horwood, Chichester, UK.

Arup, H (1977), 'Discussion on electrochemical removal of chlorides from concrete bridge decks', *Materials Performance*, 16 (11).

Bamforth, PB (2007), '*Early-age thermal crack control in concrete*', CIRIA C660, London, UK.

British Standards Institute (2004), '*BS EN 12504-4 'Testing Concrete. Determination of Ultrasonic Pulse Velocity*', European Committee for Standardization, Brussels.

Broomfield, JP (1997), '*Corrosion of Steel in Concrete*', Chapman and Hall, London, UK.

Browne, RD, Geoghegan, MP and Baker, AF (1983), '*Corrosion of Reinforcement in Concrete Construction*', ed A.P. Crane, Ellis Horwood, London, UK.

Bungey, JH and Millard, SG (1996), '*Testing of Concrete in Structures*', Blackie, London, UK.

Chess, P and Green, W (2020), '*Durability of Reinforced Concrete Structures*', CRC Press, Taylor & Francis Group, Boca Raton, FL, USA.

Concrete Institute of Australia (2011), '*Cracking in Concrete Slabs on Ground and Pavements*', Recommended Practice Z15, Sydney, Australia.

Concrete Institute of Australia (2014), '*Durability Planning*', Recommended Practice Z7/01, Concrete Durability Series, Sydney, Australia.

Concrete Institute of Australia (2015), '*Performance Tests to Assess Concrete Durability*', Recommended Practice Z7/07, Concrete Durability Series, Sydney, Australia.

Concrete Society (2010), '*Non-Structural Cracks in Concrete*', Technical Report TR22, Camberley, Surrey, UK.

Foo, M (1985), '*Study of the Potential Field Associated with a Corroding Electrode*', PhD Thesis, Monash University, Melbourne, Australia.

Ingham, J (2009), 'Forensic engineering of fire-damaged structures', *Proceedings of the Institution of Civil Engineers – Civil Engineering*, 162, May, 12–17.

Pullar-Strecker, P (1987), '*Corrosion Damaged Concrete*', Butterworths, London, UK.

Standards Australia (2018), 'AS 1012.14 Method for securing and testing cores from hardened concrete for compressive strength', Sydney, Australia.

Stark, D (1984), 'Determination of permissible chlorid.e levels in prestressed concrete', *PCI Journal*

Stern M and Geary AL (1957) 'Electrochemical polarization – 1 A theoretical analysis of the shape of polarisation curves', *Journal of the Electrochemical Society*, 104, 56–63.

Van Daveer, J R (1975), 'Techniques for evaluating reinforced concrete bridge decks', *ACI Journal*, December.

Chapter 7

Condition survey and diagnosis (B) – laboratory measurements

7.1 GENERAL

The various methods that have been described in the previous chapter and which form the basis of on-site investigations may give an indication of the extent of the corrosion of the reinforcement in the concrete or the extent of deterioration of the concrete. Such methods are mainly directed to the question of whether corrosion and deterioration is occurring and if so, how fast are they occurring. Apart from a few cases when the cause of the corrosion is obvious, (ponding of rainwater, exposed inadequately protected fittings attached to the reinforcement and the like), the next question of 'Why is it corroding?' or 'Why is it deteriorating?' will have to be determined by laboratory testing.

Even in the presence of high concentrations of aggressive agents, good concrete can prevent corrosion, by preventing the aggressive agents from penetrating through the concrete to the reinforcement. The penetration of ions to the metal surface is however facilitated by poor concrete quality and substandard construction practice. It is a very common occurrence that corrosion of reinforcement is associated with inadequate cover of concrete. Since the penetration of most aggressive agents is governed by principally diffusion kinetics then the time taken for the concentration of such an aggressive agent to build up to a critical value at any given point within the concrete is proportional to the square of the distance of that point from the surface. In general terms if 50 mm cover will give 10 years of life, 100 mm will give 100 years of life. As an example, it is recommended in AS 3600 (Standards Australia, 2018) that structures in marine environments built using 40 MPa concrete must have a minimum cover of 45 mm in fully submerged conditions. Corrosion problems are almost inevitable if the concrete cover is lower than the recommended values in AS 3600.

The corrosion of embedded steel usually takes the form either of pitting or uniform (general) corrosion. Pitting is favoured when hydrolysis reactions are sustained and a large cathode/anode area ratio exists. The products of the corrosion reaction are soluble at the low pH conditions near the anode. Therefore, considerable corrosion can occur without the concrete cracking or spalling. This, of course, is the most dangerous form of corrosion, because it can lead to considerable loss of section of the steel with no evidence of corrosion at the surface of the concrete. The possibility of an unpredicted collapse is therefore much greater when pitting corrosion obtains. Uniform (general) corrosion is favoured by carbonation or by excessive chloride content, so a large number of closely situated pits are formed (in the case of the latter). The corrosion produced is solid rust (hydrated haematite, 'red rust') which has a much greater volume than the original steel. Hence, cracking, and spalling occurs at a relatively early stage of the corrosion process, which means that an unsightly rust stain may precede by some time the possibility of structural collapse.

If the underlying cause of the corrosion or deterioration is likely to be a deficiency of the concrete, the corrosion investigator is often required to determine what particular deficiency of the concrete is causing the problem. At this stage, contractual and legal considerations start to play an increasingly large part in the overall investigation and it becomes even more vital than usual that all documentation associated with the investigation is adequately completed (dates, locations of samples, instructions to analysts) properly recorded and preserved for future reference, sometimes many months, or even years later.

Thus there are a number of tests that can be carried out to examine the quality of the concrete and these will often be directed to determining the underlying cause of the corrosion and/or deterioration that may have been detected either by electrochemical or non-destructive test methods or by visual observation. Concrete and core samples may be analysed/examined for cement (binder) content and type, original water content, and density to assess the concrete quality. Appropriately sized cores may also be used to measure compressive strength. A comprehensive account of methods for the chemical analysis of concrete is given in BS 1881-124 (2015) and some of these methods are described by Bungey and Millard (1996). Water penetrability (absorption, sorption, permeability) testing may also be necessary. Knowledge of the concrete quality may then need to be used to determine the extent to which inadequate quality was a contributing factor to reinforcement corrosion.

The depth of chloride penetration into concrete (chloride profile) is determined by laboratory measurement. Discs approximately 10–25 mm thick sliced from core samples or alternatively, powdered drillings taken at various depths from the surface with an arrangement similar to that shown previously (at Section 6.10.3), are chemically analysed for acid soluble (total) chloride content.

The mechanism(s) of chemical deterioration of elements of a concrete structure will need to be determined by laboratory testing. For example, alkali aggregate reaction (AAR) and/or delayed ettringite formation (DEF) determinations, the type of sulphate attack, acid (inorganic or organic) attack etc. as discussed at Chapter 2.

Microbial analysis may be necessary from a biological deterioration point of view. Physical deterioration assessment may be necessary. Determination of the depth of fire damage may also be required.

7.2 CEMENT (BINDER) CONTENT AND COMPOSITION

A major factor controlling the ability of a concrete to maintain the high pH environment that is necessary to preserve the passive film on the surface of the steel is the original cement content or the aggregate/cement ratio of the concrete.

For cementitious binders consisting of type GP (Portland) cement alone, common methods for the determination of the original cement content of the concrete depend upon the fact that the hydrated cement compounds in the concrete are usually much more easily decomposed by the action of dilute hydrochloric acid than are the calcium and silica compounds that are present in most common aggregates.

ASTM C1084-19 (2019) gives the details of a standard method for the determination of the cement content of a hardened concrete. The sample is crushed, dissolved in hydrochloric acid, and the calcium ion content determined by titration with a standard EDTA solution. It is then assumed that the CaO represents 63.5% of the mass of the cement. A check on the results may be obtained by determining the silica content of the solution. This determination is based upon the fact that if the hydrochloric acid solution is evaporated to dryness then on treating with boiling water and holding at temperature for about five minutes, only hydrated silica remains insoluble. If the solution is filtered and the residue held for about an hour at 1,000°C then it can be weighed as SiO_2. It is then assumed that the SiO_2 represents 21.0% of the mass of the cement.

Chemical analysis of concrete by BS 1881-124 (2015) can be used to determine the acid soluble calcium and silicon contents of the concrete, from which the approximate cement content can be determined. For cementitious binders consisting of type GP (Portland) cement alone, the calcium content alone will generally be sufficient, but if the concrete also contains shell, limestone aggregate, or other sources of calcium carbonate then the silicon content also needs to be measured.

Chemical analysis by the ASTM C1084-19 (2019) and BS 1881-124 (2015) methods may be less accurate for concrete containing SCMs unless a reference sample is available for comparison (Concrete Institute of Australia, 2015).

Petrographic examination (see Section 7.16) can determine the type of cement used (e.g. Type GP, SR, GB, white cement, calcium aluminate cement, blended cement) and the volume percent of binder (BSI, 2015 & Concrete Society, 2010).

7.3 AIR CONTENT

The entrained air content of hardened concrete can be estimated by petrographic examination (Concrete Institute of Australia, 2015), refer Section 7.16.

7.4 WATER/CEMENT (BINDER) RATIO

The original water/cement (w/c) (water/binder, w/b) ratio is often determined as an indicator of the penetrability of the concrete. The cement content of the original mix is measured as described above. The original water content is regarded as the sum of the water that originally filled the capillary pores of the concrete (the pore water), plus the water of hydration contained within the hydrated cement compounds (the bound water).

In order to determine the pore water content, a sample of the concrete is dried at 105°C until constant mass is achieved. After weighing, the sample is immersed in a low viscosity organic fluid and subjected to a vacuum, so that all the air in the capillary pores is evolved. Release of the vacuum allows the fluid to enter the pores and from the increase in mass of the sample the volume of the capillary pores can be calculated.

In order to determine the bound water, the same sample of concrete (from which the capillary water has all been removed by the original heat treatment) is crushed and heated to 1,000°C in a stream of nitrogen and the water evolved determined by passing the nitrogen over a drying agent (often magnesium perchlorate) and measuring the increase in mass of the drying agent.

The total original water content is then the sum of the pore water and the bound water (each of these may have to be corrected for the pore water and the bound water in the aggregate – determined separately) and this divided by the cement content gives an indication of the original water/cement (binder) ratio.

Bungey and Millard (1996) suggest that the final result is rarely likely to have an accuracy greater than ±0.1. The same authors suggest that sufficient accuracy is often obtained if the bound water in the cement is put equal to 23% of the mass of the cement.

The UK Concrete Society (2010) suggest an estimated accuracy (expressed as a reproducibility) of ±0.1 of the actual w/c (w/b) ratio for the range 0.4–0.8. It is less accurate for concrete that is poorly compacted, carbonated, air entrained or cracked (Concrete Institute of Australia, 2015). Results reported by Barnes and Ingham (2013) varied by up to ±0.2 from the actual value.

Petrographic examination (see Section 7.16) is typically more reliable as it also provides information about carbonation, air entrainment, and voids.

For accuracy it requires comparison with reference materials of known water/cement (binder) ratio. Provided the sample is representative of the placement (rather than of a defective area), the results may be accurate within an error of ±0.05 of the actual water/cement (binder) ratio for the range 0.4–0.8. Greater accuracy is achievable for relatively new concrete where reference materials consist of exactly the same concrete cured in the same way as the in situ concrete (Concrete Society, 2010).

The penetration of reinforced concrete by aggressive agents is dependent upon the penetrability of the concrete and the past over-emphasis on strength requirements to the detriment of durability was often associated with a failure to recognise the importance of low penetrability in obtaining durable construction in aggressive environments. The penetrability is related to the water/cement (binder) ratio. Penetrability increases as the water/cement (binder) ratio increases, and consequently the rate at which carbon dioxide can penetrate concrete increases with the water/cement (binder) ratio. The appropriate water/cement (binder) ratio for any given environment will depend upon the durability required. As an example, for concrete in a marine environment the maximum recommended water/cement (binder) ratio can be 0.4.

It is possible to achieve high strengths with a lean mix by the use of low water/cement (binder) ratios. The durability of such concretes can be poor however if the effect of the decreased penetrability resulting from the low water/cement (binder) ratio is outweighed by the deleterious effect of the reduced cement content in the mix. A reduction in the cement content of a mix reduces the chloride binding (and adsorption) ability of the mix and also provides less neutralising capacity to counteract the effect of the ingress of carbon dioxide. The minimum cement content recommended for concrete exposed to sea water can be 400 kg/m^3.

Insufficient compaction and vibration leads to low strength concrete. More importantly it leads to a high penetrability and easy access for aggressive agents. The design of the reinforcement can often impede adequate vibration by preventing the vibrator from entering into the structure. Similarly, too dense a spacing of the reinforcement can make it difficult for the concrete to penetrate adequately into the reinforcement cage.

7.5 SCM CONTENT AND COMPOSITION

The presence and nature of SCMs (e.g. fly ash, slag, silica fume, etc) in hardened concrete can be determined by petrographic examination, refer Section 7.16.

Scanning electron microscopy coupled with energy-dispersive x-ray spectroscopy (SEM-EDS) may also need to be employed.

7.6 WATER ABSORPTION, SORPTION, AND PERMEABILITY

Papworth and Green (1988) have introduced the term 'penetrability' to describe the ability of a concrete to resist the penetration of aggressive agents into concrete and up to the metal concrete interface. They have pointed out that the penetrability of e.g. water is in fact controlled by two physical phenomena, the absorption of the penetrant on to the surface of the concrete (in this case the surfaces of the capillary pores), and the permeability. The absorption is a function of the chemical attraction of the substrate for the penetrant, that is, primarily the chemical bonding within the cement and aggregate components of the concrete. The permeability is a function of the size and volume of the capillary pores. Papworth and Green (1988) have pointed out that what is normally termed the absorption as for example in the Initial Surface Absorption Test (ISAT) (BSI, 1996), is in fact a function of both variables.

Water absorption and sorptivity are measures of concrete pore volume (porosity) and structure as indicated by the volume and rate of uptake of water by capillary suction respectively (CC&AA, 2009). Solvent absorption can also be used as a measure of the capillary porosity of concrete (Peek et al., 2007).

Absorption tests measure the total weight gained by concrete immersed in water. Absorption tests focus on filling the voids in concrete without reference to the rate at which this occurs, although some tests might only partially fill the voids if the duration of the absorption period is fixed. Absorption test results are often referred to as the 'volume of permeable voids (VPV)' or the 'apparent volume of permeable voids (AVPV)'. Absorption values are expressed as percentage void space by volume of sample, or less commonly as the mass of absorbed water as a percentage of the dry mass of sample.

The simplest of the absorption tests (AS 1012.21, 1999; ASTM C462-13, 2013; BS 1881 Part 122, 2011) all involve drying the concrete to remove as much pore water as possible, weighing the sample, soaking in water under standardised conditions and re-weighing to determine the amount of water taken up. Such tests do not yield a quantitative assessment of durability but are useful for forming comparisons between different concretes (Concrete Institute of Australia, 2015).

The Road Authority of Victoria in Australia (VicRoads) use the AS 1012.21 (1999) VPV/AVPV water absorption test to assess the potential durability of structural concrete. AS 1012.21 (1999) is an Australian adaptation of the ASTM C642-06 (2006) method. In the AS 1012.21 (1999) test method, cylinders or cores are cut into four equal slices so that several results are obtained from the one test specimen. The slices are oven dried to a constant weight then immersed and boiled in water in order to saturate the concrete's permeable voids and thus give a measure of the total pore volume (Concrete Institute of Australia, 2015).

VicRoads (2020) VPV performance assessment criteria for different grades of structural concrete for bridge construction are shown in Table 7.1. VicRoads concrete grades are specified with a minimum cementitious material content, a maximum w/b ratio and a minimum compressive strength at 3, 7, and 28 days. They are designated by the letters VR (VicRoads) followed by a three-digit number indicating the minimum cementitious material (binder) content in kg/m³ and a two-digit number indicating the specified minimum compressive strength (cylinders) at 28 days. The maximum VPV values in Table 7.1 are at 28 days for each concrete grade for both test cylinders and concrete test cores cut from cast in situ and sprayed concrete and rounded down to the nearest whole number.

The minimum compressive strength requirements (from cylinders) for each VicRoads (2020) concrete grade are shown in Table 7.2 whereby the designations for the concrete grades are as above.

In contrast, sorptivity tests measure the rate of water absorption. The results can be used in formulae for assessing the depth of penetration of water with time under capillary action. Sorptivity is measured by placing concrete in contact with water with no pressure head and measuring weight gain with time to obtain a sorptivity value. The corresponding visible height rise is often also measured however this is a less accurate method for calculation of a sorptivity value. Sorptivity values are calculated as volume absorbed per unit surface area related to the square root of time, e.g. mm³/mm²/min$^{1/2}$ which is often reduced to mm/min$^{1/2}$.

Water sorptivity tests include the Taywood test (Concrete Society, 1988) and ASTM C1585-13 (2013) test. The Taywood sorptivity test, refer Figure 7.1, was originally a brick test. Recognising the significance of the rate of water ingress due to capillary action, as opposed to tests which only

Table 7.1 VPV performance assessment criteria (VicRoads Specification 610 on structural concrete)

Concrete Grade	Maximum VPV Values at 28 days (%)		
	Test Cylinders (compacted by vibration)	Test Cylinders (compacted by rodding)	Test Cores
VR330/32	14	15	17
VR400/40	13	14	16
VR450/50	12	13	15
VR470/55	11	12	14
VR520/60	11	12	14
VR535/65	10	11	13
VR550/70	10	11	13
VR580/80	9	10	12
VR610/90	9	10	12
VR640/100	9	10	12

Source: VicRoads (2020)

Table 7.2 Minimum compressive strength requirements for VicRoads
(Specification 610) structural bridge concretes

Concrete Grade	Minimum Compressive Strength (MPa)		
	3 days	*7 days*	*28 days*
VR330/32	14	20	32
VR400/40	17	26	40
VR450/50	23	35	50
VR470/55	25	40	55
VR520/60	27	45	60
VR535/65	29	43	65
VR550/70	31	52	70
VR580/30	34	60	80
VR610/90	38	67	90
VR640/100	42	75	100

Source: VicRoads (2020)

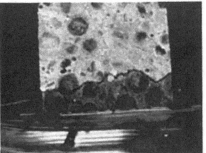

Figure 7.1 Taywood soprtivity test. A sample is stood in contact with water and
the height rise and weight gain measured with time. (Courtesy of
Concrete Institute of Australia, 2015)

measure the volume of pores, the original brick test was adapted for concrete in the late 1970s. The test is mentioned in UK Concrete Society
Technical Report TR31 (1988) but was never standardised.

The ASTM C1585-13 sorptivity test is relatively simple to undertake.
Although it takes slightly longer than the Taywood sorptivity test, it gives
results that are more representative of in situ concrete, and is an international standard (Green and Papworth, 2015).

Sorptivity tests can be used to develop an understanding of concrete quality, for concrete mix acceptance and quality assurance as well as when modelling of water ingress is required (Concrete Institute of Australia, 2015).

The Road Authority of New South Wales (NSW) in Australia (Roads and
Maritime Services, RMS) utilise a water sorptivity test to assess the

Table 7.3 Effectiveness of curing water sorptivity test criteria

Exposure classification	Maximum sorptivity penetration depth (mm)	
	Shrinkage Limited cement	Blended cement
A	35	35
B1	25	25
B2	N/A	20
C1	N/A	8
C2	N/A	8
U	In accordance with Annexure B80/A1	

Source: RMS (2012)

effectiveness of the curing of concrete for different exposure classifications for bridges. In the RMS T362 water sorptivity test (2012), 100 x 100 x 350 mm samples are pre-conditioned at 50% RH and then placed under 50 mm of water for 6–24 hours before being broken open in flexure using a beam test rig and the depth of water penetration measured. As the water head is low the results are a measure of the concrete capillary suction (Concrete Institute of Australia, 2015). The RMS T362 water sorptivity test criteria are provided at Table 7.3.

The RMS bridge concrete specification (2019) requires that the maximum sorptivity penetration depth is first verified on a trial mix using the method and duration of curing ('curing regime') proposed for use on the Works. At the trial mix stage, the curing of the sorptivity test specimen must be identical to that proposed for the concrete member. At the construction stage, the curing of the concrete member must be identical to that of the sorptivity test specimen. Provision of charts of the curing temperature and humidity versus time to verify that the required curing has been achieved are then also necessary (Roads and Maritime Services, 2019).

As advised at Section 2.3, the penetration of water into concrete under a pressure head (permeability) is an important durability performance parameter for concrete exposed to water pressure. The coefficient of permeability or hydraulic conductivity is defined, as advised at Section 2.3, by Darcy's Law, and has the units of length/time (typically $m.sec^{-1}$).

Test methods vary for measuring the coefficient of water permeability of concrete. For example, they include ASTM D5084-16a (2016); Main Roads Western Australia Test Method WA 625.1 (1998); and, the method developed by Taywood Engineering, refer Figure 7.2, and Figure 7.3, published as an in-house procedure and was included in Concrete Society Technical Report TR31 (Concrete Society, 1988; Concrete Society, 2008). The various water permeability test methods differ in the type of permeameter used, which can be either rigid or flexible wall.

Figure 7.2 Taywood water permeability test. One face of samples is kept under constant pressure by the compressed air bottle. The other side can be left open to witness time to water penetration or closed so that water flow rate can be measured. (Courtesy of Concrete Institute of Australia, 2015)

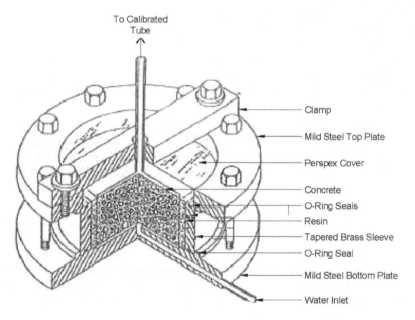

Figure 7.3 Taywood water permeability test rig. (Courtesy of Papworth & Green, 1988)

7.7 DEPTH OF CHLORIDE PENETRATION

The profile of chloride penetration into the concrete can be determined by analysing discs approximately 10–25 mm thick sliced from core samples. Alternatively, powdered drillings can be taken at various depths from the surface with an arrangement similar to that shown previously at Section 6.10.3. Samples for the determination of the profile of chloride penetration may be obtained on-site at the same time as the test for carbonation depth described in Section 6.9. However, unlike the carbonation test the chloride penetration test requires the cores or the samples of the concrete drillings to be taken back to the laboratory for analysis. The chloride profile can be fitted to distribution equations discussed at Section 5.4. These assume diffusion to be the primary mechanism of penetration and the diffusion coefficient determined. However, under wetting and drying conditions the use of diffusion equations may be inappropriate because the wetting and drying leads to a concentration of the absorbed chloride in the penetrant solution by evaporation of water which is replaced by further salt containing solution. The effect of sorptivity (capillary suction) may also need to be considered.

In structures such as those exposed to the sea it is useful to obtain a profile of chloride concentration from the concrete surface to the rebar to provide information regarding the rate of chloride penetration. This information is used to estimate the time required for the chloride level at the rebar to exceed the concentration necessary to sustain pitting propagation if this has not already occurred.

The acid soluble (total) chloride concentration can be measured using the analytical technique described in BS 1881-124 (2015) or AS 1012.20 (2016), for example. The technique involves grinding the dried concrete to a fine powder passing through a 150 µm sieve, then dissolving it in nitric acid and analysing the resultant solution for chloride either by the Mohr method or the Volhard method.

In the Volhard method a known excess of silver nitrate is added and the excess titrated with ammonium thiocyanate using ferric ammonium sulphate as the indicator. The ammonium thiocyanate initially precipitates all the silver as silver thiocyanate but as soon as the precipitation is complete, the next drop of ammonium thiocyanate produces a red brown precipitate of ferric thiocyanate, refer Figure 7.4.

In the Mohr procedure the chloride containing solution is titrated with silver nitrate in the presence of potassium chromate, as soon as there is excess of silver ions, they react to form red silver chromate.

Automatic titration is now a more common method utilised where chloride ion selective electrodes are employed and potentiometric titration (at the point of inflexion) undertaken, refer Figure 7.5.

The acid soluble (total) chloride concentration includes the chemically combined and adsorbed (Van Der Waals forces) chloride components. The water soluble, or free, chloride concentration is of greater interest than the

Figure 7.4 Volhard titration for acid soluble (total) chloride content determination. Ammonium thiocyanate addition (from burette) has produced an endpoint of red brown precipitate of ferric thiocyanate. (Courtesy of Concrete Institute of Australia, 2015)

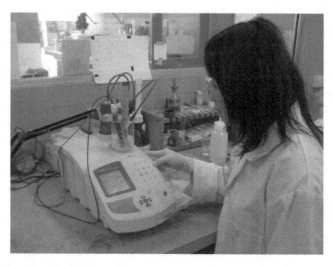

Figure 7.5 Automatic potentiometric titration for acid soluble (total) chloride content determination. (Courtesy of Nhu Nguyen, SGS Perth, Western Australia)

acid soluble since it is water soluble chlorides that cause corrosion. The concentration of chloride ions available for corrosion can be measured by the process of water extraction for 24 hours followed by the same titration technique used for the determination of acid soluble chlorides. Different water extraction techniques have been assessed by Hope et al. (1985). The free chloride concentration can also be measured by squeezing the pore solution out of concrete and analysing the solution by potentiometric titration with a chloride ion selective electrode.

However, free chloride (water soluble) testing of concrete powder samples will dissolve chemically bound and adsorbed chloride ions that are in equilibrium leading to an activation value that is higher than the true value. The extent to which free chlorides are extracted with water is also a function of the length of dissolution time. Furthermore, the binder content is often unknown and is hard to determine especially when they contain SCMs. Pore squeezing on the other hand is not easy to do. Moreover, the pore water content may vary over time, dependent on the exposure conditions.

Total chloride (acid soluble) concentration based on binder mass is therefore the preferred method used to define the chloride concentration necessary to sustain pitting propagation. Chemically bound and adsorbed chloride are in equilibrium with the free chloride concentration and as such all chloride is also relevant.

7.8 SULPHATE ANALYSIS

The sulphate and sulphoaluminate content of concrete can be analysed by the methods given in BS 1881-124 (2015) and AS 1012.20 (2016) for example. In this case the acid solution prepared as described previously is analysed for sulphate ions. This is done by adding barium chloride and gravimetrically determining the amount of barium sulphate precipitated.

Knowledge of the sulphate content can indicate whether sulphate attack has occurred or is likely to occur in the future if the concrete remains exposed to its environment. Normal Portland cements contain a concentration of sulphate arising from the original gypsum added to prevent flash set of about 5%. Sulphate in excess of this may be evidence of sulphate attack and so sulphate analyses are typically carried out in conjunction with original cement content determinations.

As indicated at Section 2.4.4, the common sources of sulphates which may attack concrete are groundwater, soils, industrial chemicals, seawater, and products of sulphur oxidising bacteria. Groundwater and soils may contain sulphates of calcium, magnesium, sodium, potassium and ammonium in various quantities depending on the location with magnesium sulphate being the most aggressive to concrete followed by ammonium sulphate, refer Section 2.4.4.

Petrographic examination, refer Section 7.16, of thin sections from core samples is most commonly employed to identify the form of sulphate attack. The smell of ammonia gas may also be present in the case of ammonium sulphate induced concrete deterioration (Collins & Green, 1990).

Laboratory methods such as x-ray fluorescence (XRF) and x-ray diffraction (XRD) on crushed and pulverised concrete core sections to identify inorganic elements and compounds may also be necessary to confirm the type of sulphate attack.

7.9 ALKALI AGGREGATE REACTION

The presence and extent of alkali silica reaction (ASR) (or alkali carbonate reaction, ACR) within a concrete structure or building can only be unequivocally established by taking core samples and performing a laboratory study, including petrographic examination (Thomas et al., 2011), refer Figure 7.6, and Section 7.16.

Laboratory studies can also be utilised to measure the residual expansion in ASR-affected concrete by accelerated exposure testing of cores. Structure and building elements may also be instrumented and monitored for future behaviour before deciding on a course of action.

Uranyl acetate staining of concrete core samples followed by examination under fluorescent light has also been undertaken to detect the presence or otherwise of ASR (Natesaiyer and Hover, 1989).

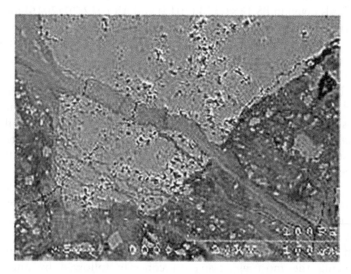

Figure 7.6 ASR gel deposition through both coarse aggregate and cement paste.

7.10 ALKALI CONTENT

In some circumstances, the risk of AAR is managed by specifying a maximum concrete alkali content. For example, in New Zealand a maximum alkali limit of 2.5 kg/m^3 (equivalent Na$_2$O) is used for many concretes when potentially reactive aggregate is used. Determination of the alkali content of hardened concrete may therefore be required to assess compliance with such specifications, or to evaluate the cause of AAR in an existing structure to help identify appropriate precautions for future construction (Concrete Institute of Australia, 2015).

BS 1881-124 (2015) provides a method for determining acid soluble sodium and potassium contents. Barnes and Ingham (2013) reported that inter-laboratory test results from this method were consistently higher than the target, which may reflect extraction of alkalis that would normally not be available for AAR, such as alkalis bound within minerals in the aggregate. Of potentially greater concern is that the range of results they reported exceeded 1.0 kg/m^3 (equivalent Na$_2$O) for all concrete tested. Thus, this method may not be accurate enough for practical use (Concrete Institute of Australia, 2015).

Other in-house methods may be more appropriate, such as methods that measure water soluble alkalis, and/or that separate the binder fraction from the aggregate and analyse only the binder fraction (Concrete Institute of Australia, 2015).

7.11 DELAYED ETTRINGITE FORMATION

Delayed ettringite formation (DEF) is identified in hardened concrete by petrographic examination, refer Section 7.16.

7.12 ACID ATTACK

Inorganic acid attack of concrete can be indirectly determined by spraying concrete breakouts, core samples, etc with phenolphthalein pH indicator solution and noting the depth of neutralisation from the inorganic acid attack rather than the depth of carbonation.

As indicated at Section 2.4.7, the effects of organic acids on concrete cannot be determined by depth of neutralisation (pH) indication or concentrations of solutions, which may reasonably be applied to inorganic acids as a whole, as virtually each organic acid must be considered individually with regard properties such as its solubility in water and, most importantly, the solubility of its calcium salt (Lea, 1998).

Laboratory methods such as XRF and XRD can be utilised on crushed and pulverised concrete core sections to identify inorganic elements and compounds of relevance in terms of the products of reaction from inorganic acid attack.

Organic based compounds can be determined using laboratory-based methods such as fourier transform infra-red (FTIR) spectroscopy, for example, and so the products of reaction of organic acids-based attack may be detectable. Consideration may need to be given however to any organic-based admixtures already present in the concrete.

7.13 CHEMICAL ATTACK

Laboratory methods such as XRF and XRD can be utilised on crushed and pulverised concrete core sections to identify inorganic elements and compounds in terms of chemical attack.

As mentioned previously, organic based compounds can be determined using laboratory-based methods such as FTIR spectroscopy, for example, remembering though that organic-based admixtures may already be present in the concrete.

7.14 MICROBIAL ANALYSIS

Where microbial attack is a potential concern or suspected, such as in water, and wastewater retaining structures, samples of biological slime, or other surface deposits can be analysed by a microbiologist to identify the presence of bacteria, algae, slime, fungi etc. that could attack the underlying concrete (Concrete Institute of Australia, 2015).

Specialist microbiological analytical services are available for this purpose. Samples must be collected, stored and transported in accordance with their recommendations (Concrete Institute of Australia, 2015).

7.15 PHYSICAL DETERIORATION DETERMINATION

Petrographic examination, refer Section 7.16, is also useful for identifying the cause and extent of physical attack such as by freezing and thawing (Concrete Institute of Australia, 2015).

Petrographic examination is also the definitive technique for determining the depth of fire damage in concrete (Ingham, 2009).

7.16 PETROGRAPHIC EXAMINATION

Examination of concrete samples by microscope techniques will reveal information about concrete composition, aggregate type, concrete quality, and the causes and extent of damage.

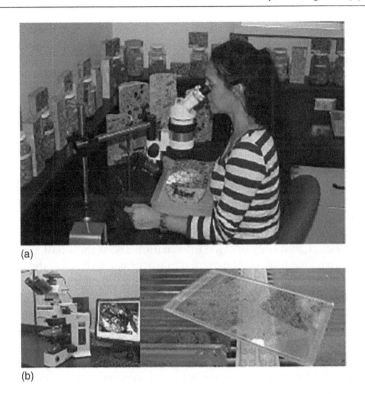

(a)

(b)

Figure 7.7 **Petrographic examination.**

Petrographic examination is a technique used by geologists to investigate rocks and aggregates. It involves visual examination of concrete surfaces under reflected light in-hand specimen or by binocular microscope, refer Figure 7.7, or of thin sections under polarised transmitted light, refer Figure 7.6.

Scanning electron microscopy (SEM) can be used to detect further detail and is usually augmented by spot chemical analyses by x-ray techniques (SEM-EDS) to identify the composition of specific features.

A geologist will be able to analyse concrete samples to identify features related to the aggregates, but if information about the concrete is required, then the examination should be carried out by a petrographer with specialist knowledge about concrete materials and technology.

Samples for petrographic examination are generally prepared from cores. Other means of sampling may cause microcracking or other damage.

In addition to being valuable to identify, as described previously, the cause and extent of chemical attack both from external aggressive agents such as sulphates and acids and from within the concrete itself such as AAR and DEF as well the cause and extent of physical attack such as by freezing and thawing and fire, it is also useful for identifying surface defects, cracking, or penetration of the concrete by materials such as surface treatments (protective or remedial) and can distinguish zones of complete and partial carbonation.

The UK Concrete Society Technical Report TR71 (2010) describes petrographic examination of concrete in more detail.

Experienced concrete petrographers use optical microscopes in accordance with ASTM C856/C856-M (2020).

7.17 COMPRESSIVE STRENGTH

Tests for concrete strength fall into two main categories, destructive tests, and non-destructive tests. Destructive tests give a definitive value directly measured from the structure, but as they damage the concrete, they are generally restricted, and only provide information about the small area tested.

Non-destructive tests can be carried out over wide areas on the structure. However, they do not definitively measure strength, and accurate interpretation of results requires calibration against results from tests on laboratory specimens or destructive in situ tests (Green & Papworth, 2015).

Measuring compressive strength of concrete in an existing structure will generally involve a combination of non-destructive tests (i.e. no concrete removal) and destructive tests (e.g. core samples extracted). It should be noted that the amount of concrete removed must not reduce the structural capacity or other performance parameters of the element.

Concrete Institute of Australia Recommended Practice Z11 (2020) gives general guidance on evaluating concrete strength from the results of testing cores taken from the structure. It describes the reasons for strength evaluation by the use of cores and how to obtain cores from concrete structures so that they will be suitable for testing and the test results will be 'significant' in the true statistical sense. Steps in the determination and evaluation of concrete strength from the results include providing for corrections to the indicated compressive strength of the core for: length to diameter ratio of the core; presence of reinforcement in the core; position of the core axis in relation to the standard cylinder axis; age of the concrete; quality of compaction of the concrete; and, the curing regime experienced.

BS EN 13791 (2007) and associated guide BS 6089 (2010) detail methods of measuring and assessing in situ compressive strength using cores, rebound hammer, ultrasonic pulse velocity, and pull-out tests.

7.18 REPORTING

7.18.1 Background

The end-result of any condition survey is the preparation of a report and it is this report that determines what action, if any, is taken. An example of the requirements of such a report is given in the 'Australian Concrete Inspection Manual' (Isaacs and Nagarajan, 1995). Whether the survey has been carried out at the behest of a concerned structure and/or building owner worried by signs of distress visible in the structure/building or on behalf of lawyers representing an aggrieved party during litigation, the report of a condition survey will generally assume the same format. Since the reader is unlikely to have a detailed technical knowledge of the mechanisms of corrosion or concrete deterioration, it is the responsibility of the writer to express the results of his work in a logical fashion so that the reader can follow the reasoning that leads to the inevitable recommendation of some or no action.

Corrosion (and concrete deterioration) can be a remarkably localised phenomenon and so it is always necessary to commence any report with the details of the commission that the inspector has been given so that the reader knows precisely what the inspector has been asked to inspect and what the report covers and does not cover along with any limitations to the investigation (e.g. cost, access, difficulty of sampling, etc.). For the same reason it is vital that the inspector identifies by means of photographs or drawings exactly where the reported measurements are made. It may be necessary to point out to the reader where problems of access have prevented measurements being taken.

A mass of photographs of concrete cracks with or without rust stains weeping from them may look impressive as a piece of journalism but is really of little value in helping the reader to understand whether the problem has been caused by (say) inadequate cover, an inappropriate concrete, contact of the reinforcement with a galvanic cathode etc. The report of the initial visual inspection (Section 6.2) must lead to the reasons why the inspector has undertaken or wishes to undertake the various on-site measurements that are described in Sections 6.3 to 6.9 and the description of the on-site measurement results must in a similar fashion justify the off-site laboratory testing that may be considered necessary. By definition it is a technical scientific report and should be set out as such.

However, the end-result of any condition survey is the production of a report and the end-result of any report must be recommendations and conclusions which are expressed in clear non-technical language and on the basis of which financial decisions can be made with as little uncertainty as possible. Managers and accountants are trained to cope with the problems of raising and disbursing money; the one thing that they hate is uncertainty and so it is the task of the inspector to remove as much of that uncertainty as is ethically possible.

Whether the report is to be delivered to an asset owner or the solicitors representing a claimant, the format is the same. It must always be remembered that the reader is not a technical expert in your discipline. A recommended lay-out is as follows.

7.18.2 Commission/scope of services

This details in non-technical language the brief that the inspector has been given. It is often a good idea to use the commissioning authority's own words.

7.18.3 Technical background

A brief introduction of the general deterioration phenomena involved. Not specific to problems being investigated.

7.18.4 Site investigation

The site investigation should be recorded in as much detail as necessary so that the scope and methodology of the investigation could be repeated based on the information presented. The location, date, and types of testing, for example, should be presented before discussing the results of the testing.

It should be noted also that unless you can produce your site notebook with detailed sketches of what has been observed, possibly photographs added later, signed, and dated your evidence is almost worthless.

7.18.5 Hypothesis

What you think might have happened and the reasons for the laboratory tests.

7.18.6 Laboratory testing

Signed reports from the laboratory staff, detailing the origin of the samples, (reference numbers that can be related back to the site notebook), and the exact instructions that they were given.

7.18.7 Commentary on laboratory results

This just discusses what the results have shown, without relating them to the background of the investigation.

7.18.8 Conclusions and recommendations

This is not a summary of all that has been done, it relates back strictly to the brief that started the whole thing off and answers (if possible) the questions that were posed. It is written in language that is as non-technical as possible and provides as much certainty as to the next stages (if any) of the process.

REFERENCES

American Society of Testing Materials (2006), 'Standard Test Method for Density, Absorption and Voids in Hardened Concrete', ASTM C642-06, West Conshohocken, Pennsylvania, USA.

American Society of Testing Materials (2016), 'Standard Test Methods for Measurement of Hydraulic Conductivity of Saturated Porous Materials Using a Flexible Wall Permeameter', ASTM D5084-16a, West Conshohocken, Pennsylvania, USA.

American Society of Testing Materials (2013), 'Standard Test Method for Measurement of Rate of Absorption of Water by Hydraulic Cement Concretes', ASTM C1585-13, West Conshohocken, Pennsylvania, USA.

American Society of Testing Materials (2019), 'Standard Test Method for Portland Cement Content of Hardened Hydraulic-Cement Concrete', ASTM C1084-19, West Conshohocken, Pennsylvania, USA.

American Society of Testing Materials (2020), 'Standard Practice for the Petrographic Examination of Hardened Concrete', ASTM C856/C856M-20, West Conshohocken, Pennsylvania, USA.

Barnes, R and Ingham, J (2013), 'The chemical analysis of hardened concrete: Results of a 'round robin' trial –parts I and II', Concrete, Vol. 47, No.8, 45–48 Oct 2013 and Vol. 48, No. 1, 48–50 Dec 2013/Jan 2014.

British Standards Institute (1996), 'Testing Concrete Part 208: Recommendations for the determination of the initial surface absorption of concrete', BS 1881-208, London, UK.

British Standards Institute (2010), '*Assessment of in-Situ Compressive Strength in Structures and Precast Concrete Components*', Complementary guidance to that given in BS EN 13791, BS 6089, British Standards Institution, London, UK.

British Standards Institute (2011), 'Testing Concrete Part 122: Method for determination of water absorption', BS 1881-122, London, UK.

British Standards Institute (2015), 'Testing Concrete Part 124: Methods for analysis of hardened concrete', BS 1881-124, London, UK.

BS EN 13791 (2007), '*Assessment of in-situ compressive strength in structures and precast concrete components*', European Committee for Standardization, Brussels.

Bungey, J P and Millard, H G (1996), '*Testing of Concrete Structures*', Chapman and Hall, London, UK.

Cement Concrete & Aggregates Australia (2009), '*Chloride Resistance of Concrete*', June, Sydney, Australia.

Collins, FG and Green, WK (1990), 'Deterioration of concrete due to exposure to ammonium sulphate', Use of Fly Ash, Slag & Silica Fume & Other Siliceous Materials in Concrete, Concrete Institute of Australia & CSIRO, Sydney, Australia, 3 September.

Concrete Institute of Australia (2015), Recommended Practice, Concrete Durability Series, 'Z7/07 Performance Tests to Assess Concrete Durability', Sydney, Australia.

Concrete Institute of Australia (2020), 'The Evaluation of Concrete Strength by Testing Cores', Recommended Practice Z11, Sydney, Australia.

Concrete Society (1988), 'Permeability Testing of Site Concrete', Technical Report TR31, Camberley, Surrey, UK.

Concrete Society (2008), 'Permeability Testing of Site Concrete', Technical Report TR31, Camberley, Surrey, UK.

Concrete Society (2010), 'Concrete Petrography – An Introductory Guide for the Non-Specialist', Technical Report TR71, Camberley, Surrey, UK.

Green, W and Papworth, F (2015), *'Concrete Durability Performance Testing – The Approach Adopted in a Concrete Institute of Australia Recommended Practice'*, Proc. Concrete 2015 Conf., Concrete Institute of Australia, Melbourne, Australia, 30 August – 2 September.

Hope, B, Page, J and Poland, J (1985), 'The determination of the chloride content of concrete', *Cement and Concrete Research*, Vol 15, 683–870.

Ingham, J (2009), 'Forensic engineering of fire-damaged structures', *Proceedings of the Institution of Civil Engineers – Civil Engineering*, 162, May, 12–17.

Isaacs, HP and Nagarajan, RR (1995), *'Australian Concrete Inspection Manual'*, UNSW Press, University of New South Wales, Australia.

Lea, FM (1998), *'The Chemistry of Cement and Concrete'*, ed Peter C Hewlitt, Edward Arnold, London, UK.

Main Roads Western Australia (1998), 'Water Permeability of Hardened Concrete', Test Method WA 625.1, Perth, Western Australia.

Natesaiyer, K and Hover, K (1989), 'Further study of an in-situ identification method for alkali-silica reaction products in concrete', *Cement and Concrete Research*, 19, 5, September, 770–778.

Papworth, F and Green W (1988), *'Life Cycle Aspects of Concrete in Buildings and Structures'*, Taywood Engineering Ltd., Leederville, Western Australia.

Peek, AM, Nguyen, N and Wong, T (2007), *'Durability Planning and Compliance Testing of Concrete in Construction Projects'*, Proc. Corrosion Control 007 Conf., Australasian Corrosion Association Inc., Sydney, Australia, 25–28 November.

Roads & Maritime Services (2012), 'Interim test for verification of curing regime – Sorptivity', Test Method T362, October, Sydney, Australia.

Roads & Maritime Services (2019), 'Specification B80 Concrete Work for Bridges', Edition 7/Revision 2, November, Sydney, Australia.

Standards Australia (1999, R2014), 'AS 1012.21 Methods of testing concrete Method 21: Determination of water absorption and apparent volume of permeable voids in hardened concrete', Sydney, Australia.

Standards Australia (2016), 'AS 1012.20 Methods testing concrete Method 20: Determination of chloride and sulfate in hardened concrete and concrete aggregates', Sydney, Australia.

Standards Australia (2018), 'AS 3600 Concrete Structures', Sydney, Australia.

Thomas, MDA, Fournier, B, Folliard, KJ and Resendez, YA (2011), 'Alkali-Silica Reactivity Field Identification Handbook', Federal Highway Administration, Report No. FHWA-HIF-12-022, Washington DC, USA, December.

VicRoads (2020), 'Section 610 – Structural Concrete', February, Melbourne, Australia.

Chapter 8

Repair and protection (A) – mechanical methods

8.1 INTRODUCTION

Although steel-reinforced concrete structures are expected to have a service life of many, many decades, early deterioration all too often occurs and the most common cause of impact on serviceability is corrosion of the reinforcing steel. The problem is not limited to old structures but is frequently diagnosed in structures less than ten years old. Investigations into likely causes found that failures could commonly be attributed to shortcomings on the part of designers and builders that could have been avoided. Lack of sufficient cover and poor-quality concrete are still among the most frequent sources of durability problems involving corrosion. Although the causes of durability problems have been widely publicised, it appears that lack of durability-focussed design, competent supervision, and poor-quality workmanship are still a frequent cause of defects and deterioration.

The repair methods that are employed to restore a deteriorated reinforced concrete structure depend on the nature and the extent of the deterioration. In general repair techniques may be 'mechanical' or 'electrochemical', though often a combination of the two is necessary to ensure a long-lasting solution to the problem. Cathodic protection and other electrochemical techniques will be considered in the following chapters and this chapter will be concerned with 'mechanical' techniques. These include the simple sealing of cracks, local patching and surface rendering with plain concrete or polymer-modified cementitious based mortars, recasting with concrete, even complete demolition of building elements and their replacement by new structural elements. Inhibitors may also be an option and coating and penetrant treatment of concrete surfaces may be necessary. Structural strengthening of elements is also possible.

An important consideration with all repair techniques is the capacity of the repaired structure to bear load. Unless load is taken off structural elements by jacking onto props before the repairs are carried out, the repaired section will not be able to share loads that are applied later. They will not carry any of the original load and this will have to be borne by the remaining structure with possible consequent damage in the remaining structure. If a substantial amount of damaged concrete or corroded bars have to be replaced, attention will have to be given to the effects of such removals on the load-carrying capacity of the affected members, and the stability of the structure will have to be checked. Substantial amounts of concrete or steel must never be removed without the load bearing capacity of the structure being checked by a structural engineer. Sometimes the repairs can be carried out safely if they are done in small sections, but in some cases it will be necessary to provide substantial temporary bracing.

8.2 CRACK REPAIR

Apart from those cracks that appear within the first few months of a structure's existence the major causes of cracking are stresses on the structure and corrosion of the reinforcement. Crack repairs may be necessary to restore the load-bearing capacity and possibly the safety of a structure, to restore durability where the cracks permit air (carbon dioxide gas), chlorides, and moisture to reach the reinforcement and cause corrosion, to restore the watertightness of a structure which is leaking or to improve the appearance of a structure which has become unsightly because of cracking.

Before an appropriate repair method can be selected it is necessary to identify the type of crack and the cause of cracking. If a crack is discovered soon enough so that one is reasonably sure that no aggressive contaminants have penetrated the crack sufficiently far to attack any reinforcement, then such a crack can be sealed.

Crack repair methods available include:

- Injection (epoxies, polyurethanes, micro-cement grouts).
- Filling (rout & caulk, rout & pour) (polyurethanes, silicones, polysulphides, epoxies, epoxy-urethanes, bitumen modified urethanes).
- Coatings (acrylics, polyurethanes, epoxies, co-polymers, cementitious, elastomeric systems, fibre reinforced).
- Membranes (acrylics, polyurethanes, bituminous, fibre reinforced, torch-on).
- Impregnants (silicates, silanes, siloxanes, silane/siloxane).
- Locally bandaged (coatings and membranes + fibre reinforced).

In terms of crack injection, for example, the material that is injected into the crack has to be of sufficiently low viscosity that it will be able to penetrate as far as possible down the crack. Mortar material is usually injected as a microfine cement slurry which aids the restoration of the high pH environment around the reinforcement that inhibits corrosion. Rigid resins, typically a two-pack epoxy or a polyurethane with a pot life of a few hours can be injected into the crack and after it has solidified to a hard relatively inflexible material it will effectively seal the crack against any ingress of corrosive agents.

If a 'live' crack is filled with a rigid material, the crack may re-open, or a new crack may appear nearby. Consequently, the repair in such a case should either be flexible enough to accommodate future movement, or the movement should be prevented. If it is not possible to prevent continued movement, a flexible sealant must be used. The width of a joint that can be sealed by a flexible sealant is controlled by the width of the crack and the anticipated amount of movement in the joint. The strain capacity of the sealant is unlikely to exceed 25% of the joint width and so in order to accommodate crack movements which are comparable with the width of the crack a wide sealing groove may have to be made as in Figure 8.1.

The width of the new joint and the strain capacity of the sealant determine the magnitude of the movement that can be absorbed. The sealant must be separated from the bottom of the joint by a bond breaker (backer rod, backer tape) otherwise continued movement of the crack surfaces can

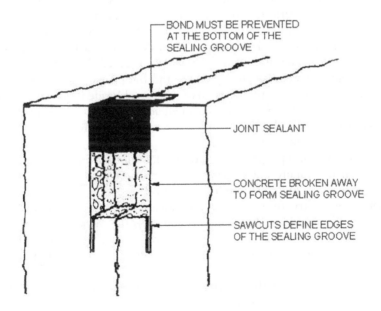

Figure 8.1 Repair of a 'live' crack. (Courtesy of Pullar-Strecker, 1987, p.43)

set up fatigue cracks in the sealant. What is required is that when the crack moves, debonding needs to take place from the backer rod/tape and not from the sides of the joint, viz weakest bond needs to be to the bond breaker. Where the use of a flexible sealant in the joint will still not accommodate all the movement, the best solution may be to fill the crack with a rigid material and to form a straight-movement joint nearby.

Width:depth ratios are also important when it comes to repair of 'live' cracks, namely:

- For joint widths up to 12 mm → 1:1 width to depth ratio.
- For joint widths 12 mm to 50 mm → 2:1 width to depth ratio.
- For joint widths >50 mm → sealant cannot bridge.

If the cracks were either caused by corrosion of the bars, or have penetrated so far into the structure that aggressive agents can attack the reinforcement there will be no point in filling the cracks since localised corrosion will continue as a result of the action of the aggressive agents that have caused the original corrosion and which are still present at the metal surface.

8.3 REPAIR AND PROTECTION OPTIONS

As indicated at Section 5.3, during the lifetime of a reinforced concrete structure there are periods with respect to corrosion initiation and corrosion propagation of reinforcement that can be defined. The initiation period, when critical or threshold amounts of chlorides or the carbonation front or the extent of leaching of the concrete has not reached the reinforcement so that the reinforcement remains passive, and the propagation period when reinforcement corrosion proceeds. In the case of chloride-induced corrosion, as indicated at Section 5.3, there is also an active corrosion (metastable pitting/pit growth) period subsequent to corrosion initiation.

Various technologies and approaches are possible during the initiation, corrosion onset, and propagation periods of a reinforced concrete structure or its elements. Figure 8.2 presents a summary of some technologies and approaches, excluding the 'do nothing' option.

Whenever a structure is being investigated all the available options (including 'do nothing') should be presented to the owner/stakeholder for consideration. In short, each option has its pros and cons and the structure owner/stakeholder cannot make a considered decision unless all options are presented (Chess & Green, 2020).

Chess and Green (2020) note that the 'do nothing' option has led to successful long service lives of various reinforced concrete structures (or for large sections of structures) where the structure owners or stakeholders have

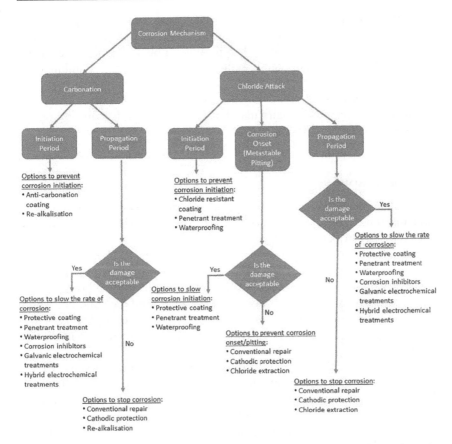

Figure 8.2 Flowchart of concrete repair and protection options (excluding the 'do nothing' option).

not needed to undertake any repair and protection works to the structures for their 50 years of service life achieved thus far.

8.4 PATCH REPAIR

8.4.1 Stages in the process

Concrete patch repair, conventional concrete repair, conventional patch repair, patch repair, patching, etc. are all descriptions of a common repair technique for reinforced, precast, and prestressed concrete structure elements suffering from chloride-induced, carbonation-induced, and/or leaching-induced reinforcing steel corrosion. Patch repair involves removal (or breakout) of physically deteriorated concrete (by mechanical tools or ultra-high water pressure), cleaning the steel within the patch, application of a coating system to the steel, priming of repair area surfaces and finally restoring the concrete profile with typically a polymer-modified cementitious based repair mortar (hand applied, poured, sprayed or combinations thereof) (Green et al., 2013).

In general, as a result of patching the serviceability of the structure will be prolonged, but ongoing repairs may be required (dependent mostly on the rate of reinforcement corrosion).

8.4.2 Breakout

Removal or breakout of concrete prior to patch repair is achieved using mechanical tools (e.g. jackhammers), refer Figure 8.3, or by ultra-high pressure water (hydrodemolition), refer Figure 8.4.

Hydrodemolition is usually faster and less likely to damage adjacent sound concrete than percussion methods.

The space that is created behind the bars during breakout should be sufficiently wide to permit removal of rust from all around the bars. The length of bar and the amount of the circumference exposed during breakout will depend on the extent and type of corrosion and other forms of remediation to be employed.

If cathodic protection (CP) is also to be applied, then the repair is largely aesthetic as the CP will mitigate the corrosion. The repair must still be made durable since it may support the CP anode or its loss would simply re-expose the reinforcement to the corrosive environment.

If only patch repairs are being used, then the repair should extend all around the bar (usually extending 15–40 mm behind the bar) and at least 50–100 mm beyond the point where any corrosion damage or rust can be observed along the length of the bar, refer Figure 8.5 as an example. A mirror may be used to ensure that all the rust has been removed from the back of the bars.

If chloride contamination levels are high in the as yet undamaged parent concrete, then further concrete removal may be required in order to ensure good service from the freshly patched area.

It is also important to note that the surrounds of each breakout need to be square edged or 'delineated' by saw cutting so as to avoid feather edging, refer Figure 8.5.

The soundness of the concrete breakout areas should also be checked before any patch repair material reinstatement (hand applied, poured, sprayed or combinations thereof) is undertaken, refer Figure 8.5.

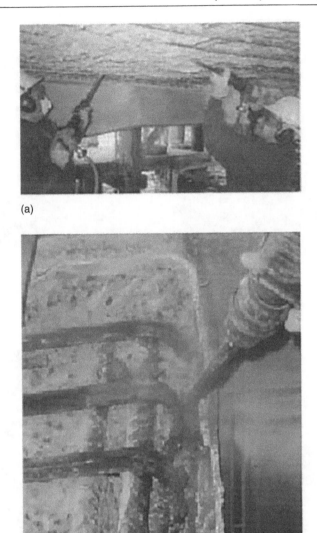

(a)

(b)

Figure 8.3 Concrete breakout by jackhammer (air-powered).

(a)

(b)

Figure 8.4 Concrete breakout by hydrodemolition.

8.4.3 Rebar coatings

Cleaning to a Class 2½ finish as specified in AS 1627 (2017) is necessary to ensure the removal of chloride contaminated products on the surface (and in the pits) of the reinforcement (rebars). However, in indoor situations etc power tool cleaning may be all that is practically achievable. Often removal of concrete and corrosion product from the back of a bar proves difficult and the lack of adequate cleaning is one of the most frequent causes of repair failures. The time interval between cleaning and repairs should also

(a)

(b)

Figure 8.5 (a) Concrete breakout to behind reinforcement, chased until uncor-
roded and with the repair area delineated to avoid feather edging and
(b) concrete breakout area soundness checking example.

be as short as possible, particularly if the structure is located in an aggressive environment. In such an environment, aggressive ions can easily be
trapped in the oxide film and can re-initiate corrosion after the repair. In a
marine environment flash rusting of freshly cleaned bars can frequently
occur.

If there is considerable reduction in the cross-sectional area of the bars
then further steel will have to be added/augmented, by welding, and lapping
new bars to adjacent sound reinforcement. The concrete will have to be cut
back beyond the defective zone into sound material to provide sufficient lap
distances.

Proprietary coating systems can then be applied by hand to prepared existing, new or augmented reinforcement (rebar) surfaces. Coating systems include but are not limited to: zinc-rich epoxy; epoxy; and resin modified cementitious. Method of application is by brush ensuring complete envelopment of the bars. Figure 8.6 shows an example of zinc-rich epoxy rebar coating application and an example of a resin modified cementitious rebar coating system.

(a)

(b)

Figure 8.6 (a) Zinc-rich epoxy rebar coating application by brush and (b) resin modified cementitious rebar coating system application.

8.4.4 Bonding agents

It is essential that there should be good adhesion (bond) between the repair mortar (patching material) and the concrete. This is achievable by making the repair surface saturated surface dry (SSD) by saturating with water, for as long as necessary so that the water can penetrate well into the structure and then allowed to dry until it is damp/surface dry (with no free standing water). Very wet surfaces will reduce the bond due to too high a w/c ratio of the repair mortar at the bond line. Dry surfaces will mean that water is sucked out of the repair mortar such that the w/c ratio of the repair mortar at the bond line is too low.

Bonding agents may also be employed to ensure good adhesion between the repair mortar and the concrete. Proprietary bonding agents include: acrylic; styrene butadiene rubber (SBR); epoxy; and polymer-modified cementitious based. The method of application is by brush, refer Figure 8.7, and in the case of the SBR and epoxy-based systems they must remain 'tacky' prior to the application of the repair mortar, otherwise they will act as 'debonding agents'. The 'tackiness' of the acrylic based bonding agents can be restored by a further application of bonding agent so as to achieve re-emulsification, refer Figure 8.7.

8.4.5 Patching materials

Patching materials are most commonly pre-blended, pre-packaged, ready-to-use, polymer-modified cementitious mortars with a wide range of properties including: lightweight for overhead use; cohesive, flowable, or pourable;

Figure 8.7 Bonding agent application to patch repair area (subsequent to rebar coating application).

structural grade/high strength; fairing, thin-build, levelling; fast setting; and then some with additional properties such as chloride resistance, carbonation resistance, low shrinkage, high bond strength, and high electrical resistivity.

Supplier provided mixing paddles are necessary and application can be by hand, form and pour, or sprayed, refer Figure 8.8.

Polymer-modified mortars consist of Portland based cements including in some instances blended cements to which a suspension of polymeric particles in water has been added. These suspensions include styrene butadiene and acrylic lattices. The addition of these polymers to Portland cement mortars improves workability yet permits the lowering of the water/cement (binder) ratio, reduces shrinkage and penetrability, and also improves adhesion, and tensile strength.

In the mid-late 1970s patch repair was commonly undertaken using epoxy mortars or simply sand and cement mortars. In the mid-1980s polymer-based mortars, using epoxy resin as the binder, began to crack, and fall off, as the thermal, elastic modulus, and tensile strength differences created high tensile strains in the concrete around the patch. The simple sand/cement mortars were also found to crack and fall off, due to high shrinkage, and lack of bond. Polymer-modified cementitious mortars were then developed

(a)   (b)

(c)   (d)

Figure 8.8 Patching material mixing and application (by hand, form and pour, and sprayed).

to improve bond, reduce shrinkage, minimise incompatibility problems, and provide other specialist performance such as chloride resistance, high electrical resistivity, carbonation resistance, etc.

Although patching is widely practised throughout the industry, repair methods that involve the removal of part but not all the chloride-contaminated concrete followed by patching with chloride-free concrete could be self-defeating. In such a case, cathodic areas are created in the newly alkaline, chloride free repair areas, but increased anodic activity is initiated elsewhere as a result of chloride contamination in the area around the patch. This phenomenon has been termed the 'incipient anode', 'ring anode' or 'halo' effect (Concrete Society, 2011).

Page and Treadaway (1982) were the first to suggest that the mechanism of incipient anode formation in chloride-contaminated concrete is the concept of macrocell development (the formation of spatially separated anodes and cathodes with the anodes being the areas adjacent the repair and the cathodes being the repair itself).

When it comes to 'incipient anode', 'ring anode' or 'halo' effect management, Green et al. (2013) report that the there is a perception in the industry that the only way to do this is by the insertion of galvanic anodes within patch repairs, refer Section 10.1. However, Green et al. (2013) point out that it should be remembered that conventional concrete patch repair utilising a reinforcement coating system (zinc-rich epoxy, epoxy or resin modified cementitious), a bonding agent (acrylic, SBR, epoxy or polymer-modified cementitious) and a polymer-modified cementitious repair mortar (hand applied, poured, sprayed or combinations thereof), effectively 'isolate' the patch such that macrocell activity is minimised and incipient anode formation managed. By the use of reinforcement coating systems, cathodic reactions are severely restricted, on the steel surfaces within the patch. Bonding agents provide a restriction to ionic current flow out of the patch into the surrounding parent concrete. Furthermore, polymer-modified cementitious repair mortars have electrical resistivities which restrict ionic current flow between the patch and surrounding parent concrete.

Green et al. (2013) further advise that if all three aspects, reinforcement coating/bonding agent/repair mortar, are effectively combined, then macrocell activity is minimised. Even if just a reinforcement coating (polymeric-dominated and isolating) and a polymer-modified cementitious repair mortar are used, without the use of a bonding agent but with the method of surface preparation of repair surfaces being SSD, then Green et al. (2013) indicate that macrocell activity will still be minimised.

Green et al. (2013) also report on research from others (Morgan, 1996) done in the 1980s and 1990s, for example, involving polymer-modified Portland cement mortars with high electrical resistivity achieving corrosion protection and in addition not affecting corrosion of steel in adjacent unrepaired areas (i.e. not leading to incipient anode development). Also, Morgan (1996) studied in the mid-1980s the influence of a variety of different reinforcement coating systems on corrosion activity. They concluded that a low permeability polymer-modified repair mortar used in conjunction with a zinc-rich epoxy rebar coating provided the best protection to reinforcing bars both in the repair zone and in the adjacent concrete.

Furthermore, Green et al. (2013) report of the work of Berndt (2013), for example, on the long term performance (in excess of 35 years) of various repair techniques to the reinforced concrete piles (500 mm diameter) of two large (260 m long and 48 m wide) reinforced concrete sugar storage sheds in Cairns, Queensland, Australia. Deterioration to the 2,118 reinforced concrete piles included a combination of reinforcement corrosion, alkali-silica reaction and salt hydration distress (or physical salt attack). Berndt (2013) states that patch repairs with shotcrete up to 39 years old were generally successful and devoid of incipient anode problems except where the removal of chloride contaminated concrete around repair areas was inadequate.

8.4.6 Equipment and workmanship

Patch repair equipment and workmanship has advanced since the 1970s. Equipment advances include more compact and increased capacity (air-powered and electric) jack hammers (refer Figure 8.3), saws (diamond tipped blades), grinders, etc. Hydrodemolition (high pressure water blasting), refer Figure 8.4, as a concrete breakout method for large area repairs including overhead. Water blasting as a surface preparation technique. Power tool/equipment mixing on-site of mortars to provide more homogeneous product, refer Figure 8.8.

Workmanship advances have included the ability to undertake overhead repairs by hand packing (because of high cohesive strength mortars), refer Figure 8.8, rather than just form and pour; improved workmanship quality because of better and tailored equipment and materials; as well as increased availability to information, knowledge, and training.

8.5 SPRAYED CONCRETE (SHOTCRETE/GUNITE)

There is a growing use of shotcreting (guniting) for concrete repairs. One of its main advantages is that large area and large volume repairs can be undertaken including overhead without the need for formwork. However, the same care is required in removing defective concrete, providing a clean substrate, and cleaning the bars as with other types of repairs. The defective or contaminated concrete is removed with percussive tools or hydrodemolition. The steel is cleaned with high pressure water or wet abrasive blasting.

In the wet method of shotcrete application, all the ingredients are mixed together before they are applied. In the dry method, the solid materials are mixed together first and water is added at the nozzle. The water/cement (binder) ratio in the wet process tends to be higher than in the dry, as it is necessary to achieve sufficient workability for pumping. Consequently, the concrete placed in the dry process has a higher compressive strength and density than in the wet one. Fibres (steel or synthetic) and mesh reinforcement may be used with shotcreting repairs and can be attached to pins fired into the concrete.

Silica fume is commonly incorporated into sprayed mortars (shotcrete or gunite). Whereas preferred thicknesses in the past were in the range of 25–60 mm, using silica fume, layers with a thickness of up to 400 mm can be placed readily without any adhesion problems. Silica fume mixes are characterised by very low penetrability and very high electrical resistivity. Sometimes the adhesion is improved, and the penetrability reduced, by the addition of polymeric-based additives to the shotcrete mix. Blended cements can be utilised in sprayed mortars. Control of line and thickness of placed material is by timber profiles or tensioned wires.

The performance of shotcrete as a repair medium is very much influenced by the quality of the workmanship, in particular the skill of the nozzle operator.

8.6 RECASTING WITH NEW CONCRETE

When areas affected by corrosion become too large for repairs to be effected by simple patching techniques or sprayed mortars it may be necessary to remove all the old concrete and re-cast fresh concrete around the reinforcing bars. When doing this it is important to ensure that the environment of the bars is returned to an alkaline condition. This may require removal of carbonated or chloride-contaminated concrete and its replacement with low-penetrability concrete or a cementitious based mortar that will resist renewed contamination.

A major requirement for a repair mix is low penetrability, and this is usually achieved by using fairly rich mixes with low water/cement (binder) ratios. The water/cement (binder) ratio should not exceed 0.40 (maybe even as low as 0.35). Such a mix will normally have a low workability and so be difficult to place in the confined space that may be present behind the exposed reinforcement. The use of a superplasticiser may be necessary in such circumstances. The attainment of a low penetrability can be facilitated by taking care that prolonged moisture curing is carried out and by the addition of SCMs (e.g. silica fume) to a repair mix. After the formwork has been removed, curing can be continued by covering the exposed concrete with polyethylene sheet or with a curing compound.

It is essential that there should be good adhesion between the new and old concrete. The contact surface should be saturated with water for as long as necessary so that the water can penetrate well into the structure and then be allowed to dry until it is damp (SSD). Very wet surfaces will reduce the bond. Possible adhesion problems may also be avoided by the application of a

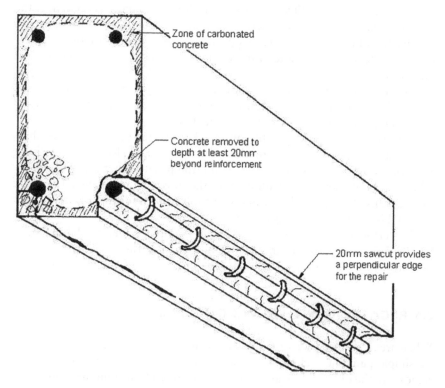

Zone of carbonated concrete

Concrete removed to depth at least 20mm beyond reinforcement

20mm sawcut provides a perpendicular edge for the repair

Figure 8.9 Replacement of concrete cover. (Courtesy of Pullar-Strecker, 1987, p.46)

bonding agents. Proprietary mixes, refer Section 8.4.5, are available for this purpose. The join between the new and the old concrete should always be perpendicular to the face of the concrete. This prevents the natural compressive forces that are set up by expansion and contraction of the structure from 'levering off' either the new concrete or the old. An example is shown in Figure 8.9.

The load-carrying capacity of a repair will depend not only on its strength, but also on the elastic modulus, bond strength, shrinkage, and creep properties. It is desirable that the properties of the repair material should be similar to those of the original concrete. Differential shrinkage and creep between the two materials will reduce the load-bearing contribution of the repair and similarly a large difference between the coefficients of thermal expansion of the two materials could be harmful.

8.7 INHIBITORS

Inhibitors may be added to a corrosive environment to lower the corrosion rate by retarding the anodic process, the cathodic process, or both. An

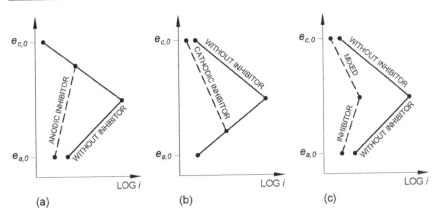

Figure 8.10 Inhibitor actions. (Courtesy of Wranglen, 1985, p.168)

anodic inhibitor increases the anode polarisation and hence moves the corrosion potential in the positive direction, refer Figure 8.10, whereas a cathodic inhibitor moves the corrosion potential in the negative direction, refer Figure 8.10 and impacts the cathodic process (cathodic polarisation). With so called mixed inhibitors, the corrosion potential change is smaller and its direction is determined by the relative size of the anodic and cathodic effects, refer Figure 8.10.

Two techniques of application of chemical inhibitors to reinforced concrete are used and these may be termed admixture and surface applied or topical. In the case of admixture inhibitors, the chemical compound is added to the concrete batch (in new structures), to concrete repair materials (hand applied, poured and sprayed) and are even incorporated into some penetrant and coating treatment systems. Chemicals that have been used include inorganic nitrites, amphipathic molecules that develop hydrophobic properties in the concrete and organic amine compounds. The latter are believed to be migratory inhibitors so that they are able to diffuse through the concrete towards the steel and inhibit the corrosion process.

Surface (topically) applied inhibitors rely for their action on their ability to diffuse through the concrete to the surface of the reinforcement after they have been applied to the surface of the concrete. They are usually used in remedial situations either after the appearance of surface cracking or in situations in which a corroding structure needs repair. In the case of surface applied inhibitors, the compound is applied to the surface of the concrete either as a spray or as a gel or paste. Chemicals that have been used include organic amines and carboxylate compounds.

Admixture corrosion inhibitors have been in use for some time to extend the life of reinforced concrete infrastructures. The most common among these are the inhibitors based on inorganic nitrites (Berke, 1991). These inhibitors generally provide protection by reinforcing the protective ferric oxide passive film around the steel by oxidising ferrous ions:

$$2Fe^{2+} + 2OH^- + 2NO_2^- \rightarrow 2NO + Fe_2O_3 + H_2O$$

$$Fe^{2+} + OH^- + NO_2^- \rightarrow NO + \gamma - FeOOH$$

The nitrite is not adsorbed into the passive film, but some of it is consumed as part of the inhibitive process. Sodium nitrite is reported (Elsener, 2001) to cause moderate to severe loss of compressive strength and similar results were found with sodium benzoate and potassium chromate. Sodium nitrite is also reported to enhance the risk of alkali aggregate reaction (AAR). Calcium nitrite on the other hand may increase the compressive strength and act as a moderate accelerator.

Although anodic inhibitors are very effective and widely used, they usually have an undesirable property: if the content of inhibitor is or gradually becomes so low that it does not suffice to cover all the surface of the metal acting as an anode, the dangerous combination small anode and large cathode is obtained, which may lead to pitting. In such cases the inhibitor may do more harm than good. In general, if nitrites are added in recommended amounts there is enough present to avoid such problems; attempts to economise on the amount added could, however, be dangerous.

In recent years inhibitors based on organic amines and alkanolamines have become available with the claimed advantage that, as well as being capable of application as admixtures, they may be applied as remedial agents to the surface of the concrete. The effect of a number of these amine based inhibitors is shown in Figure 8.11.

Such inhibitors are adsorbed on the surface of the metal and provide a protective shield against the attack of aggressive agents. The adsorption occurs through the formation of a metal-amine bond and the organic 'tail'

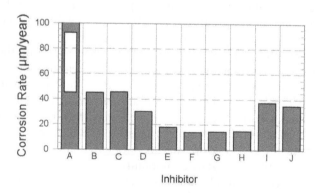

Figure 8.11 Effectiveness of some amine-based inhibitors. (Courtesy of Elsener, 2001, p22)

of the molecule provides the surface coverage. It is proposed that these inhibitors, when applied to the surface of an already corroding structure, can diffuse through concrete (Bjegovic et al., 1993) and once they reach the steel, they form a protective barrier over the metal surface inhibiting further corrosion. They can also be used in admixture applications.

There is, however, little published work on the mechanisms of the inhibition, the long term performance of the inhibitors or the effect of their concentration on the level of inhibition. The inhibitors are generally proprietary products and so most of the information available is that given by suppliers.

Green et al. (2013) advise of some the performance concerns relating to organic migratory inhibitors including:

- Extent of migration?
  (Concrete moisture content; concrete porosity; exposure effects; ambient conditions)
- Can they migrate in a suitable timeframe?
- Can you readily measure their locale?
- Longevity?
  (Ongoing migration; back diffusion to the environment; evaporation; rinsing or leaching of the surface)
- Stability?
  (How long will they continue to perform; are they consumed)
- Accelerate corrosion?
  (Insufficient concentration; partly-cover rebar; different concentrations due to concrete porosity; macrocell corrosion due to inhibitor concentration differences; large cathode/small anode effects)
- What effect on localised pitting corrosion?
  (Pits in chloride contaminated conventional reinforcement; pits in chloride contaminated prestressing steel)
- Excessive applications can cause chemical attack of concrete surfaces?

As for the inorganic inhibitors, if the content of an organic migratory inhibitor is not sufficient to cover all the surface of the metal acting as an anode, the dangerous combination of small anode and large cathode is obtained, which may lead to pitting. For example, Nairn et al. (2003) advises that 'thorough penetration of inhibitor into carbonated concrete may be particularly important, lest uneven penetration leads to the formation of macrocells and worsening of the corrosion problem'.

Holloway et al. (2003) found for a commercially available migrating organic inhibitor that at low concentrations chloride-induced corrosion of steel was enhanced and they postulated that this was due to incomplete monolayer film formation decreasing the relative areas of anodic to cathodic sites and, hence driving the anodic reactions.

Holloway & Forsyth (2006) report on the investigation and testing on some migrating organic inhibitor treated steel reinforced concrete test panels mounted on the concrete piles of a jetty in Western Port, Victoria, Australia. Duplicate panels were placed in the 'splash zone' and 'high tide region' relative to the water level. The splash zone panels were only directly exposed to seawater in the event of waves greater than 1–2 metres, while the high tide panels were positioned at the mean high tide mark and were typically submerged twice daily. They found that the migrating organic inhibitor was found to increase the amount of corrosion relative to the controls.

The variable and questionable performance of inhibitors for hardened concrete has led Broomfield (2007) to conclude 'that it may prove that inhibitors for hardened concrete are most effective for carbonated and low chloride level concretes, with low cover, and reasonable permeability'. He also points out that, 'however, any application of inhibitors for repair of existing structures should be done on a trial basis with proposed long term monitoring of the treatment'.

The effects of inhibitors on any subsequent CP and electrochemical treatments may also need consideration.

8.8 COATINGS AND PENETRANTS

8.8.1 Anti-carbonation coatings

Anti-carbonation coatings may be applied after carbonation repairs to stop further atmospheric carbon dioxide ingress. The carbonation resistance should always be checked rather than relying upon coating formulation as testing has shown that not all acrylics (for instance) have good carbonation resistance (Green, 1988).

Penetrants (e.g. silanes, siloxanes, silicones or blends thereof) cannot be used for carbonation resistance as they have been known to accelerate carbonation in laboratory tests (Broomfield, 2007).

Much has been published about anti-carbonation coatings and their effectiveness, but it is not easy to find independent information about specific products. Careful comparison of product information may be required. This may include the requirements for independent testing. The principles of anti-carbonation coatings are that they are porous enough to let water vapour move in and out of the concrete, but the pores are too small for the large carbon dioxide molecule to pass through (Broomfield, 2007).

Klopfer (1978) states that an anti-carbonation coating must maintain an R-value (equivalent air layer thickness with respect to CO_2 diffusion) >50m for it to be deemed effective and an S_D-value (equivalent air layer thickness with respect to water vapour diffusion) <4 m to allow an acceptable level of water vapour to pass. A coating applied at 200 μm would thus need to achieve a CO_2 diffusion coefficient of the order of 5×10^{-7}cm²/sec to

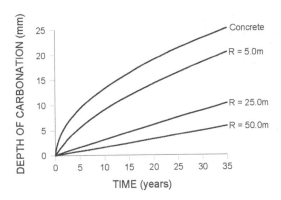

Figure 8.12 Effect of coatings on the progress of carbonation. (Courtesy of Green, 1988)

produce an R-value = 50m, and a water vapour diffusion coefficient of 1.3 × 10^{-5}cm^2/sec to achieve an S_D-value = 4 m (Green, 1988).

AS/NZS 4548.5 (1999), for example, provides guidelines on testing for carbon dioxide diffusion and water vapour transmission.

A theoretical consideration of the effect of coatings on the progress of carbonation is shown in Figure 8.12, which is based on a simple square-root of time model. This was developed from Fick's first law of diffusion and assumes that a known quantity of carbon dioxide will be required to neutralise an element of concrete. By applying coatings of known thickness and CO_2 diffusion resistance to the concrete, the rate of access of CO_2 into the concrete will be reduced, from which the reduced rates of carbonation can be calculated.

8.8.2 Chloride-resistant coatings

Coating systems that restrict the ingress of chloride ions are available for application to concrete as a preventative measure or for application after chloride repairs.

As for anti-carbonation coatings, it is not easy to find independent information about specific products. Careful comparison of product information may be required. As previous, this may include the requirements for independent testing.

AS/NZS 4548.5 (1999), for example, provides guidelines on testing for chloride ion diffusion resistance.

Andrews-Phaedonos et al. (1997) note for coastal bridge structures for example, that protective coatings for each environmental zone need to fulfil the following criteria:

a. Penetrability performance to enhance durability: resistance to ingress of waterborne chloride ions, O_2, CO_2, and water sorption (uptake of water and waterborne ions by capillarity).

b. Resistance to physical and chemical degradation mechanisms: resistance to ultraviolet light, mechanical damage, biological growth, and compatibility with the alkaline concrete substrate.
c. Practical application to produce a defect free film.
d. Compliance with appropriate health and safety and environmental requirements.

As an example of the variation in the performance of different generic types of coatings for concrete, chloride ingress data from laboratory exposure and field exposure are shown at Table 8.1 and Figure 8.13.

8.8.3 Penetrants

Penetrants (e.g. silanes, siloxanes, silicones or blends thereof) and pore blocking materials (e.g. silicates) may be applied after concrete repairs to reduce the penetration of moisture and chloride ions. However, as mentioned previously, penetrants cannot be used for carbonation resistance as they have been known to accelerate carbonation in laboratory tests (Broomfield, 2007).

As for anti-carbonation and chloride-resistant coatings, it is not easy to find independent information about specific penetrant and pore blocking products. Careful comparison of product information may be required. As previous, this may include the requirements for independent testing.

Table 8.1 Laboratory chloride ingress results

Coating Type	Mortar Substrate	Accelerated Weathering	Cumulative % Chloride	Efficiency Index %
None	AS3799	No	2.17	0
Cementitious	AS3799	Yes	0.04	98
		No	0.06	97
Epoxy	AS3799	Yes	0.11	95
		No	0.04	98
Acrylic	AS3799	Yes	0.86	60
		No	0.87	60
Silane	AS3799	Yes	0.36	83
		No	1.33	39
None	C311	No	2.01	0
Cementitious	C311	Yes	0.08	96
		No	0.06	97
Epoxy	C311	Yes	0.11	94
		No	0.23	88
Acrylic	C311	Yes	1.06	47
		No	1.44	28
Silane	C311	Yes	0.24	88
		No	0.52	74

Source: Andrews-Phaedonos et al. (1997)

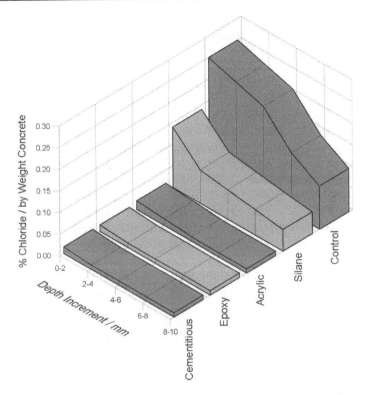

Figure 8.13 Chloride profiles from site test slabs after 16 months' exposure. (Courtesy of Andrews-Phaedonos et al., 1997)

APAS Specification 0168 (2001), for example, provides guidelines on testing of some performance properties for penetrant treatments.

Performance assessment of pore blocking materials tends to be by specific laboratory and field based testing in terms of most commonly water penetrability but may include resistance to ingress of waterborne chloride ions, O_2 and CO_2.

8.9 STRUCTURAL STRENGTHENING

Steel plate bonding and steel or concrete column jacketing are the traditional methods of external reinforcing of concrete structures. Steel plates bonded to the tension zones of concrete members have shown to be increasing the flexural capacity of the members (Fleming & King, 1967). This traditional method has been used over the world to strengthen bridges and buildings. However, the corrosion of the steel plates, deterioration of the bond between steel and concrete, installation difficulties such as the necessity of heavy equipment in installing, have been identified as major drawbacks of this technique (Nezamian & Setunge, 2002). As a result, fibre-reinforced polymer (FRP) composite system strengthening has been developed as an alternative to this method.

FRP systems can be used for the strengthening or retrofitting of existing concrete structures to resist higher design loads, correct deterioration-related damage or increase ductility (Nezamian & Setunge, 2002). Externally bonded FRP systems have been used to strengthen and retrofit existing concrete structures overseas and in Australia (Karla & Neubauer, 2003). The FRP composites combine the strength of the fibres with the stability of the polymer resins. They are defined as polymer matrices that are reinforced with fibres or other reinforcing material with a sufficient aspect ratio (length to thickness) to provide a desirable reinforcing function in one or more directions. The FRP composite materials are different from traditional construction materials such as steel, aluminium, and concrete because they are anisotropic, i.e. the properties differ depending on the direction of the fibres (Nezamian & Setunge, 2002).

The characteristics of FRP composites depends on many factors such as type of fibre, its orientation, and volume, type of resin, and quality control used during the manufacturing process. FRP systems come in a variety of forms, including wet lay-up systems, and procured systems. There are several constituent materials in commercially available FRP repair systems such as resins, primers, putties, saturants, adhesives, and fillers as well as protective coatings. FRP composites gain their strength largely from the fibres, which are usually glass, carbon, or aramid fibre. FRP materials are lightweight, non-corrosive, non-magnetic and exhibit high tensile strength (Nezamian & Setunge, 2002).

Externally bonded FRP systems have also been applied to strengthen masonry, timber, steel, and cast-iron structures. They have been used in structural elements such as beams, slabs, columns, walls, joints/connections, chimneys, and smokestacks, vaults, domes, tunnels, silos, pipes and trusses (Nezamian & Setunge, 2002).

Figure 8.14 shows an example of application of carbon FRP fabric systems application to strengthen some reinforced concrete substructure bridge beam elements.

Figure 8.15 provides an example of carbon FRP laminates application to reinforced concrete bridge substructure elements.

8.10 PILE JACKETING

Jacketing of reinforced or prestressed concrete piles will not be discussed in detail suffice to advise that in marine conditions Florida Department of Transport has found that encasing bridge piles in concrete is not effective in stopping corrosion. Prestressing strands in a marine pile were found to have

(a)

(b)

Figure 8.14 Carbon FRP fabric application example, including epoxy bonding agent, to bridge substructure elements.

corroded through shortly (three years) after repair, see Figure 8.16 (Broomfield, 2007).

Similarly, it would not have mattered whether the jacket was say FRP, fibreglass, etc. instead of concrete and nor would it have mattered whether the pile encapsulation system within the jacket was resinous instead of concrete, the same dramatic chloride-induced pitting corrosion would have occurred. As is evident from the previously discussed mechanisms of chloride-induced pitting corrosion of steel in concrete, refer Section 4.7, chloride ion-induced pit propagation does not ultimately need oxygen to sustain itself because within the pits resides steel (iron) in pH 4 or so hydrochloric acid. Macrocell corrosion involving a large cathode (at which the oxygen reduction reaction occurs) and small anodes/pits (at which iron oxidation

(a)

(b)

Figure 8.15 Carbon FRP laminate application example to bridge substructure elements.

Figure 8.16 Pretensioned tendons corroded through three years after concrete pile jacketing repair. (Courtesy of Broomfield, 2007)

occurs) can lead to more rapid pitting but should the oxygen concentration become limited then the pitting propagation corrosion mechanism reverts to steel (iron) oxidation in acid solution (where the reduction reaction becomes hydrogen evolution within the pits).

The installation of jackets to already corroding reinforced concrete piles in a saline environment will also hide any corrosion.

REFERENCES

Andrews-Phaedonos, F, Collins, FG, Green, WK and Peek, AM (1997), '*Assessment of protective coatings for concrete bridges in marine or saline environments*', Proc. *Corrosion & Prevention 1997 Conf.*, Australasian Corrosion Association Inc., November.

APAS (2001), 'Silane-based water repellents for concrete and masonry', Australian Paint Approval Scheme Specification 0168, Rev No 6.

Berke, N (1991), 'Corrosion Inhibitors in Concrete', *Concrete International*, 13, No. 7, 24–27, July.

Bjegovic, L, Sipos, V, Ukrainczyk, P and Miksic, B (1993), 'Corrosion and Corrosion Protection in Concrete', ed. Narayan Swamy, 2, 865–877.

Berndt, M (2013), 'Long-Term performance of concrete repairs in complex conditions', Proc. Corrosion 2013 Conf., NACE International, Orlando, USA, March, *Paper No.* 2214.

Broomfield, JP (2007), '*Corrosion of Steel in Concrete: Understanding, Investigation and Repair*', second edition, Taylor & Francis, London, UK.

Chess, P and Green, W (2020), '*Durability of Reinforced Concrete Structures*', CRC Press, Taylor & Francis Group, Boca Raton, FL, USA.

Concrete Society (2011), 'Cathodic protection of steel in concrete', Technical Report 73, Surrey, UK.

Elsener, B (2001), '*Corrosion Inhibitors for Steel in Concrete*', European Federation of Corrosion, Publications No.35, Institute of Materials, London, UK.

Fleming, CJ and King GEM (1967), 'The development of structural adhesives for three original uses in South Africa', RILEM Int. Symp. on Synthetic Resins in Build. Constr., RILEM, Paris, 75–92.

Green, WK (1988), '*Evaluation of coatings as carbonation barriers for reinforced concrete*', *Proc. Corrosion and Prevention 1988 Conf.*, Australasian Corrosion Association Inc., November, Paper 1–4.

Green, W, Dockrill, B and Eliasson, B (2013), 'Concrete repair and protection – overlooked issues', Proc. Corrosion and Prevention 2013 Conf., Australasian Corrosion Association Inc., November, Paper 020.

Holloway, L, Nairn, K and Forsyth, M (2003), 'Concentration effects in concrete penetrating corrosion inhibitor performance', Proc. Corrosion Control & NDT Conf., Australasian Corrosion Association Inc., November, Paper 094.

Holloway, L and Forsyth, M (2006), 'Identifying and understanding the inhibition mechanisms and performance of two organic concrete reinforcement corrosion inhibitors', Proc. Corrosion & Prevention 2006 Conf., Australasian Corrosion Association Inc., November, Paper 053.

Karla, R and Neubauer, U (2003), 'Strengthening of the Westgate Bridge with carbon fibre composites – a proof engineer's perspective', 21st Biennial Conference of the Concrete Institute of Australia, 245–254.

Klopfer, H (1978), 'The carbonation of external concrete and how to combat it', Bautenshutz and Rausanierung, 1, No 3, 86–97.

Morgan, DR (1996), 'Compatibility of concrete repair materials and systems', *Construction and Building Materials*, Vol. 10, No. 1, 57–67.

Nairn, KM, Holloway, L, Cherry, B and Forsyth, M (2003), 'A review of surface applied corrosion inhibitors', Proc. Corrosion Control & NDT Conf., Australasian Corrosion Association Inc., November, Paper 093.

Nezamian, A and Setunge, S (2002), 'User friendly guide for rehabilitation or strengthening of bridge structures using fiber reinforced polymer composites', Report 2002-005-C-04, CRC Construction Innovation, Research Program C: Delivery and Management of Built Assets, Project 2002-005-C:Decision Support Tools for Concrete Infrastructure Rehabilitation.

Page, CL and Treadaway, KWJ (1982), 'Aspects of the electrochemistry of steel in concrete', *Nature*, 297, 109–115.

Pullar-Strecker, P (1987), '*Corrosion Damaged Concrete*', Butterworths, London, UK.

Standards Australia (1999, Reconfirmed 2013), 'AS/NZS 4548.5 Guide to long-life coatings for concrete and masonry – Part 5: Guidelines to methods of test', Sydney, Australia.

Standards Australia (2005, Reconfirmed 2017), 'AS 1627 Metal finishing – Preparation and pretreatment of surfaces. Part 4 – Abrasive blast cleaning of steel', Sydney, Australia.

Wranglen, G (1985), '*Corrosion and Protection of Metals*', Chapman, and Hall, London, UK.

Chapter 9

Repair and protection (B) – cathodic protection

9.1 INTRODUCTION

One of the most effective means of reducing the corrosion of reinforcing bars is the application of CP. This is defined in AS 2832.5 (2008) as 'The prevention or reduction of corrosion of metal by making the metal the cathode in a galvanic or electrolytic cell'. A greatly simplified explanation of the action of CP is shown in Figures 9.1 and 9.2.

A corroding reinforcing bar has anodic areas where iron is dissolving from the surface by means of the reaction:

$$Fe(s) \rightarrow Fe^{2+}(aq) + 2e^-$$

In these areas, positive current leaves the bar and flows into the concrete. It also has cathodic areas where the reaction:

$$\tfrac{1}{2}O_2(g) + H_2O(l) + 2e \rightarrow 2OH^-(aq)$$

proceeds and positive current flows on to the bar from the concrete. The electrical circuit is shown in Figure 9.1 and the corrosion current (i_o) is given by:

$$i_o = (E_a - E_c)/(R_c + R_a)$$

Where E_a and E_c are the single electrode potentials of the anodic and cathodic reactions and R_c and R_a are the resistances associated with the cathodic and anodic areas respectively, refer also Figure 9.1.

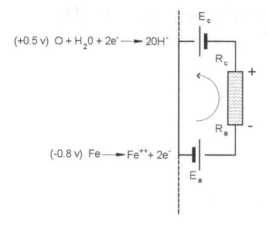

Figure 9.1 Currents associated with the corrosion process.

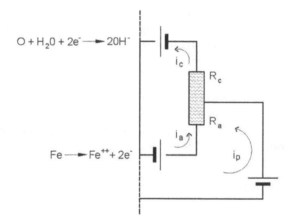

Figure 9.2 The corrosion process with a polarising current applied.

CP consists of passing a positive electrical current on to the reinforcement from an electrode that is usually buried in the concrete close to the reinforcement. This flow of current is brought about by connecting the reinforcing bar to the negative terminal of a source of electrical potential and the electrode to the positive terminal. The reinforcing bar thus becomes the cathode and the electrode an anode and the current (ionic) flows through the concrete electrolyte. A direct current flows on to the bar as shown diagrammatically in Figure 9.2. The anodic current is reduced and in principle if the applied potential is large enough, it is reversed.

Since the currents flowing into a junction must equal those flowing out, then:

$$i_c = i_p + i_a$$

summing potential drops around a circuit, then:

$$E_a - E_c = i_a R_a + i_c R_c = i_a R_a + \left(i_p + i_a\right) R_c$$

so that the anodic current is given by:

$$i_a = i_o - i_p R_c / \left(R_a + R_c\right)$$

It can thus be seen that the application of a current on to the surface of the steel to be protected reduces the corrosion current. Because positive current is flowing on to the bar the anodic reaction is suppressed and the cathodic reaction is enhanced.

Attaching the reinforcing bar to the negative terminal of a source of potential must lower the potential of the bar with respect to the electrolyte (concrete) in which it is immersed. The attainment of a sufficiently negative potential while the current is flowing is sometimes used as a criterion that enough current is flowing on to the bar to reduce the corrosion rate to a desired level. However, the cathodic reaction that takes place:

$$\frac{1}{2}O_2(g) + H_2O(l) + 2e^- \rightarrow 2OH^-(aq)$$

generates hydroxyl ions and these serve to replace the alkalinity that must have been reduced to allow the corrosion to take place. This allows re-passivation by the reaction:

$$2Fe^{++}(aq) + 6OH^-(aq) \rightarrow \gamma - Fe_2O_3(s) + 3H_2O(l) + 2e^-$$

It was seen at Chapters 5 and 6 that the corrosion potential of a passivated surface is more positive than that of an active surface. An alternative criterion for the effectiveness of a CP system is therefore the attainment of a sufficiently positive potential when the current is turned off. There are thus two complimentary criteria of the adequacy of a CP system, the attainment of a sufficiently negative potential when the current is turned on and a sufficiently positive potential when the current is turned off.

Since the suppression of the anodic process takes place independently of the environment in which the metal is situated, CP is a method by which a deteriorating reinforced concrete structure can be protected without removing the aggressive agent from the surface of the steel. It is primarily for this reason that within the United States Federal Highways Authority has stated that it regards 'The only way of stopping corrosion in bridge decks suffering de-icing salt corrosion' as being impressed current CP.

CP can be applied by galvanic means as well as impressed current means. The CP anode (electrode connected to positive terminal) may be a metal which is more active in the Galvanic Series of metals, with the current flowing due to the driving voltage from the galvanic cell formed, and the more electronegative metal corroding. This is known as galvanic cathodic protection (GCP). Alternatively, the driving voltage may be derived from an external source, such as a transformer/rectifier, in conjunction with a CP anode which corrodes only at a very slow rate. This is impressed current cathodic protection (ICCP).

It should also be noted that the electrical continuity of all the reinforcement to be protected is vital to the successful implementation of CP. If there is a break in the reinforcement, then current will flow between the pieces of reinforcing bar through the concrete. Where the current flows onto the bar, the steel will be protected, but where the current leaves the bar it will be accompanied by dissolution of the iron.

9.2 HISTORY OF CATHODIC PROTECTION

Sir Humphry Davy was the founder of CP. Davy postulated in January 1824 (Davy, 1824a), that it would be possible to prevent corrosion of copper sheathing on ships by connecting it to zinc, tin, or iron (Ackland, 2005). This was based on his observations and conclusions from over two decades of working with galvanic couples and he certainly considered the beneficial effects for the more positive metal in the couple at least as early as 1812 (Ackland, 2005). The copper sheathing on the ships was needed to reduce fouling and prevent attack on the timber by worms. Davy then reported on full scale trials in June 1824 (Davy, 1824b) and showed the complete effectiveness of zinc and iron in protecting the copper (Ackland, 2005). He also published in June 1825 (Davy, 1825) the influence of ship movements on the efficiency of the protection action and the effects on fouling (Ackland, 2005).

Davy's work defined the two ways in which we apply CP to this very day, by using galvanic anodes or by imposing impressed current where current is forced from an anode through the electrolyte onto a structure (cathode) using a DC power supply (Ackland, 2005). Both methods of CP act to shift the potential of the structure to be protected in the negative direction.

The anode is consumed while the cathode is protected if there is sufficient current to provide the requisite polarisation (Ackland, 2005).

9.3 IMPRESSED CURRENT CATHODIC PROTECTION

In terms of CP of reinforced and prestressed concrete structures and buildings, it is accepted that a properly designed, installed, operated and maintained impressed current system will halt reinforcement and prestressing steel corrosion in concrete regardless of the chloride content or extent of carbonation of the concrete as well as the extent and rate of reinforcement or prestressing steel corrosion (Holloway et al., 2012).

CP has been applied to reinforced concrete structures and buried prestressed concrete pipelines since the 1950s (Unz, 1955; Unz, 1956). Richard Stratfull first applied impressed current CP to atmospherically exposed steel in concrete in a successful trial in the US in 1959 (Stratfull, 1959). Stratfull continued to investigate and experiment with impressed current CP and in 1973 installed the first full scale installation to a bridge deck in California (Stratfull, 1974). Direct experience with impressed current CP of concrete structures in Australia, for example, dates back to the 1980s (Grapiglia & Green, 1995).

There are many types of impressed current anode, and the selection of the correct one is important in determining the overall cost and durability of the system. The variability in impressed current anode types presents the CP designer with great flexibility to provide the optimum protection current at the correct locations for the desired structure life. Combinations of impressed current anodes can also be utilised to provide the most whole-of-life (life cycle) cost effective solution. Table 9.1 provides an outline of the various impressed current anode types by category (Green, 2014). These are further discussed in Section 9.6.

9.4 GALVANIC CATHODIC PROTECTION

There are various galvanic anode types for the provision of cathodic current to reinforcing steel (or prestressing steel) in concrete. Galvanic anodes based on metallised coatings (electric arc sprayed) of pure zinc (>99% purity) have been applied to a limited extent to structures in Australia, for example, since the mid-1990s and those based on aluminium-zinc-indium to a limited extent since 2000 (Green, 2001). Galvanic CP systems for concrete based on encased discrete zinc (Zn) anodes or Zn sheet anodes on the other hand are a more recent development in Australia (Holloway et al., 2012). The discrete Zn anodes are typically encased in a medium that is formulated to maintain the electrochemical activity of the Zn (since Zn is prone to passivation from oxides from its own corrosion products) (Holloway et al., 2012). Table 9.2 provides an outline of the various

Table 9.1 Concrete impressed current anode types by category

Surface applied	Encapsulated	Immersed/buried
Organic conductive coating	Electrocatalytically coated titanium mesh or grid, surface installed and embedded into cementitious overlay	High-silicon iron (with chrome for chloride environments)
Arc sprayed zinc		Mixed metal oxide coated titanium
Thermally sprayed titanium	Electrocatalytically coated titanium ribbon, ribbon mesh, or grid, embedded into cementitious grout in recesses (chases, slots, or grooves) cut into the cover concrete	Platinised titanium
Conductive cementitious overlay		Platinised niobium
	Electrocatalytically coated titanium strip, mesh, grid, or tubes, embedded into cementitious grout in holes drilled into the concrete	
	Platinum-coated titanium rods in conductive graphite-based backfill in holes drilled into the concrete	
	Conductive titanium oxide ceramic tubes embedded in cementitious grout in holes drilled into the concrete	

Source: Green (2014)

Table 9.2 Concrete galvanic anode types by category

Surface applied	Encapsulated	Immersed/buried
Arc sprayed zinc	Zinc mesh within jackets	Cast zinc alloys
Arc sprayed Al-Zn	Zinc mesh in cementitious overlay	Cast aluminium alloys
Arc sprayed Al-Zn-In		Cast magnesium alloys
Adhesive zinc sheet	Discrete zinc anode in patch repairs	
	Discrete zinc cylinders	
	Discrete rolled zinc sheet	
	Discrete zinc strips	

Source: Green (2014)

galvanic anode types by category (Green, 2014). These are discussed further in Section 9.7.

The encapsulated discrete zinc anode types are discussed at Section 10.1 as electrochemical treatments for reinforced concrete as they typically do not output sufficient current to effect CP per se but rather a reduction in reinforcement corrosion rate (associated with cathodic current receipt).

9.5 THE APPLICATION OF CATHODIC PROTECTION

The choice of the source of the applied potential depends upon the nature of the installation to be protected and both impressed current and galvanic anode techniques have been used for the CP of reinforced concrete structures and buildings. A most important consideration affecting the choice of anode system is the capacity of the system to provide an even distribution of current over the whole surface of the reinforcing bar to be protected and to provide sufficient current to afford the level of protection required. Since concrete provides a moderately high resistivity environment for the metal most of the current entering the concrete will reach the metal surface close to the entry point of the current into the concrete. This implies that the current density must be more or less constant over the surface of the concrete.

This can be achieved either by the immersion of the reinforced concrete structure in a high conductivity environment such as seawater (which would ensure that current from a remote anode would be evenly distributed across the surface of the concrete) or the source of current must itself be distributed more or less evenly over the surface. The choice of anode system is therefore primarily determined by the accessibility of the surface of the concrete. When CP is applied to a reinforced concrete structure that is exposed above the ground, the anode must be placed in the surface of the concrete. It cannot be placed further away from the metal to be protected than the surface of the concrete. Consequently, it is usually necessary, because of the proximity of the anode to the metal to be protected and the fact that the electrical resistivity of the concrete prevents the spread of current, to use impressed current, and an anode system which is distributed across the surface of the structure.

For such a structure no part of the reinforcement is more than about 500 mm from a source of current and this is termed a 'close anode system'. For structures that are immersed either in water or in the ground, the surface of the concrete is rarely accessible and so the current distribution relies on the conductivity of the surrounding medium – ground or water to distribute the current from an anode situated some way away from the structure to be protected. If the source of current can be situated a considerable distance away from the structure to be protected the electrical resistance between the anode and all parts of the structure is effectively constant. Such a system is termed a 'remote anode system'.

The two types of system, close anodes, and remote anodes will be considered separately as will the different impressed current anodes and galvanic anodes.

9.6 IMPRESSED CURRENT ANODES

9.6.1 Historic

The earliest close anode systems for use on exposed reinforced concrete structures has been described by Wyatt and Irvine (1987). The anodes have generally consisted either of a large number of metallic conductors which have been set in the concrete in order to provide as even a distribution of current as is possible or of a conductive layer which is laid over the reinforcing grid to be protected. Impressed current systems are commonly employed and the difference between the various systems is primarily concerned with the nature of the anode.

Considerable problems have been experienced in finding a stable anode material for close anodes. The earliest systems consisted of discrete 'pan-cake' anodes distributed within a conductive asphalt overlay and it is under-stood that systems based upon this arrangement have continued to provide satisfactory service for up to 15 years. Systems involving often 50 mm thick coke breeze overlays to which the current has been supplied by 'anode flex' were developed where the 'anode flex' is an electrical cable of which the protective polymer sheath is so highly filled with graphite that it can con-duct current into the surroundings throughout its length. The whole system is then protected by a concrete or shotcrete (gunite) screed.

These systems have worked well, but for application to beams or the underside of bridge decks where the additional weight that would be added by such a system might be undesirable, other anode systems have been developed which can be applied over the whole surface to be protected.

9.6.2 Soil/water anodes

Impressed current anode systems for the cathodic protection of in soil and in water reinforced and prestressed concrete elements are remote anode sys-tems comprising traditional anodes from impressed current CP of steel in soil and water.

Impressed current anodes may be constructed of any conductive material but are usually constructed of relatively noble alloys or other materials that have low consumption rates. The range of materials used for such anodes is much wider than that available for galvanic systems. It ranges from low cost scrap steel which suffers large losses to platinised (Pt) or mixed metal oxide (MMO) coated titanium which exhibit low consumption rates and are effi-cient but expensive.

Steel, graphite, and silicon iron can be used for underground applications. Graphite and silicon iron are brittle and must be handled with care at all times. The anode is usually surrounded by backfill such as coke breeze to improve electrical contact between the anode and surrounding soil and to substantially reduce the consumption rate.

For marine work MMO coated titanium, platinised titanium (Pt-Ti) or platinised niobium (Pt-Nb) are used.

Impressed current anodes can operate at significantly higher driving potentials than galvanic anodes. The maximum values are often deter-mined by safety considerations or the acceptable levels of gases that can be generated. Similarly, current output is significantly higher than from galvanic anodes. Table 9.3 lists some typical impressed current anode

Table 9.3 Impressed current anode materials and their consumption rates

Material	Consumption rate (kg/Amp year)	Typical current density (amps/m²)
MMO coated titanium	10^{-6}	100–600
Platinised titanium	10^{-5}	100–1,000
Platinised niobium	10^{-5}	100–1,000
Silicon/chromium/iron	0.4	10–40
Magnetite	0.1	3–60
Steel	7–9	0.1–1
Lead/silver alloy	0.05–0.1	100–200
Graphite	0.2–0.5	10–40

materials and their properties. AS 2832.1 (2015), for example, also provides details of anode materials.

9.6.3 Mesh anodes

A widely used anode system consists of a mesh of expanded titanium that is coated with mixed rare earth oxides (such as platinum, iridium, ruthenium etc.) (MMO), refer Figure 9.3. In a typical repair the contaminated concrete is broken-out and the reinforcing bar cleaned. After the concrete has been made good, usually with shotcrete, the mesh is pinned to the outside surface and the appropriate connections made.

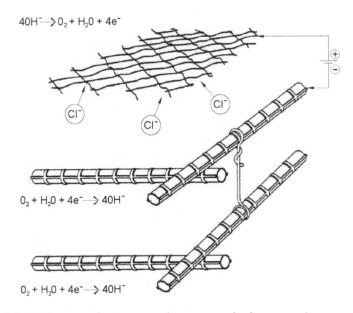

Figure 9.3 Mesh system for impressed current cathodic protection.

The whole mesh is then covered with a shotcrete overlay and the resistivity of the overlay is restricted to typically 30,000 ohm.cm or less. The low resistivity of the overlay is specified in order to improve the current distribution. The rare earth oxides act catalytically on the anodic reaction and currents of 100mA/m² of the anode surface yield only minimal decreases in the pH of the surrounding shotcrete/concrete.

9.6.4 Ribbon/grid anodes

An alternative to the use of titanium mesh anodes is to apply impressed current CP by means of embedded or slotted ribbon MMO anodes backfilled with a proprietary cementitious and CP compatible grout with known electrical resistivity characteristics and increased alkalinity (buffering capacity) to resist acidification (since the electrochemical reactions at the anode to grout/shotcrete interface are oxidising, producing acidity), refer Figure 9.4. The ribbon is usually between 6 mm to 25 mm in width and approximately 1 mm thick. The anode slots are typically spaced 150 to 300 mm apart and the depth of cover to the reinforcing steel dictates their depth and orientation. In some applications, ribbon anodes can be fixed to the reinforcing steel with insulating clips before placing the repair mortar or new concrete.

9.6.5 Discrete anodes

Another system is the 'internal anode' system. Such a system consists of discrete anodes distributed in the surface of the concrete. It is one of the most cost effective systems for beams, piles, and columns. The anodes are relatively easy to install and do not require extensive saw cutting or use of concrete overlays. The discrete anodes are typically inserted into drilled holes that are 20–25 mm diameter and backfilled with proprietary,

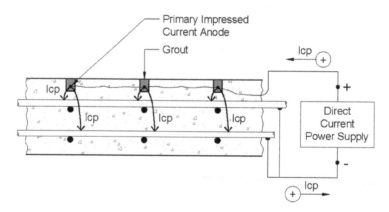

Figure 9.4 Slotted anode system for impressed current cathodic protection. (Courtesy of Strategic Highway Research Program, 1993, p.89).

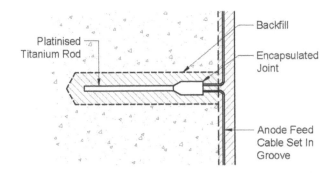

Figure 9.5 The 'platinised titanium rod/carbon rich backfill' discrete anode system. (Courtesy of Morgan, 1986, p.267)

non-shrink, cementitious and CP compatible grout with known electrical resistivity characteristics and increased alkalinity (buffering capacity) to resist acidification (since the electrochemical reactions at the anode to grout interface are oxidising, producing acidity). The length and spacing of the anodes is dependent on the steel density and protection requirements for CP. Several systems are available. These include discrete MMO titanium ribbon systems, MMO titanium cylindrical systems and conductive ceramic/titanium composite anodes. A platinised titanium rod in a carbon rich backfill system is also available, refer Figure 9.5.

9.6.6 Arc sprayed zinc

It is possible to spray metallic zinc onto concrete surfaces and operate it as an impressed current anode system (Apostolos et al., 1987), refer Figure 9.6. Zinc can be flame or arc sprayed over the surface of the concrete and a connection made to it by embedding a plate connected to the power source in the surface of the concrete and spraying over it.

A good contact can be made with the whole surface in this way, but it should be noted that an anodic current may involve dissolution of some of the zinc.

9.6.7 Conductive organic coatings

Conductive organic coatings have been developed which can be applied over the whole surface to be protected. Like the conductive asphalt overlay, 'primary anodes' are used to connect the source of current to the conductive organic coating, refer Figure 9.7.

9.6.8 Remote (soil/water) anodes

The earliest CP systems applied to structures which were buried underground were applied to prestressed concrete pipes. Such a system was

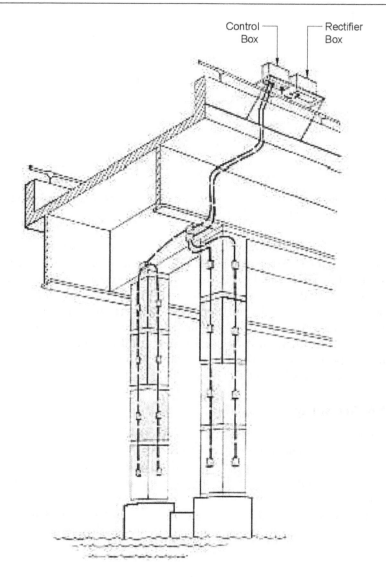

Figure 9.6 System utilising sprayed zinc as the anode. (Courtesy of Apostolos et al., 1987)

designed by Spector (1962) for application to the already existing Tel Aviv Municipal Area Pipeline located in southern Israel, refer Figure 9.8. Application of CP was subsequent to the observation of distress on similar pipelines. The scheme uses impressed current and deep well groundbeds. The pipes range from 500 mm to 1220 mm diameter and are buried between 1 and 15 m below ground level.

The design was based upon separating the pipeline network into insulated sections with a transformer/rectifier (DC power supply) station for each

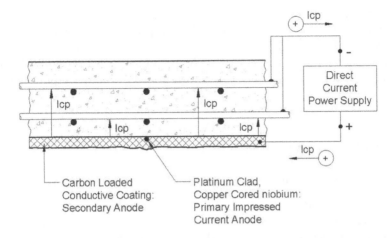

Figure 9.7 Conductive coating for impressed current cathodic protection. (Courtesy of Strategic Highway Research Program, 1993, p.107)

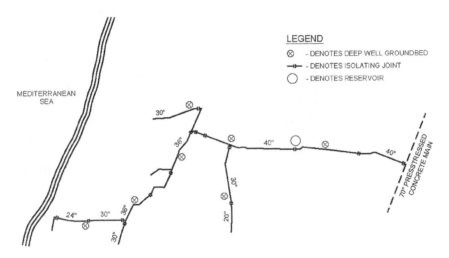

Figure 9.8 The Tel Aviv water supply system. (Courtesy of Spector, 1962)

section. The groundbeds were situated about 1.5 to 2 km apart and the location of the transformer/rectifiers was determined by the availability of a suitable AC power supply, access to the site for a drilling rig and the ability of the chosen site to ensure an adequate distribution of current to the line. In order to determine the current density necessary to bring about protection, preliminary tests were carried out using temporary anodes and the current necessary to bring about a 300 mV negative shift was measured. The current densities found to be necessary were of the order of 3 to 4 mA/m², but it should be noted that a certain amount of bonding of foreign structures had to be carried out to prevent interference with these structures.

A standard groundbed consisting of a string of 20 silicon iron anodes (150 mm × 75 mm) in a borehole 100 m deep so that the string was below the water table was used throughout this project. The current drawn from each anode would be about 1 to 15 Amp and for such a groundbed and a standard transformer/rectifier (T/R) with a 30 V/30 Amp output could be used.

9.7 GALVANIC ANODES

9.7.1 Remote (soil/water) anodes

Traditional anodes to AS 2239 (2003), for example, are used for galvanic CP of reinforced and prestressed concrete elements in-ground and in-water. Three materials are commonly used for galvanic anodes, viz:

- Zinc, with minimum amounts of other metals alloyed.
- Magnesium, with minimum amounts of other metals alloyed.
- Aluminium, with minimum amounts of other metals alloyed.

The consumption rates and application are shown in Table 9.4 below. Anode selection is based on economic and engineering considerations. The consumption rate in kilograms per ampere-year is the weight of anode material that would be consumed from an anode discharging one ampere for one year. This figure is useful in calculating the expected life of a galvanic anode installation. In seawater applications, sometimes the capacity of the anode in ampere-hours per kilogram is used.

Magnesium is widely used in underground applications. Although its efficiency is low (about 50%), this is more than offset by its very negative potential which provides high current output. Magnesium can provide protection in high resistivity soils (less than about 5,000 ohm.cm) whereas zinc

Table 9.4 Galvanic anode materials

Anode Type	Potential (V) versus Copper/Copper Sulphate	Consumption Rate (kg/Amp year)	Application
Zinc	−1.1 −1.55 (std)	12	Marine and low-to-medium resistivity soil s = soils or waters
Magnesium	−1.75 (high potential)	74	Buried and non-saline waters
Aluminium	−1.1	3.2 (Al-Zn-Hg) 3–4 (Al-Zn-In) 3–9 (Al-Zn-In)	Marine

Source: (AS 2239, 2003)

anodes are limited to use in soils of about 1,000 ohm.cm or less. In seawater, aluminium, and zinc anodes are used. Special alloys of these metals are usually used to ensure the best efficiency of operation. Anodes are commercially available in various shapes and sizes depending on the use. Galvanic anodes are usually cast around a central steel core (wire, rod, or strip), which provides an attachment for cable or for connection of the anode directly to a structure.

Galvanic anode systems for prestressed concrete pipelines have been used in situations where the anodes could be buried at the same time as the pipeline was laid. The Ross River Pipeline designed by Gourley (1978) was installed at Townsville in Queensland, Australia. The soil resistivity along the route of the line was in many places less than 3,000 ohm.cm and so CP was installed during the original construction of the pipeline. The line consists of a 1,220 mm precast concrete cylinder around which is wound the prestressing wire and which is fitted with a copper contact strip in the surface of the precast cylinder to reduce the overall resistance of the coil of the prestressing tendon. All the pipes in the line were linked by joining the copper strips.

The current density for a galvanic anode system was of the order of 2 mA/m^2 for the initial polarisation and this decreased subsequently to as low as 0.5 mA/m^2. The required size of the anode banks was calculated using a current criterion. The calculation was carried out on the basis of the total area of reinforcing steel surface using the assumption that the current density which would be required to yield a 300 mV negative potential swing would be 1 mA/m^2. In order to calculate the number and size of anodes used, the voltage drop across the mortar coating of resistivity 50,000 ohm.cm was calculated for a current density of 1 mA/m^2 to be 6 mV. Typically, in a soil of resistivity 1,000 ohm.cm it was found that six anodes each 35 mm × 35 mm × 1.5 m with a mass of 13.4 kg each connected in parallel would be sufficient to provide the requisite current for 20 years, refer Figure 9.9. With this system the initial current delivered was about 1.2 mA/m^2 and the potential at every point along the line was moved to the range −450 to −900 mV.

Benedict (1990) has described a wide variety of such systems and reported that the current densities on mortar or concrete coated steel were all in the range 0.2 to 2mA/m^2.

9.7.2 Thermally sprayed metals

Zinc thermal spray anode has been used as an anode material for galvanic CP of bridge members in warmer climates (Green, 2001; Ip and Pianca, 2002). The anode material is a zinc wire that is fed into special application equipment that melts (typically by a high amperage arc), and then sprays the molten zinc directly onto the surface of the reinforced concrete structure with compressed air. Wires are first connected to the reinforcement steel, and to a galvanised steel plate that is applied to the surface of the concrete.

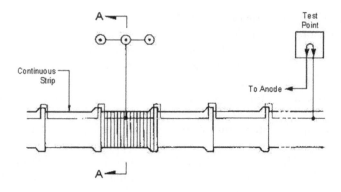

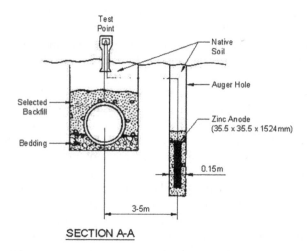

SECTION A-A

Figure 9.9 Galvanic anode system for prestressed concrete cylinder pipe. (Courtesy of Berkeley & Pathmanaban, 1990, p.79)

The thermally sprayed zinc coats the steel plate and the concrete surface. The CP system is activated by connecting the reinforcement wires to the anode wires.

Application temperature, zinc-to-concrete bond strength, concrete moisture, and anode life have been identified as the primary concerns surrounding this type of system. The major drawback identified with this anode is the loss of current output in dry areas. The use of a brush or spray applied humectant has been reported to have overcome the problems with the concrete drying out. The zinc anode thickness is typically of the order of 450 to 550 microns and requires surface preparation with an abrasive blast to achieve a minimum anode adhesion. The zinc thermal spray has a high level of porosity and absorbs water when in the splash zone. The absorption of moisture causes a decrease in circuit resistance and increases the current output of the CP system. The use of a supplemental topcoat can decrease the flow of current into the environment from the exposed side of the anode and could assure that all of the current flow would be through the concrete directly to the reinforcement steel. This would reduce the self-consumption of the anode, thereby increasing the CP system life. The zinc thermal spray anode system is most effective in areas where the anodes are subject to periodic immersion or splashing. The system is less effective in dry areas or climates where the anodes are not periodically wetted (Rothman et al., 2004).

In the US a research programme conducted during the 1990s developed a new Al-Zn-In alloy for thermal spray applications (NACE, 2005). This anode system has a higher driving potential than thermally sprayed zinc. The relatively high driving potential of this type of anode was proposed to provide better levels of protection of reinforcement steel in dryer areas.

9.7.3 Zinc mesh with fibreglass jacket

There is quite a degree of experience describing the zinc (Zn) mesh in fibreglass jacket type system (Kessler et al., 1996; Allan et al, 2000; Leng, et al., 2000; Tinnea et al., 2004; Baily et al., 2009; Tehada, 2009), which has been most typically used for concrete piles in marine environments. Many of the papers reported on the project installation but did not provide performance data and longer-term findings.

This system relies on there being sufficient moisture in the mortar and concrete within the fibreglass jacket to ensure low enough resistivity for the anode to output current. This is considered a limiting factor for this system, particularly above the tidal zone.

It has been reported that for zinc in concrete jackets, normal concrete can be used to encase the anode within the jacket if it is to be installed in a marine environment whereas activated concrete can be used for non-marine applications (Whitmore, 2004).

There are case studies of zinc mesh used with composite planks which are used to hold the zinc mesh in contact with the concrete. The composite plank is also required to hold sufficient moisture to ensure continued operation of the anode (Rothman et al., 2004).

9.7.4 Zinc sheet anodes

Zinc sheet anodes with a conductive hydrogel adhesive have been proposed for the CP of atmospherically exposed reinforced concrete. The anode is attached to the surface of the concrete by the pressure sensitive hydrogel adhesive that is on one side of the anode material and a direct electrical bond is established between the zinc sheet and steel reinforcement. The hydrogel adhesive is also formulated to maintain the activity of the zinc while also allowing ionic conduction between the zinc and concrete. The paper by Dykstra and Wehling (2002) describes the system and its installation in more detail. The paper by Bennett and Firlotte (1996) presents some of the initial research and development works aimed at identifying alloy and hydrogel adhesive combinations that would provide suitable adhesion while sustaining current densities able to afford protection to the structure.

More recently, a paper by Giorgini and Papworth (2011) presents some of the outcomes from the early work and difficulties related to zinc sheet anodes in the late 1990s before discussing the formulation of an alternate system. It is noted that changes were made to the manufacturing and adhesive pH. The improved output (higher induced potential shift) of the alternate system compared to those previously is in part attributed to the adhesive having a slightly acidic pH (Giorgini & Papworth, 2011).

9.8 THE ACTIONS OF CATHODIC PROTECTION

9.8.1 General

In order to examine the way in which CP reduces the corrosion rate of the reinforcement it is necessary to examine the corrosion reaction of steel in concrete in a little more detail. It may be assumed that steel immersed in sound concrete is exposed to the liquid that is contained within the pores and capillaries of the hardened cement paste. This is primarily a saturated solution of calcium hydroxide with a pH of about 12.8. The iron oxidation reaction at pH 12.8 may be written:

$$Fe(s) + 2H_2O(l) \rightarrow Fe(OH)_2(s) + 2H^+(aq) + 2e^-$$

And the single electrode potential for this is about -0.80 V (vs standard hydrogen electrode, H) (Pourbaix, 1966), equivalent to -1.12 V (vs Cu/satd $CuSO_4$, copper sulphate electrode (CSE)). ASTM Standard C786 (2015) proposes that the possibility of corrosion is less than 10% when the potential of the reinforcing bar is more positive than -200 mV (CSE), and so it can be seen that the object of CP is, in a broad-sense, twofold. It must be to re-create such an alkaline environment that the reinforcing bars are put into the passive state and it must drive the chloride ions away from the metal surface so that the possibility of pitting corrosion is diminished.

9.8.2 Thermodynamics

CP of steel reinforcement in concrete therefore differs from the CP of (say) steel in seawater. In seawater the object of the application of a current is to polarise the metal into the immune region whereas in concrete the object of the application of the current is to change the environment so the conditions

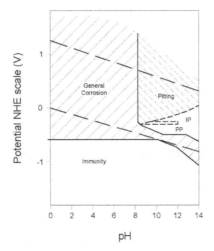

Figure 9.10 Pourbaix diagram showing region of 'perfect passivity (PP)'. (Courtesy of Pourbaix, 1973)

for passivity are re-established. This is 'CP by perfect passivity' as put forward by Pourbaix (1973), refer Figure 9.10.

9.8.3 Kinetics

Evans Diagrams are relevant to highlight the kinetic effects of CP of steel in concrete, refer Figure 9.11, namely:

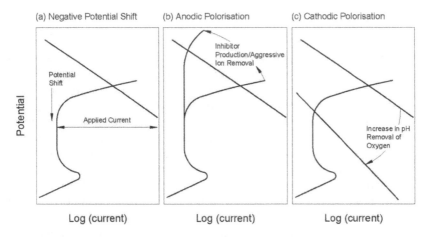

Figure 9.11 Schematic Evans Diagrams summarising the kinetic effects of cathodic protection.

a. Negative Potential Shift – shifting the potential from the pitting into the passive region.
b. Anodic Polarisation – increasing the region of passivity by producing inhibitors (OH^-) and removing aggressive ions (Cl^-).
c. Cathodic Polarisation – polarising the cathodic kinetics by increasing the concentration of products (OH^-) and reducing the concentration of reactants (oxygen, O_2).

9.9 CRITERIA FOR CATHODIC PROTECTION

9.9.1 Background

A vital part of the design of a system of CP is a method of determining whether the scheme is in fact applying an appropriate level of protection. It is no use waiting to see if visible damage occurs or if the structure fails due to pitting corrosion to see whether the level of cathodic current that has been applied is adequate to protect the structure. A number of criteria have therefore been developed that should enable a corrosion engineer to decide whether the structure is being protected. The criteria to be discussed include:

a. The potential criterion.
b. The 300 mV shift criterion.
c. The 100 mV potential decay criterion.

Terminology relevant to these criteria are discussed. The criteria proposed in the 'concrete CP standards', AS 2832.5 (2008) and ISO 12696 (BSI, 2016), for example, will conclude this section.

9.9.2 Potential criterion

This criterion has been long used for the protection of coated steel structures which are buried underground. For steel exposed to soil conditions it was found many years ago that polarisation to a potential more negative than –0.85 V CSE renders the metal immune from corrosion and this has long been used as a CP criterion in the pipeline industry.

In an attempt to examine the protection of reinforcing bar in an environment similar to that which it would experience when immersed in concrete, Hausmann (1969) examined the corrosion of steel bars exposed to a saturated calcium hydroxide environment containing a range of chloride ion concentrations. His results are shown in Figure 9.12. He found that a potential more negative than –0.71 V CSE was sufficient to prevent corrosion of the bars. Hausmann (1969) also concluded that 'corrosion of steel can be arrested in chloride contaminated concrete if sufficient current is applied to shift the polarisation potential to a minimum value of –0.71 V to a copper/

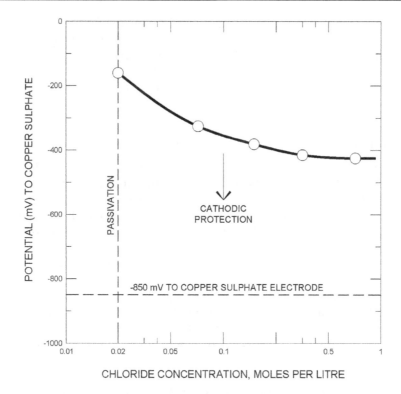

Figure 9.12 Corrosion and its relation to potential in simulated pore water. (Courtesy of Hausmann, 1969)

satd copper sulphate electrode (CSE)'. Under these circumstances it is probable that the cathodic current would be concentrated on those regions of the metal surface where the passive film had broken down and would generate a sufficient concentration of hydroxyl ions to cause the passive film to be reformed.

Hausmann (1969) further suggested that it would be dangerous to adopt the –0.85 V (CSE) criterion used for buried pipelines since the overprotection so obtained would lead to the evolution of hydrogen and the disruption of the bond between the concrete and the steel. He was also concerned that with the high strength steel used for prestressed structures hydrogen evolution could lead to embrittlement.

Vrable (1976) examined the electrochemical behaviour of rebar steel in high pH solutions and found that hydrogen evolution does not occur at potentials more positive than about –1.17 V (CSE) and that a potential, measured at the surface of the metal to be protected of –0.77 V (CSE) was sufficient to prevent corrosion in all circumstances.

Overprotection is, however, more serious in the case of prestressed concrete structures reinforced by high tensile steel. The susceptibility of steel to hydrogen embrittlement increases as the yield strength of the steel increases. In general, high tensile steel with a strength greater than 1,000 MPa is considered susceptible to hydrogen embrittlement in severe environments. For such structures the NACE Recommended Practice (2005) for their CP limits the potential of the tendons to values more positive than –900 mV (vs Ag/AgCl/0.5 M KCl).

9.9.3 Instantaneous off measurements

In the case of a buried steel structure such as a pipeline, a reference electrode is placed on the surface of the ground above the pipeline and the potential measured by means of a high impedance voltmeter connected to the pipeline. In the case of a reinforced concrete structure potentials can only be measured at the surface of the concrete and thus errors will be introduced because the passage of the CP current through the concrete will be associated with a voltage gradient in the concrete. The concrete adjacent to the steel will be negative to concrete at the surface. The magnitude of this potential drop will of course depend upon the magnitude of the protective current that is flowing to the metal to be protected. Ward (1977) has, however, shown that for a variety of concretes, if the potential of the steel is of the order of –0.750 V (CSE) then the voltage drop in the concrete is of the order of 2 mV/mm.

In order to eliminate errors due to the IR drop in the concrete an 'instantaneous off' technique is used (Cherry, 2004). The technique relies upon the principle that the IR drop between the reference electrode and the structure (which is part of the resistance overpotential of the structure) disappears 'instantaneously' when the polarising circuit is broken, whereas the diffusion overpotential takes a much longer time to decay (Cherry, 2004). On switching off the polarising current, the potential decay will look like the curve shown in Figure 9.13.

Cherry (2004) calculated that most of the activation polarisation is lost within the first 0.1 s in most circumstances and further suggested there may be such an overlap between the different decay processes (ohmic (aka resistance), activation and diffusion (aka concentration)) at switch off that differentiation between them is not possible.

It should be noted that Ackland and Dylejko (2019) have recently shown that 'instant off potentials' do not represent the true polarised potential, but rather the corrosion potential at the moment of switch off.

It should be noted also that Angst (2019) has recently examined that instant off potentials are not IR-free potentials (or polarised potentials). He notes that the time needed for both activation polarisation and ohmic drops to vanish is shorter than 0.1 s. The time constants of activation polarisation and ohmic effects have been reported to be of the order of 10^{-4} s … 10^{-1} s, and 10^{-7} s, respectively. Thus, by the time one takes instant off potential readings, activation polarisation has also most likely decayed to a

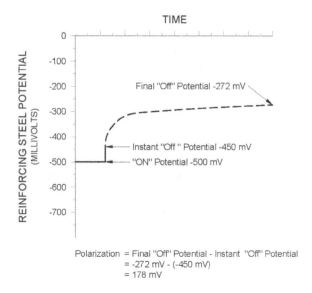

Figure 9.13 Instantaneous off potential measurements. (Courtesy of NACE, 2005)

considerable extent, while significant however, hardly quantifiable-concentration polarisation is still present. Angst (2019) concludes that instant off potentials are essentially a measure of some residual concentration polarisation, which may explain why instant off potentials allow in some cases to assess the corrosion state, but that the conceptual understanding of instant off potential measurements as an adequate approach to remove IR drop while maintaining activation polarisation may be questioned.

9.9.4 300 mV shift criterion

This criterion was also adopted from the pipeline industry. It suggests that a shift in the potential of the reinforcing bar of at least 300 mV (in the negative direction) from that which it demonstrated immediately prior to the application of the CP current indicates that the current density that has been applied is adequate to prevent further corrosion. This criterion takes no account of the IR drop in the concrete.

During free corrosion, current is flowing off the anodic areas of the steel surface and so that the actual interface potential will be negative to that measured at the concrete surface, whereas when CP is applied the current will be flowing on to the metal surface and so the interface potential may be considerably positive to that measured at the concrete surface. The actual swing at the interface may therefore be less than that measured at the surface by an unknown amount.

The negative 300 mV shift criterion of the surface potential is consistent with the values indicated in ASTM C876 (2015) and the results of Hausmann

(1969), in that whereas an initial potential of –400 mV to CSE would indicate a high probability of corrosion of the steel beneath, polarisation to –700 mV CSE (the actual swing at the metal surface would as described above be less) could place the metal in the range described by Hausmann (1969) as adequate to stop the corrosion.

9.9.5 100 mV potential decay (polarisation) criterion

A major shortcoming of the 300mV shift criterion is that it fails to recognise that the potential measured at the concrete surface is a function of the relative areas of cathodic and anodic activity beneath the reference electrode. As CP proceeds, the areas of anodic activity decrease so that the rest potential of the structure becomes more positive (less negative) and the application of 300 mV to the originally more negative rest potential may represent gross overprotection.

The 100mV decay criterion attempts to overcome the disadvantage of the 300mV criterion by referring the shift always to the instantaneous (instant) value of the rest potential. It suggests that a minimum negative polarisation of 100mV is necessary to ensure that the current density applied is adequate to prevent further corrosion. In this case the polarisation is measured as the immediate rise in potential when the current is switched off. Since 'this shift must be unaffected by any IR drops in the concrete due to the CP current', this implies that it must be measured by an instantaneous off technique which excludes the IR drop in the concrete. This criterion effectively allows for the change in pH which has arisen around the cathode (reinforcement) as a result of the application of the CP.

The 100 mV decay criterion is consistent with the results of Cherry and Kashmirian (1983) who suggested that a 100 mV shift represents the application of sufficient current density to a reinforcing bar to move the bar into the perfectly passive region of the potential-pH (Pourbaix) diagram. The criterion also allows for an initial high current period while the pH in the region of the anodic areas is building up to protective values and then a lower current period when the sole function of the CP is to maintain the whole of the surface in the perfectly passive range appropriate to the level of chloride which is still present adjacent to the interface.

Glass (1999) indicates that the 100 mV potential decay criterion can be used to monitor development of steel passivity induced by a cathodic current and that the achievement of 100 mV of negative potential shift by the low current densities normally applied when cathodically protecting steel in concrete indicates that, in theory, the steel corrosion rate is negligible.

Broomfield (1992) reports that in a large survey of impressed CP systems in North America, no failures appear to have been uncovered when the 100 mV potential decay criterion was met.

Bennett and Broomfield (1997) propose that for cathodically protected, atmospherically exposed reinforced concrete, a polarisation decay or development of 150 mV or more in anodic areas is needed for 'worst case conditions'. However, they note that given that most structures are not worst case, that complete depolarisation is not usually observed, and that depolarisation values usually exceed the minimum required, a polarisation (potential) decay of at least 100 mV is a reasonable, practical objective.

9.9.6 AS 2832.5 criteria

Australian Standard AS 2832.5 (2008) 'Cathodic protection of metals – Part 5: Steel in concrete structures', Section 2, for example, details criteria for establishing whether a CP system is providing adequate protection to a reinforced concrete structure. The standard does not rank the applicable criteria and each is equally relevant.

There is an overriding safety criterion in the standard namely that no 'instantaneous off' steel/concrete potential (measured between 0.1 s and 1.0 s after switching the d.c. circuit open) shall be more negative than –1100mV for plain reinforcing steel or more negative than –900mV for prestressing steel with respect to Ag/AgCl/0.5 M KCl. This is to ensure no compromising of bond in the case of plain reinforcing steel and no hydrogen embrittlement risk in the case of prestressing steel.

The standard then requires that initial and continuous adjustment of the CP system shall then be based on meeting one of the following criteria:

a. **Potential decay criterion.** A potential decay over a maximum of 24 h of at least 100 mV from the instant off potential.
b. **Extended potential decay criterion.** A potential decay over a maximum of 72 h of at least 100 mV from the instantaneous off potential subject to a continuing decay and the use of reference electrodes (not potential decay sensors) for the measurement extended beyond 24 h.
c. **Absolute potential criterion.** An instantaneous off potential (measured between 0.1 s and 1.0 s after switching the d.c. circuit open) and more negative than –720 mV with respect to Ag/AgCl/0.5 M KCl.
d. **Absolute passive criterion.** A fully depolarised potential, or a potential which is continuing to depolarise over a maximum of 72 h after the CP system has been switched off, which is consistently less negative than –150 mV with respect to Ag/AgCl/0.5 M KCl.

Compliance with at least one of the of the above criteria shall be achieved within 6 months or alternatively, within a longer period as agreed with the structure owner.

Compliance with at least one of the above criteria shall also be maintained on a continuous basis for the life of the system.

The standard notes that the criteria given in a), b), c) and d) have been found to indicate adequate polarisation which will lead to the maintenance or re-establishment of protective conditions for the steel within the concrete (such a note being normative like the criteria and the above other requirements).

Within part of the informative (non-normative) Appendix A section of AS 2832.5 (2008) is noted the following:

- If environmental conditions, which favour corrosion of reinforcement, are likely to occur during the lifetime of the structure, or during service, CP is one method of preventing corrosion of steel in concrete.
- Sufficient corrosion protection will be achieved if specific criteria of protection are met at representative points on the structure.
- The criteria of protection in this Standard are based on electrochemical considerations regarding corrosion processes and on practical experience.

9.9.7 ISO 12696 criteria

The international standard for CP of steel in concrete is ISO 12696 (BSI, 2016). ISO 12696 states 'Prestressing steel may be sensitive to hydrogen embrittlement and, due to the high tensile loading, failure can be catastrophic. It is essential that caution is exercised in any application of cathodic protection to prestressed elements'. It further states 'No instantaneous off steel/concrete potential more negative than –1,100 mV (with respect to Ag/AgCl/0.5 M KCl) shall be permitted for plain reinforcing steel or -900 mV (with respect to Ag/AgCl/0.5 M KCl) for prestressing steel'.

In terms of criteria of protection, ISO 12696 states for any structure, any representative steel in concrete location shall meet any one of the criteria given in items a to c namely:

a. An instantaneous off potential more negative than –720 mV with respect to Ag/AgCl/0.5 M KCl.
b. A potential decay over a maximum of 24 h of at least 100 mV from instantaneous off.
c. A potential decay over an extended period (typically 24 h or longer) of at least 150 mV from instantaneous off subject to continuing decay and the use of reference electrodes (not potential decay probes) for the measurement extended beyond 24 h.

9.9.8 Other standards

Broomfield (2017) indicates that the first major international standard for CP of steel in concrete was originally published by NACE International in 1990 (NACE, 1990). This was followed by the European CEN standard BS EN 12696 in 2000. He advises that although the NACE standard has been reapproved and republished several times it has not changed significantly. However, the heavily revised and expanded BS EN ISO 12696 (BSI, 2016) now covers the CP of steel embedded in concrete in atmospherically exposed, buried, and submerged service, as well as both GCP and ICCP.

Broomfield (2017) also notes that NACE has published standards for concrete pipes (NACE, 2014a) and buried and submerged RC structures generally (NACE, 2014b). There are also several state-of-the-art reports on evaluating RC cathodic protection systems (NACE, 2016), CP of prestressing steel in concrete (NACE, 2002) and CP of masonry buildings incorporating structural steel frames (NACE, 2010).

9.9.9 CP criteria are proven

Concrete CP criteria have been in existence for many decades and have been proved time and time again to show that corrosion to steel reinforcement, whether it be conventional reinforcement or prestressing steel, as well as whether the cause of corrosion is chlorides, carbonation, or leaching, is arrested by CP.

It is not possible, given the numerous variables within steel in concrete, exposure environments, element type, treatment period, etc. that an applied potential or required current could be determined to predict theoretical criteria for CP achievement (Solomon, 2005). Due to this limitation, protection criteria being applied for RC CP systems are largely based on accumulated practical experience and are criteria which have repeatedly demonstrated reliability (Solomon, 2005).

Broomfield (2019) in a recent overview of CP criteria for steel in atmospherically exposed concrete concluded and recommended the following:

1. There is a basis in theory and experiment for the absolute and potential shift criteria in the standards.
2. There are also extensive field data supporting the criteria.
3. Important laboratory tests supporting the 100 mV depolarisation criterion have not been followed up to confirm them or to see how well 'damp sand' experiments replicate steel in concrete.
4. There are many values for the Tafel constant and polarisation curves for steel in concrete in the literature. More information on the anodic polarisation curve of steel in concrete, pore solutions and sand with and without chloride could help inform the basis and limitations of the potential shift criteria.
5. With the desire for ICCP anode design life of up to 100 years on structures around the world, there is a need to use criteria that minimise current flowing from the anode. The 100 mV depolarisation criterion seems superior to the E-log I and absolute potential criteria in achieving this. However, caution should be applied to any attempt to reduce the depolarisation from 100 mV as it is shown that this criterion can require the movement of chloride away from the steel to achieve protection in the most aggressive conditions ($>1.6\%$ Cl⁻ by mass of

cement) and accurate and complete determination of the chloride level in less aggressive conditions (<0.3% Cl⁻ by mass of cement) is subject to error and practical sampling limitations.

6. The issue of stability and ability to calibrate reference electrodes needs further investigation if absolute potential criteria and potential limits are to be controlled in the field in a meaningful way.

9.10 SELECTION AND DESIGN OF CATHODIC PROTECTION SYSTEMS

9.10.1 General considerations

Whether using ICCP or GCP, a monitoring system is essential to ensure that the system is being fully effective in providing the desired levels of protection. It is therefore necessary to install as a minimum, a number of permanent reference electrodes scattered across the structure in order to obtain meaningful data in relation to levels of protection being provided and system output adjustments necessary.

CP is primarily used to control corrosion on reinforced concrete structures due to the ingress of chlorides. It has also been used on historic steel framed stone structures where the stonework has cracked due to expansive corrosion of the steel framework. CP has also been applied to bridges and prestigious new structures either where chloride ingress is expected or where chlorides have been used in the mix water.

CP arrests the corrosion process across the entire structure where the anodes are applied. Provided the anodes are installed correctly and the passage of sufficient protective current is possible through the concrete on to the affected steel surfaces then the system will be effective and provide a long service life. The passage of protective current will be interrupted by the presence of any air voids caused by cracks or delaminations. However, the amount of repair works needed on a reinforced concrete structure suffering distress will be minimal when CP is to be applied. Concrete containing high percentages of chlorides does not need removal provided it is still adherent within the parent concrete. Furthermore, it must be ensured that all the steel reinforcement is electrically continuous such that all of it is part of the circuit and receives protection. Any electrically isolated portions of structure or with high resistance circuit contacts will not be protected and thus corrosion can still propagate undetected until external rust staining becomes evident.

In most instances ICCP anode systems need replacement every 25 to 100 years, dependent upon the anode type selected. GCP systems will normally require anodes replaced every 5 to 25 years, again dependent upon the type selected.

Life cycle cost analysis has frequently demonstrated that ICCP systems are one of the most cost effective repairs for structures with 20 years or more residual life. Once applied, the system stops the cycle of regular concrete repair associated with structures having high chloride levels.

ICCP systems require ongoing monitoring throughout the life of the system in order to ensure effectiveness and good service. This is normally performed by a number of monitoring reference electrodes being installed across the structure and wired back either to central test boxes or the transformer/rectifier (T/R) cabinet. More sophisticated systems are available where the entire CP and monitoring systems can be monitored and adjusted remotely.

ICCP is not easily applied to prestressing steel due to its sensitivity to hydrogen embrittlement. Hydrogen can be evolved as part of the cathodic processes if excessive protection is applied to the steel. Thus, it must be ensured that protection applied to these types of structures is carefully controlled to avoid overprotection. A good monitoring and control system would be mandatory.

Installation of anodes within the structure must be undertaken with care to ensure that short circuits with embedded steel work are avoided. This can be a particular problem due to the presence of tie wire and tramp steel. The passage of current can be affected if old patch repairs have been performed using epoxy grouts or if insulating membranes exist between the anodes and steel reinforcement.

Gases are evolved by the anodes and need to permeate away to avoid the build-up of a gas insulating layer at the anode/concrete interface resulting in an increasing circuit resistance often referred to as 'gas blockage'. Most ICCP anode systems have a maximum output rating due to the need for evolved gases to be able to permeate away from the surface and permit current to flow continuously unimpeded.

Oxidation of water can occur at concrete CP anode surfaces resulting in the production of hydrogen ions and consequent generation of acid. Oxidation of hydroxyl ions can also occur at concrete CP system anode surfaces, resulting in consumption of alkalinity and consequent reduction in pH. If chloride ions are present, then oxidation of chloride ions to hypochlorous acid (HOCl) and hydrochloric acid (HCl) can also occur. Excessive acid generation can lead to acid attack (neutralisation) of the alkaline cementitious grout and mortar and then concrete surrounding the anode.

It has been shown that CP generates hydroxyl ions on the reinforcement surface and will also attract positive ions such as calcium, sodium, and potassium. This will increase the alkalinity around the reinforcing or prestressing steel. In principle this could cause alkali silica reaction (ASR) or accelerate ASR in susceptible concretes. Broomfield (2007) advises that this has been demonstrated in the laboratory, however, there are no recorded cases of ASR being caused or accelerated by CP of in-field structures.

9.10.2 General design considerations

The design of a CP system consists primarily of determining the appropriate array of anodes that will enable the potential source to deliver a sufficient current density to reduce the rate of corrosion of the protected structure to an acceptable level. Transformer/rectifiers (T/Rs) for impressed current systems can be rated up to 50 volts without exceeding safety requirements, but are often limited to about 15 volts (8 volts at the anode) in order to avoid the damage that may be caused to the anode circuit by highly anodic potentials in the presence of chloride ions.

From a knowledge of the driving potential available, the total allowable resistance of the circuit can be calculated and then the anode layout and the sizing of the connecting cables can be carried out to ensure that the current can be provided.

CP systems are usually divided into a number of separate 'zones' that are independently powered. This division of the system into zones (and sub-zones for some systems) enables the applied voltage to be kept down to reasonably small values and also enables the number and size of the anodes in a given zone (sub-zone) to be varied to suit the density of reinforcement in that zone (sub-zone) and the state of corrosion of that reinforcement.

9.10.3 Current density

The particular current density needed for protection of a structure depends on the particular structure, the service environment, and nature of the corrosion damage.

While current densities for protection can vary in the range of 2 to 20mA/m^2 or greater, based on experience and literature reviews the following are typically recommended:

- Old structures exhibiting chloride-induced corrosion: initial applied current density during polarisation period of 20 mA/m^2 reducing probably in 1 to 4 months to approximately 10 mA/m^2 and with time this would be expected to reduce further.
- New reinforced concrete structure: 0.2 to 2 mA/m^2 for atmospherically exposed concrete and 0.1 to 5 mA/m^2 for steel in water saturated buried/submerged structures.

Higher current densities are applied during the initial polarisation period which typically could be anything from 12 to 24 weeks under normal situations.

At the end of the polarisation period the potentials reach stable values where they are not gradually shifting towards higher protection levels (polarising). Once full polarisation (stability) has been achieved then it is normal to be able to reduce applied current densities to avoid overprotection and preserve the systems' operational life.

9.10.4 Anode layout

The design layout for anodes is dependent on the extent and density of steel reinforcement together with factors affecting the corrosion rate.

Total steel area and layout can be calculated from as-built drawings where available. The total area calculated should allow for variations in surface roughness plus all interconnecting steel such as tie wire, braces, etc. Many structures will have higher steel density areas such as corners and laps.

9.10.5 Power requirements

9.10.5.1 General

The primary requirement of a CP circuit is that it must be capable of delivering sufficient current to the steel/concrete interface to reduce the rate of corrosion to an acceptable value. The design of a CP circuit therefore involves a calculation of the voltage of the power supply that is required to overcome the various resistive components that are present in the circuit. If the voltage drops that occur when a given current is passed through each of the resistive components is calculated then the total voltage that must be delivered by the power source is the sum of the individual voltage drops plus the voltage swing that it is hoped to induce in the conductor that is to be protected. It is a matter of common experience that the initial current requirement to achieve adequate polarisation is considerably greater than that which is necessary in the long term, but the design must be capable of delivering this initial current.

Although commonly described as a protection criterion, the current density criterion is more accurately described as a design criterion. It represents the current density that the designer of the CP system strives to apply to the structure to be protected in the fairly certain knowledge (it is not always certain and when it fails, there is often considerable controversy) that it will be a sufficiently large current density to enable the structure to achieve one of the potential criteria described previously. The range of values of current density that may be necessary to reduce the rate of corrosion of reinforcing steel in concrete to an acceptable value corresponds to about 5 to 25 mA/m^2 of the metal surface. For conventionally designed structures this has often been put equal to 20 mA/m^2 of the concrete surface.

The two major sources of voltage drop (in addition to the polarisation voltage at the cathode) are associated with those points in the circuit where the current density is comparatively high, that is in the cabling that connects the reinforcement to the potential source and in the immediate environment of the anode. For most reinforced concrete structures, the resistance between the reinforcement that is being protected and the region a little way away from the anode is small. Even though the concrete resistance may be high, there is such a large volume between the anode and the cathode through which the current can pass that its resistance can nearly always be neglected. Similarly, because the current density in the region where the current enters the cathode the current density is low the voltage drop in the cathodic environment can be neglected.

9.10.5.2 Anode resistance

Because the anode is nearly always vastly smaller than the cathode the current density immediately adjacent to the anode is inevitably high and so the anode resistance is always considerable. The anode resistance has been calculated for a number of electrodes of different shapes by Dwight (1936) and a more recent version is given by Sunde (1949).

The expression for a number of electrode configurations has been given by von Baeckman et al (1997). The simplest of these is the formula for the resistance to infinity of a long thin rod situated vertically in the electrolyte, buried to a depth that is at least equal to its length:

$$R = \frac{\rho}{2\pi l} \ln\left(\frac{2l}{d} \sqrt{\frac{4t+3l}{4t+l}} \right)$$

where ρ is the resistivity of the electrolyte usually measured in ohm.m, l is the length of the anode in m, d is the diameter of the anode in m and, t is the depth to which the anode is immersed in the electrolyte in m. For $t \gg d$, writing $t \approx l$ this equation can be approximated as:

$$R = \frac{\rho}{2\pi l}\left(\ln \frac{3l}{d} \right),$$

which is a commonly used expression.

It can thus be seen that an internal anode of the form shown in Figure 9.5 of 2 mm in diameter, 200 mm in length inserted to a depth of 50 mm in concrete of resistivity 20,000 ohm.cm would be 95 ohm.

Morgan, (1986) lists the resistance for a number of differently shaped anodes, but great care must be taken noting Morgan's definitions of l (length of the anode) and a (the radius of the anode).

9.10.5.3 Circuit resistance

The resistance of the cabling between the drain point on the structure and the potential source (and the potential source and the anode if this is significant) can usually be easily calculated from the length and cross sectional area of the wire (cable) using a figure of, for example, 1.8×10^{-11} ohm.m for the resistivity of copper.

9.10.5.4 Cathodic polarisation (back emf)

This is often put at two (2) volts so that the sum total of the various voltage drops has a safety margin that will ensure that the appropriate current can be passed.

9.10.5.5 Power supply

Once the total resistance of the circuit has been established, the voltage required to drive a current density of 20 mA/m² to the metal surface can be calculated and the appropriate transformer/rectifier (T/R) chosen.

9.11 STRAY CURRENT AND INTERFERENCE CORROSION

9.11.1 General

Interference effects from CP systems are a result of stray DC being picked up by a foreign structure, flowing along the structure and discharging at another point back from the structure to the source sometimes via the primary protected structure. Where current is picked up by the foreign structure CP is afforded. However, where current discharges from the foreign structure adverse effects are noted on the potentials and metal loss occurs in accordance with Faraday's Law where approximately 8 kg of steel is lost per annum if one Amp of DC continuously discharges.

The possibility of stray current effects from reinforced concrete CP systems is relatively small when compared to traditional CP systems applied to pipelines and other metallic structures. However, stray current effects could occur if portions of the structure are not adequately electrically continuous with the portions of reinforcement connected into the negative circuit. Also, any other structures such as metallic pipes cast into the concrete could also experience pick up and discharge of DC. Immersed/buried reinforced concrete structures often will be protected using immersed or buried anodes. These anode systems could readily cause adverse effects in foreign structures as sometimes the current outputs involved are fairly large and a few milliamps of stray current could be disastrous on thin sections of structures particularly pipes and lead sheath cables for example.

9.11.2 Regulatory requirements

Prior to designing and installing a concrete CP system it is important to confirm whether or not there are foreign structures buried or submerged nearby which could possibly suffer stray current effects from the CP system. This includes metallic pipes, conduits etc. mounted in direct contact with the concrete.

Often Electrolysis Committees are established which require the registration of proposed and operating CP systems in order to mitigate stray current effects. It is essential to discuss with any interested asset owner who could possibly be affected by the system to ensure that suitable testing is undertaken cooperatively and any noted stray and interference currents can be acceptably minimised.

9.12 COMMISSIONING

Whether the CP system is impressed current or galvanic it needs to be commissioned.

In the case of an ICCP system, the commissioning procedure includes the 'Pre-Commissioning Testing' of the continuity of all the electrical connections in the circuits. The anode to reinforcement resistance in each zone is measured to ensure that there are no short circuits between the anodes and the reinforcement. The power supplies are examined to ensure that they are to the appropriate rating and that the polarity of the circuit wiring is correct. Checking has to be carried out to ensure that all cables are wired and labelled correctly.

The following steps have been described in Concrete Society Technical Report 37 (1991) as forming part of the commissioning procedure for ICCP systems:

a. Circuit verification. This process involves measuring the potential shift when a polarising current is applied instantaneously. The objective of this test is to ensure that all the leads (cables) are connected the correct way around so that the reinforcement is not made the anode. This of course would be potentially disastrous as it would rapidly accelerate the corrosion.

b. Establishing the conditions before polarisation. This involves recording the base potential, that is the rest potential before any CP current is applied. The results are used to monitor the subsequent performance of the system.

c. Adjustment of current output. A system that is sometimes used is to apply an initial polarising current of (say) 20 mA/m^2 (of reinforcement surface). This is subsequently reduced as the conditions at the reinforcing bar surface develop so that the performance criteria are exceeded.

   An alternative procedure is to set the potential of the reinforcement to a potential that is 200 mV more negative than the base potential that the system demonstrated before CP was applied. The current in this case is adjusted automatically to maintain this set potential.

Often also a current limit is set on the transformer/rectifier so that the initial current is not to great and it builds up slowly.

d. Interaction testing. It is necessary to ensure that the current that is put into the concrete is all flowing to the reinforcement and is not flowing to any adjacent structures where it might be causing accelerated corrosion. In order to do this, the potential of adjacent structures is measured with respect to a remote standard reference electrode in the ground and the potential shift that occurs when the CP current is switched on and off is determined. If the potential shift is greater than 20 mV (in some places a limit of 10 mV is applied), then the system has to be modified to reduce the shift to less than the specified value.

e. Performance verification. The performance of the CP system must then be monitored to see if the system is working in accordance with the specification. If the performance criteria are not met by the system, then the system has to be adjusted to bring it up to specification.

f. Duration of commissioning. The commissioning process may take up to three months or even longer depending upon the techniques involved. Commissioning can usually be considered complete when the system demonstrates that the protection criteria (i.e. Section 9.9.9) will be obtained.

A commissioning procedure for a CP system is detailed in AS 2832.5 (2008), for example, and includes:

- Pre-Energising Inspection and Testing.
- Initial Energising.
- Initial Adjustment.
- Initial Performance Assessment.
- Adjustment of Current Output.

9.13 SYSTEM DOCUMENTATION

9.13.1 Quality and test records

In accordance with AS 2832.5 (2008), the Project Quality Plan (PQP), the quality documents arising therefrom and the inspection and test results shall form part of the permanent records of the installation of the system.

9.13.2 Installation and commissioning report

An installation and commissioning report for the system shall be prepared. The report shall incorporate as a minimum the requirements of AS 2832.5 (2008) Clause 8.2, for example.

9.13.3 Operation and maintenance manual

An operation and maintenance (O&M) manual for the system shall be pre-pared. The manual shall incorporate as a minimum the requirements of AS 2832.5 (2008) Clause 8.3, for example.

9.14 OPERATIONAL

9.14.1 Warranty period

Regardless of the CP system type(s), it is not unusual to have a 12 month or 24-month warranty (defects liability) period.

9.14.2 Monitoring

Whether the CP system is impressed current or galvanic it needs to be moni-tored during its life. Typical periods for monitoring (testing) will be (AS 2832.5, 2008) for example:

- Three monthly testing and reports in the first 12 month period.
- Six monthly tests and reports in the second year once the system potentials have polarised to stable levels and current outputs have been adjusted downwards with the achievement of full polarisation and passivation of the steel.
- Twelve monthly tests and reports thereafter.

In addition to the above, separate reports should be obtained should any special investigations be required or if any interference tests are performed with foreign structure owners.

9.14.3 System registration

In most instances where electrolysis permits are required, a compliance per-mit should be obtained from the relevant electrolysis committee in order to legally operate the CP system.

From time to time the electrolysis committee could request special tests be performed with foreign structure owners to confirm that stray current effects are being satisfactorily mitigated.

9.15 CATHODIC PREVENTION

Whereas CP is applied to an already corroding structure in order to reduce the rate of corrosion to an acceptable level, cathodic prevention is applied to

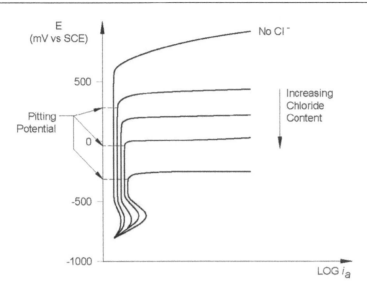

Figure 9.14 Schematic behaviour of steel in chloride contaminated concrete. (Courtesy of Bertolini et al., 1998)

a new structure to stop the corrosion initiating. The techniques of cathodic prevention are very similar to ICCP but the currents and therefore the voltages that are applied are very much smaller. Bertolini et al. (1998) has proposed a schematic illustration of the behaviour of steel in concrete in the presence of chlorides as can be seen in Figure 9.14.

For a structure exposed to a chloride aggressive environment the chloride content of the concrete will increase with time and so Figure 9.14 can be re-plotted as Figure 9.15.

The x-axis may now represent time as the chloride content increases.

CP in Figure 9.15 may be represented by the path 1-4-6-5. The metal remains perfectly passive until corrosion is initiated and the potential falls to 6. If CP is now applied the potential is made more negative and the metal is moved again into the perfectly passive (PP) zone.

Cathodic prevention in Figure 9.15 may be represented by the path 1-2-3. At the beginning of its life the potential of the metal is shifted in the negative direction to 2. Because pitting has not initiated when the chloride content increases the metal remains passive in the 'imperfectly passive' (IP) region, possibly for the life of the structure.

The equipment for the application of cathodic prevention is the same as that for ICCP. However, an initial polarising current density of up to 2 mA/m^2 is usually enough to produce a potential shift of 100 to 150 mV and the long term polarising current density necessary to maintain a 24 hour (or 72 hour) depolarisation (decay) of 100 mV may be of the order of 0.2 mA/m^2.

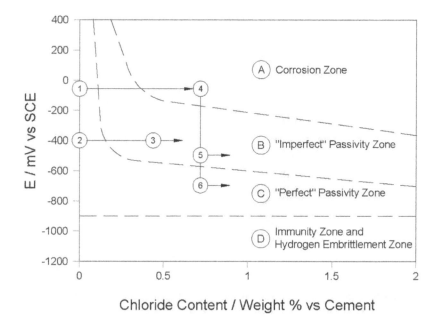

Figure 9.15 Cathodic protection and prevention. (Courtesy of Bertolini et al., 1998)

Cabling and connections can be provided during construction so that ICCP can easily and cheaply be retrofitted if it is found to be necessary at a later date. A complete impressed current cathodic prevention system can also be installed during construction and to commission it immediately so that the heavier duty equipment for the larger currents never becomes necessary.

Installation of just the anode (e.g. MMO titanium ribbon or MMO titanium mesh) together with permanent corrosion monitoring, refer Section 11.6, has also been undertaken to the most aggressive sections of elements of some long-life or critical reinforced concrete structures.

Cathodic prevention has three other long term benefits, namely:

i. The initial polarising current reduces the (already low) corrosion rate.
ii. The generation of hydroxyl ions at the cathodic surface helps to maintain and repair the passive film.
iii. The negative potential of the steel reinforcement surface may repel the negatively charged chloride ions and reduce their rate of entry into the concrete and the electro-endosmotic flow of water molecules away from the cathode may actually remove those chloride ions that have penetrated to the metal/concrete interface.

REFERENCES

Ackland, B (2005), '*Cathodic protection – black box technology?*', *Proc. Corrosion & Prevention 2005 Conf.*, Australasian Corrosion Association Inc., November, Paper 004.

Ackland, BG and Dylejko, KP (2019), 'Critical questions and answers about cathodic protection', *Corrosion Engineering, Science and Technology (CES&T)*, Vol. 54, No. 8, 688–697.

Allan, A G, Leng, D L and Davison, N (2000), '*Sacrificial anode systems for repair of reinforced concrete*', *Proc. Corrosion and Prevention 2000 Conference*, Paper 116, Auckland, New Zealand, November.

Angst, UM (2019), 'A critical review of the science and engineering of cathodic protection of steel in soil and concrete', *Corrosion*, 75, 12, 1,420–1,433.

Apostolos, JA, Parks, DM and Carello, RA (1987), 'Cathodic Protection using metallised zinc' *Materials Performance*, 26, 22.

American Society of Testing Materials (2015), 'Standard Test Method for Corrosion Potentials of Uncoated Reinforcing Steel in Concrete', C876-15, West Conshohocken, Pennsylvania, USA.

Baily, DM, Lampo, RG, Costa, J and Miltenberger, M (2009) 'Impact and Corrosion Prevention Using Polymer Composite Wrapping and Galvanic Cathodic Protection System For Pilings At Kawaihae Harbor', NACE, Corrosion 2009, Paper No. 09513.

Benedict, RL (1990), 'Corrosion Protecton of Concrtete Cylinder Pipe', *Materials Performance*, 29, 2, 22.

Bennett, J and Broomfield, JP (1997), Analysis of studies on cathodic protection criteria for steel in concrete', *Materials Performance*, 36, 12, 16–21.

Bennett, J and Firlotte, C (1996) 'A Zinc/Hydrogel System for Cathodic Protection of Reinforced Concrete Structures', NACE, Corrosion 1996, Paper No. 316.

Berkeley, KGG and Pathmanaban, S (1990), '*Cathodic Protection of Reinforcement Steel in Concrete*', Butterworths, London, UK.

Bertolini, L, Bolzoni, F, Pedeferri, P, Lazzari, L and Pastore, T (1998), 'Cathodic protection and cathodic prevention in concrete: principles and applications', *Journal of Applied Electrochemistry*, 28, 1,321–1,331.

Broomfield, JP (1992), 'Field survey of cathodic protection on North American bridges', *Materials Performance*, 31, 9, 28–33.

Broomfield, JP (2007), '*Corrosion of Steel in Concrete: Understanding, Investigation and Repair*', second edition, Taylor & Francis, London, UK.

Broomfield, JP (2017), 'Up-to-date overview of repair & protection aspects for reinforced concrete', in *Reinforced Concrete Corrosion, Protection, Repair and Durability*, Eds W K Green, F G Collins and M A Forsyth, ISBN 978-0-646-97456-9, Australasian Corrosion Association Inc, Melbourne, Australia.

Broomfield, JP (2019), 'An overview of cathodic protection criteria for steel in atmospherically exposed concrete', *Corrosion Engineering, Science and Technology (CES&T)*, Vol. 55, No. 4, 303–310.

BSI (2016), BS EN ISO 12696, '*Cathodic Protection of Steel in Concrete*', British Standards Institute, London, UK.

Cherry, BW and Kashmirian, AS (1983), 'Cathodic protection of steel embedded in porous concrete', *British Corrosion Journal*, 18, 194.

Cherry, BW (2004), 'How instant is instant?', *Journal of Corrosion Science and Engineering*, 9, 6, 1–9.

Concrete Society (1991), 'Cathodic protection of steel in concrete', Technical Report 73, Surrey, UK.

Davy, H (1824a), 'On the corrosion of copper sheeting by sea water, and on the methods of preventing this effect; and on their application to ships of war and other ships', *Philosophical Transactions*, 114, 151–158.

Davy, H (1824b), 'Additional experiments and observations on the application of electrical combinations to the preservation of the copper sheathing of ships, and to other purposes', *Philosophical Transactions*, 114, 242–246.

Davy, H (1825), 'Further researches on the preservation of metals by electro-chemical means', *Philosophical Transactions*, 115, 328–346.

Dwight, HB (1936), 'Calculations of resistance to ground', *Electrical Engineering*, 55, 1,319.

Dykstra, BG and Wehling JE (2002), 'Cathodic Protection of Steel Reinforced Concrete Structures Using a Galvanic Zinc Hydrogel System', NACE, Corrosion 2002, Paper No. 02269.

Giorgini, R and Papworth, F (2011), 'Galvanic cathodic protection system complying with code based protection criteria', *Concrete Institute of Australia Biannual Conference*, Perth, Western Australia, October.

Glass, GK (1999), 'Technical Note: The 100-mV Potential Decay Cathodic Protection Criterion', Corrosion, March, 286–290.

Gourley, JT (1978), 'The cathodic protection of prestressed concrete pipelines', Corrosion Australasia, 3, No. 1, 4–7.

Grapiglia, JP and Green, WK (1995), 'Recent developments in cathodic protection systems for reinforced concrete', Proc. Australasian Corrosion Association Conference '95, November.

Green, WK (2001), '*Australasian experiences with cathodic protection of concrete marine structures*', Proc. 15th Australasian Coastal and Ocean Engineering Conference and 8th Australasian Port and Harbour Conference, 25–28 September.

Green, W (2014), '*Electrochemistry and its relevance to reinforced concrete durability, repair and protection*', Proc. Corrosion and Prevention 2014 Conf., Australasian Corrosion Association Inc., September, Paper 001.

Hausmann, DA (1969), '*Criteria or Cathodic Protection of Steel in Concrete Structures*', Materials Protection, 23, October.

Holloway, L, Green, W, Karajayli, P and Birbilis, N (2012), '*A review of sacrificial and hybrid cathodic protection systems for the mitigation of concrete reinforcement corrosion*', Proc. Corrosion & Prevention 2012 Conf., Australasian Corrosion Association Inc., November, Paper 062.

Ip, A and Pianca, F (2002), 'Applications of Sacrificial Anode Cathodic Protection Systems for Highway bridges – Ontario Experience', NACE, Corrosion 2002, *Paper* 02267.

Kessler, R, Powers, RG and Lasa, IR (1996) 'Zinc Mesh Anodes Cast into Concrete Jackets', NACE, Corrosion 1996, Paper No. 327.

Leng, D, Powers, RG and Lasa, I (2000), 'Zinc Mesh Cathodic Protection Systems', NACE, Corrosion 2000, Paper No. 00795.

Morgan, JH (1986), '*Cathodic Protection*', second edition, NACE, Houston.

NACE (1990), '*Impressed Current Cathodic Protection of Reinforcing Steel in Atmospherically Exposed Concrete Structures*', Recommended Practice 0290-2000, NACE International, Houston.

NACE (2002), '*State-of-the-Art Report: Criteria for Cathodic Protection of Prestressed Concrete Structures*', Publication SP01102, NACE International, Houston, USA.

NACE (2005), 'Sacrificial Cathodic Protection of Reinforced Concrete Elements – A State of the Art Report', NACE International, Publication 01105, Task Group 047.

NACE (2010), '*Cathodic Protection of Masonry Buildings Incorporating Structural Steel Frames*', Publication 01210, NACE International, Houston, USA.

NACE (2014a), '*Cathodic Protection to Control External Corrosion of Concrete Pressure Pipelines and Mortar-Coated Steel Pipelines for Water or Wastewater Service*', Standard Practice 0100 (formerly RP0100), NACE International, Houston, USA.

NACE (2014b), '*Cathodic Protection of Reinforcing Steel in Buried or Submerged Concrete Structures*', Standard Practice 0408, NACE International, Houston, USA.

NACE (2016), '*State-of-the-Art Report on Evaluating Cathodic Protection Systems on Existing Reinforced Concrete Structures*', Publication 01116, NACE International, Houston, USA.

Pourbaix, M (1966), '*Atlas of Electrochemical Equilibria in Aqueous Solutions*', Pergamon Press, Oxford, UK.

Pourbaix, M (1973), '*Sections on Electrochemical Corrosion*', Plenum Press, New York, USA.

Rothman, P, Szeliga, MJ and Nikolakakos, S (2004), 'Galvanic Cathodic Protection of Reinforced Concrete Structures in a Marine Environment', Corrosion 2004, Paper No. 04308, NACE International.

Solomon, I (2005), 'Corrosion and Electrochemical Protection of Reinforced Concrete Structures', Australasian Corrosion Association, Training Course Notes, Version 1.1, Melbourne, Australia.

Spector, D (1962), Corrosion Technology, 9, 257.

Standards Australia (2003), AS 2239, 'Galvanic (sacrificial) anodes for cathodic protection', Sydney, Australia.

Standards Australia (2008), AS 2832.5, 'Cathodic protection of metals Part 5: Steel in concrete structures', Sydney, Australia.

Standards Australia (2015), AS 2832.1, 'Cathodic protection of metals Part 1: Pipes and cables', Sydney, Australia.

Strategic Highway Research Program (1993), '*Cathodic Protection of Concrete Bridges: A Manual of Practice*', SHRP-S-372, National Research Council, Washington, DC, USA.

Stratfull, RF (1959), 'Progress report on inhibiting the corrosion of steel in a reinforced concrete bridge', Corrosion, 15, 6, June, 65–68.

Stratfull, RF (1974), Transportation Research Record 500, 1–15 TRB, Washington, DC, USA.

Sunde, ED (1949), '*Earth Conduction Effects in Transmission Systems*', Van Nostrand, New York, USA, 71.

Tehada, T (2009) '*Integrated Concrete Pier Piling Repair and Corrosion Protection System*', NACE, Corrosion 2009, Paper No. 09514.

Tinnea, J, Howell, KM and Figley, M (2004) 'Triple System Galvanic Protection of Reinforced Concrete Energizing and Operation', NACE, Corrosion 2004, Paper No. 04337.

Unz, M (1955), 'Proposed methods for cathodic protection of composite structures', Corrosion, 11, 2, February, 40–43.

Unz, M (1956), 'Intrinsic protection of water mains', Corrosion, 12, 10, October, 66.

von Baeckmann, W, Schwenk, W and Prinz, W (1997), '*Handbook of Cathodic Corrosion Protection*', third edition, Gulf Publishing Coy, Houston, 538.

Vrable, JB (1976), 'Chloride corrosion of steel in concrete', ASTM STP 629, eds D.E. Tonini and S.W. Dean, *American Society for Testing and Materials*, 124.

Ward, PM (1977), 'Chloride corrosion of steel in concrete', ASTM STP 629 eds D.E. Tonini and S.W. Dean, *American Society for Testing and Materials*, 150.

Whitmore, D (2004), 'New Developments in the Galvanic Cathodic Protection of Concrete Structures', NACE, Corrosion 2004, Paper No. 04333.

Wyatt, BS and Irvine, DJ (1987), 'A Review of Cathodic Protection of Reinforced Concrete' Materials Performance, 26, No 12, 12.

Chapter 10

Repair and protection (C) – electrochemical methods

10.1 GALVANIC ELECTROCHEMICAL TREATMENTS

10.1.1 Background

Galvanic electrochemical treatment/application of reinforced concrete consists most commonly of various configurations of discrete zinc anodes including but not limited to discrete zinc 'pucks', zinc cylinder arrays, or zinc strips. These types of anodes are typically encased in a medium that is formulated to maintain the electrochemical activity of the zinc (since Zn is prone to passivation from oxides from its own corrosion products) (Holloway et al., 2012).

Galvanic discrete zinc anodes typically do not provide CP in accordance with the criteria of Standards such as discussed at Section 9.9. They do not output sufficient cathodic current, for reasons as discussed below, to effect CP per se but rather provide electrochemical treatment and a reduction in corrosion rate.

It should also be noted that some galvanic anode suppliers, contractors, and conflicted consultants are misleadingly using the acronym 'CP' to describe 'corrosion protection' that may be offered by these systems rather than cathodic protection. CP of reinforced concrete is achieved when the protection criteria of Standards such as AS 2832.5 (Standards Australia, 2008) or ISO 12696 (BSI, 2016) are met on a continuous basis. If the protection criteria of concrete CP Standards are not met, then CP is not afforded by discrete zinc anodes.

10.1.2 Discrete zinc anodes in patch repairs

One of the first of the galvanic electrochemical treatment systems developed consists of Zn encased in a high alkalinity (pH >14) lithium hydroxide-based mortar so as to maintain the activity of the Zn (Sergi & Page, 1999), see Figure 10.1.

This type of system is installed by directly connecting the zinc anode to the reinforcing steel within a concrete patch repair, see Figure 10.1. When

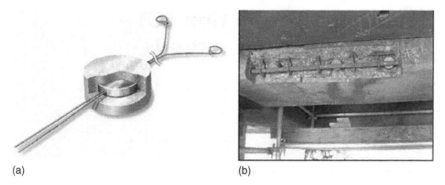

(a)   (b)

Figure 10.1 Examples of discrete zinc anodes encapsulated in lithium hydroxide-based mortar for installation in patch repairs. (Courtesy of Vector)

installed as part of a patch repair system it is claimed that the primary function of the anode is to delay the onset of incipient anode formation at the reinforcing steel immediately adjacent to the patch repair.

The phenomenon of 'incipient anode', 'ring anode' or 'halo' effect (Concrete Society, 2011) within concrete patch repairs has been previously discussed at Section 8.4.5. Page and Treadaway (1982) were the first to suggest that the mechanism of incipient anode formation in chloride contaminated concrete is the concept of macro-cell development (the formation of spatially separated anodes and cathodes with the anodes being the areas adjacent the repair and the cathodes being the repair itself).

As advised at Section 8.4.5, when it comes to incipient anode management, Green et al. (2013) report that the there is a perception in the industry that the only way to manage such is by the insertion of galvanic anodes within patch repairs. However, that is not the case. Green et al. (2013) point out that it should be remembered that conventional concrete patch repair utilising a reinforcement coating system (zinc-rich epoxy, epoxy or resin modified cementitious), a bonding agent (acrylic, styrene butadiene rubber, epoxy or polymer modified cementitious) and a polymer modified cementitious repair mortar (hand-applied, poured, sprayed or combinations thereof), effectively 'isolate' the patch such that macro-cell activity is minimised and incipient anode formation managed. By the use of reinforcement coating systems, cathodic reactions are severely restricted on the steel surfaces within the patch. Bonding agents provide a restriction to ionic current flow out of the patch into the surrounding parent concrete and polymer modified cementitious repair mortars have electrical resistivities which further restrict ionic current flow between the patch and surrounding parent concrete.

Manufacturers of electrochemical treatments for patch repairs continue to modify the size and configuration of the anode systems in an attempt to improve their current output and lifetime (e.g. Allan et al., 2000; Ball & Whitmore, 2008; Sergi, 2009) and refer to Figure 10.2.

(a)

(b)

Figure 10.2 Examples of different configurations of discrete zinc anodes encapsulated in formulated mortar (to maintain zinc activity) for installation in patch repairs. ((a) Courtesy of Vector) and ((b) Courtesy of BASF)

(a)

(b)

Figure 10.3 Examples of a 'rolled Zn sheet system encased in a slightly acidic pH environment' for installation in patch repairs. (Courtesy of SRCP)

More recently, a paper by Giorgini and Papworth (2011) introduces a rolled Zn sheet system encased in a slightly acidic pH environment, refer to Figure 10.3.

Discrete zinc anode systems that are installed within the parent concrete around the perimeter of patch repair areas are shown at Figure 10.4. These systems include an integral connecting titanium wire and are installed into drill holes and embedded in 'a specially formulated backfill mortar' (SPA, 2020).

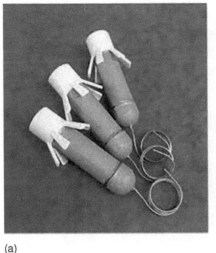
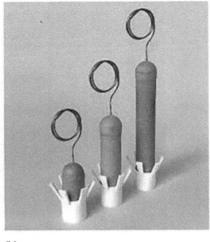

(a)

(b)

Figure 10.4 Example of a discrete galvanic anode system embedded within drill holes using a pliable viscous backfill mortar of pH <12.4. (Courtesy of Duoguard)

10.1.3 Distributed discrete zinc anodes

Encapsulated discrete zinc anodes have also been developed for installation in sound concrete to provide some electrochemical treatment.

An example of a cylinder-shaped system is shown at Figure 10.5 where installation is into drill or core holes arranged in an array with linking up of the anodes by a conductor (often a single insulated cable) which is then connected to the reinforcement. As previous, there are various papers that have been published on the application of these systems (e.g. Allan et al., 2000; Ball & Whitmore, 2008).

'Sausage'-shaped distributed discrete zinc anode systems have also been developed, refer to Figure 10.6. It is understood that the encasement material for these systems is lithium hydroxide (pH>14) based so as to maintain the activity of the Zn and to stop it passivating.

Distributed discrete zinc anode systems of the form of a rolled Zn sheet system encased in a slightly acidic pH environment (see Figure 10.7) have been developed (Giorgini & Papworth, 2011). An array of these rolled zinc sheet anodes are inserted into drill or core holes, which are then filled with zinc activator paste and connected to a conductor cable which is then connected to the reinforcement.

Distributed discrete zinc anodes that are installed in an array form into drilled holes containing a zinc activating backfill mortar and connected to a titanium wire which is connected to the reinforcement are shown at Figure 10.8.

(a)

(b)

Figure 10.5 Example of a cylinder-shaped encased distributed discrete galvanic zinc anode system. (Courtesy of Vector)

(a)

(b)

Figure 10.6 Example of a 'sausage'-shaped encased distributed discrete galvanic zinc anode system. (Courtesy of Vector)

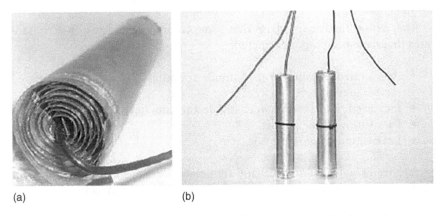

(a)   (b)

Figure 10.7 Example of a rolled zinc layer sheet discrete galvanic anode system. (Courtesy of SRCP)

Figure 10.8 Example of a discrete galvanic anode system embedded within drill holes using a pliable viscous backfill mortar of pH <12.4. (Courtesy of Savcor Products Australia)

10.1.4 Performance limitations

Holloway et al. (2012) advise that the performance of galvanic anodes in concrete depends largely on various factors including:

1. Humidity and moisture in the concrete.
2. Temperature.
3. Anode geometry and spacing.
4. pH level around anode.
5. Chemicals added around anode to maintain activated state.
6. Current density respective to anode surface area.
7. Resistance of anode to concrete.
8. Build-up of corrosion product(s) around anode.
9. Level of chloride contamination.
10. Level of corrosion activity of the reinforcement.

They go on further to advise that some common problems or early failures that have been reported include:

- Accelerated consumption of anode resulting in lower than expected service life.
- Increased resistance between anode and concrete.
- Passivation of the anode.
- Lack of moisture in the concrete.

As a result, galvanic discrete zinc anodes do not typically output sufficient cathodic current to effect CP per se but rather provide electrochemical treatment and a reduction in corrosion rate.

Preston et al. (2017) note that there is no standard or test method that defines what constitutes a galvanic anode for use in concrete. Current best practice for the specification of products is considered to be presented in the model specification in Technical Report 73 (Concrete Society, 2011). This requires an anode to meet several key performance characteristics:

i. It shall have proven technical merit.
ii. The galvanic anode system, including the galvanic metal element, the activating agent, and the backfill where needed, shall not present a corrosion risk to the steel at the end of the service life of the system.
iii. The consumption of a galvanic anode in an anode system embedded in cavities in the concrete shall not give rise to expansive forces that disrupt the concrete cover.

Preston et al. (2017) then advise that it is necessary for a client or specifier to determine if a product meets these characteristics, for which there are no defined criteria or test methods, along with a need for them to assess if the products can deliver what is needed by the structure. If this is a fully effective CP system, meeting the protection criteria in ISO 1296 (BSI, 2016), this is unlikely to be fulfilled. If it is a system able to provide some lesser degree of corrosion protection for more than a decade or two this is also unlikely to be fulfilled.

Furthermore, they note that increasingly, new products will be brought to market and prior to use it will be necessary to assess their merit. All zinc anodes in concrete require some form of activation agent to ensure that they do not passivate, which will prevent their operation. It is also necessary to ensure anodes do not passivate in storage or prior to installation, present procedures do not specify the importance of ensuring that this does not occur (Preston et al., 2017).

Chess (2019) notes the limited surface area of the galvanic discrete anodes, e.g. 0.1 m² of zinc surface per m2 of concrete for the anode shown in Figure 10.1, would give immediate concerns at the current output level and hence the level of CP which can be achieved.

Chess (2019) also notes that zinc is temperature sensitive for its output. Pure zinc anode output drops to zero below 15°C and above 40°C pure zinc will become passivated, whatever its environment.

The life expectancies forecast for galvanic discrete zinc anodes should also be questioned according to Chess (2019). He advises that most of the life expectancies of galvanic anodes are formulated using the calculation which is used for consumption of galvanic anodes in seawater. This has an amp-hour capacity for the zinc inserted along with a utilisation factor and a current density to give a life in years. Chess (2019) comments that this equation is not valid for zinc in concrete as the current level is likely to be severely restricted by the anode circuit resistance and the utilisation factor is likely to be hugely influenced by the anti-passivation measures taken and the immediate environment around the anode system, for example, if the surrounding concrete is dry.

10.1.5 Performance assessment

Preston et al. (2017) advise that it is recognised that reinforced concrete CP systems utilising galvanic anodes embedded within concrete frequently do not have the required ability to collect the data required to assess if the system meets the performance criteria in ISO 12696 (BSI, 2016). As such, these systems cannot be considered compliant with ISO 12696; although some manufacturers of these systems claim otherwise. In the informative (non-normative) notes to the criteria of ISO 12696 an assessment of corrosion risk to reinforcement is permitted when, specifically, with galvanic anodes, CP performance criteria are not met (Preston et al., 2017).

Preston et al. (2017) note further that it is clear within the ISO 12696 Standard that this assessment of corrosion risk is not an additional performance criterion, it is a reference to a possible additional assessment of corrosion rate where galvanic anode systems (and hybrid treatment systems, see Section 10.2) do not meet the CP criteria. If a galvanic anode system does not meet the criteria and the structure is found to be at risk of corrosion, there is then a requirement to supplement the galvanic anode system.

Another of the informative (non-normative) notes to the criteria of ISO 12696 (BSI, 2016) then provides actions to take if a corrosion risk is identified with a galvanic anode system for concrete, namely:

i. Turning the galvanic system into an impressed current system (if this is possible).
ii. Installing more galvanic anodes.
iii. Applying a temporary electrochemical treatment (using the galvanic anodes, if possible) and a temporary power supply.

The recommended method of determining corrosion risk presented in ISO 12696 (BSI, 2016) where a galvanic anode system is used is corrosion rate assessment using the Butler Volmer (BV) equation (Preston et al., 2017):

$$i_{appl} = i_{corr}\left(\exp\left(\frac{2.3\Delta E}{\beta_c}\right) - \exp\left(-\frac{2.3\Delta E}{\beta_a}\right)\right)$$

Where:

- i_{appl} is the applied current density (Note: it is necessary to measure current and calculate the current density using the steel surface area under consideration, which in practice requires a considerable assumption).
- i_{corr} is the corrosion rate.
- ΔE is the electrode potential shift.
- β_a and β_c are constants.

Practical field experience has shown that using this approach is not straightforward. Many galvanic anode systems are not designed with any monitoring provisions and there is no ability to isolate the anodes from the reinforcement, no method to measure the anode currents or no reference electrodes embedded in order to measure potential shift (Preston et al., 2017).

When systems have been designed with the necessary 'anode segments' to permit such measurements the sources of error in the measurement and analysis are many. When using hand-held instrumentation to measure currents from galvanic anodes in concrete there can be considerable variances in the measurements obtained, in particular as current can rapidly vary when cables are disconnected and a meter then connected in series. The selected meter range itself can also give rise to a significant error as the shunt resistance increases with decreasing range (Preston et al., 2017).

Aside from the practical difficulty in taking such measurements a particular source of error is then how the steel area under test is determined when the distribution of polarisation from the anodes may vary. It is necessary for records of the steel area associated with each monitoring segment to be documented and stated in the project operation and maintenance manual, something that in practice unfortunately does not happen. Even if records are available though, this would also be an approximation only (Preston et al., 2017).

Angst & Büchler (2015) describe further uncertainties in the results of corrosion rate measurements in galvanic corrosion cells, in relation to concrete, arising from electrode location, and errors within presently used standard formulae including the BV equation.

Flitt et al. (2015) comment that the Wagner and Traud equation has been erroneously called the BV equation and that the use of the BV equation to estimate reinforcement corrosion rate as a means of corrosion risk assessment in the case of CP systems that do not meet protection criteria, is not correct as it eliminates important electrokinetic and corrosion kinetic information.

Chess (2019) notes that achieving any of the criteria of CP standards with galvanic anodes could be very difficult, particularly if they have only a small surface area as their individual current outputs will be so much less than an impressed current system, so an additional non-normative note was added specifically for galvanic systems whose current can be measured. He discusses various concerns regarding the use of the BV equation and it making some significant assumptions. The first and most dubious according to Chess (2019) is that steel in concrete behaves in the same way that a metallic element in an aqueous environment does. The second is it ignores the fact that micro and macro corrosion cells exists in concrete, so an average corrosion rate might not actually represent the situation. There are also other assumptions such as reaction rate increasing with temperature when it is known that CP currents in aqueous environments do not.

Chess (2019) goes on to say that what in effect is being stated is that if the corrosion rate is very low, you will need less CP, which while being probably true, begs the question that if the corrosion rate is very low, why do you need CP at all?

Preston et al. (2017) comment that if there are deficiencies in using published methods of corrosion rate monitoring where galvanic anodes are used, either due to theoretical, or practical measurement aspects, how should performance of anodes be assessed? One option they propose is to design galvanic anode systems such that all anodes can be easily isolated from the reinforcement, in the manner of an impressed current system. This methodology has been used at one significant bridge installation in the UK. This enables routine monitoring of the system for depolarisation testing. Another approach, which is gaining more prevalence is to have sections of galvanic anodes that can be switched to enable monitoring at localised areas, again for standard depolarisation tests and measurements of anode currents, see Section 9.9.

Holmes et al. (2013), have discussed the benefits of using steel/concrete/portable reference electrode (potential mapping) surveys to provide an indication of the benefit achieved using galvanic anodes. Preston et al. (2017) point out though, that there is at present no method to assess the performance of galvanic anode systems against standard criteria unless it is designed to enable isolation of all anodes from the reinforcement, and this is not usually done.

The lack of performance data for galvanic anode systems is now something that needs to be considered by all parties, clients, designers, contractors, and manufacturers (Preston et al., 2017). Clarity as to the claims of whether such systems can deliver 'full' CP to the criteria in ISO 12696 (BSI, 2016) and the claims that manufacturers make to possible customers in the concrete repair contracting industry or the facility owners, who are often not CP experts, also needs to be clarified. Clear and updated guidance is required as to what can or cannot be achieved using galvanic anodes embedded in concrete (Preston et al., 2017).

Chess (2019) notes that if the anode is simply attached to the reinforcement and mortared in, its output current, and voltage cannot be monitored. In this case, this anode system is not in compliance with the code of practice (ISO 12696, BSI, 2016) for the CP of reinforced concrete. He notes that certain manufacturers are plainly showing a direct connection detail for their anode while stating on the same page it is in compliance with ISO 12696.

Broomfield (2017) advises that if a galvanic system is designed and installed to the requirements of ISO 12696 (BSI, 2016) then monitoring is a requirement. He states this means that a monitoring system should be built-in at least providing the ability to monitor current and steel potentials at selected locations. Broomfield (2017) then further comments that it is easier to ignore this requirement on a galvanic system then on an ICCP system, but it is important to ensure that structures are being protected and that anodes and connections are still functioning as designed.

Furthermore, Broomfield (2017) advises that there are no test methods for galvanic anodes which means that designers must rely on their own experience, the reported experience of others, or carry out trials before applying novel anodes or applying them in novel situations. He states that BS EN ISO 12696 (2016) gives guidance on applying galvanic anodes to create a full distributed and monitored system and that there is also a NACE Standard (2016) and a NACE state of the art report (NACE, 2005).

10.2 HYBRID ELECTROCHEMICAL TREATMENTS

10.2.1 Background

Hybrid electrochemical treatment/application of reinforced concrete involves an array of embedded discrete zinc (Zn) anodes that are operated in impressed current mode initially and then in galvanic mode. The concept of hybrid anode electrochemical treatment systems differs from the purely galvanic anode electrochemical treatment systems described above. While still incorporating an array of embedded Zn anodes, the premise of the hybrid systems is that they are initially operated in an impressed current mode, before being disconnected from the DC power source and maintained as a galvanic anode system.

Two generations of systems have been developed namely:

- **First Generation System:** developed primarily in the UK (Glass et al., 2007) and whereby the impressed current mode of operation involves a transformer/rectifier as the DC power source.
- **Second Generation System:** developed primarily in Canada (Whitmore et al., 2019) and whereby the impressed current mode of operation uses a self-powered impressed current anode within each single anode unit.

Hybrid electrochemical treatment systems, like galvanic discrete zinc anodes, typically do not provide CP in accordance with the criteria for CP of Standards such as discussed at Section 9.9. Use of the BV equation seems to be the only method of performance evaluation of hybrid electrochemical treatment systems. Assessment to CP criteria, as per Section 9.9, is typically not undertaken.

It should also be noted that suppliers, contractors, and conflicted consultants for hybrid electrochemical treatment systems are also misleadingly using the acronym 'CP' to describe 'corrosion protection' that may be offered by these systems rather than cathodic protection. CP of reinforced concrete is achieved when the protection criteria of Standards such as AS

2832.5 (Standards Australia, 2012) or ISO 12696 (BSI, 2016) are met on a continuous basis. If the protection criteria of concrete CP Standards are not met, then CP is not afforded by hybrid anode systems.

10.2.2 First generation system

The first generation hybrid electrochemical treatment system involves an array of embedded Zn anodes that are operated in impressed current mode initially (for a number of weeks) and then in galvanic mode (Glass et al., 2008). The premise of the first generation hybrid system is that it is initially operated in an impressed current mode, before being disconnected from the DC power source (transformer/rectifier) and maintained as a galvanic anode system. The UK patent holders (Glass et al., 2007) for the technology have published a paper describing the system and its principles for providing protection (Glass et al., 2008). In the paper it is proposed that during the initial impressed current phase of the treatment any active pits are re-alkalised effectively stopping active corrosion and returning the steel to a passive state. Following the application of the impressed current phase for a predetermined period the re-established passivity of the steel is maintained by the galvanic anode system.

Dodds et al. (2017) describe the hybrid corrosion protection (HCP) system as a discrete zinc anode system that is installed into pre-drilled cavities within reinforced concrete, see Figure 10.9. They describe it as combining both an impressed and galvanic system whereby the hybrid anodes are manufactured using 18 mm zinc cylinders, ranging from 42 mm to 220 mm long, with an integrated titanium connector wire. A 'coating' is applied to the surface of the zinc anode in order to keep the anode 'active' throughout its design life. Initially the system is connected to a temporary, constant 9 V DC power supply, typically for a period of at least one week, depending on the type of reinforcement (e.g. mild steel or prestressed), to deliver a charge to the reinforcing steel.

Christodoulou and Kilgour, (2013) describe an initial impressed current phase of 'typically 8 to 12 weeks' rather than one week. Christodoulou et al. (2016) on the other hand advise of 'an initial temporary energisation phase' of 7 to 28 days. Thompson et al. (2012) advise between 7 and 14 days from an external power supply of between 6 and 12 V DC.

A minimum of 50 kC/m^2 of steel surface area is passed during the initial impressed current phase according to Christodoulou & Kilgour (2013) and Christodoulou et al. (2016). The hybrid anode is then disconnected from the temporary DC power supply and connected directly to the steel reinforcement. This latter treatment phase continues for the remainder of the anodes' working life and provides a 'relatively low current to the steel reinforcement' (Dodds et al., 2017).

Dodds et al. (2017) advise that the first-generation hybrid anode system is installed at a spacing dependent on steel density (per m^2 of concrete),

Figure 10.9 Installation of the first-generation hybrid anode system to a reinforced concrete pier. (Courtesy of Dodds et al., 2017)

and the individual anodes are connected in series with 'insulated' titanium wire. Different sizes of anode are available and are named based on
the expected charge output during the anode's life (e.g. D350 and D500
anodes relate to an output of 350 and 500 kC, respectively). The size of
the anode is chosen based on the design life required, average current
density and steel density within the concrete structure (Dodds et al.,
2017).

According to Christodoulou et al. (2016), first generation hybrid anodes
are usually designed to provide a residual service life of 30–50 years 'based
on an expected current demand from the surrounding reinforcing steel in a
structure'.

In terms of terminology for this treatment method, Holloway et al. (2012)
indicate that the UK patent holders for the technology (Glass et al., 2007)

published a paper in 2008 describing the system and its principles for providing protection (Glass et al., 2008). In the Glass et al. (2008) paper the system is referred to as 'hybrid electrochemical treatment', 'hybrid treatment' or 'hybrid corrosion treatment'. Terminology of HCP (Christodoulou et al., 2016; Glass et al., 2016) has also been used.

On the other hand, Christodoulou and Kilgour (2013) call it HCP and do so even though there is no data and assessment in the paper where the HCP system to prestressed concrete substructure elements of two bridges complies with the CP criteria of Standards, i.e. see Section 9.9.

To add further confusion, Thompson et al. (2012) term it hybrid electrochemical protection (HEP).

It should be noted that CP has for nearly 200 years meant cathodic protection. The use of the acronym CP for 'corrosion protection' for steel in concrete (Christodoulou et al., 2016; Glass et al., 2016), HCP (Christodoulou et al., 2016; Glass et al., 2016) and HCP (Christodoulou & Kilgour, 2013) to describe this electrochemical treatment method are therefore considered misleading to say the least.

10.2.3 Second generation system

Recently a two-stage, self-powered modular electrochemical treatment system based on embedded discrete Zn anodes has been developed (Whitmore et al., 2019). This second generation hybrid electrochemical treatment system provides a two-stage current application as follows (Whitmore et al., 2019):

- During Stage 1, the single anode unit uses a self-powered impressed current anode to perform an electrochemical treatment aimed at passivating the reinforcing steel.
- The two-stage anode then automatically switches over to Stage 2 and provides galvanic cathodic current aimed at maintaining steel passivity for the life of the system. This switch occurs automatically when the voltage of the impressed current element drops to that of the galvanic element.

These anodes are designed to be installed in drilled holes. Each anode includes both the impressed current and galvanic components (Figure 10.10) and is connected to the steel using a single wire (Whitmore et al., 2019).

10.2.4 Performance assessment and limitations

Use of the BV equation seems to also be the only method of performance evaluation of the hybrid electrochemical treatment systems (e.g. Thompson et al. (2012); Christodoulou & Kilgour, 2013; Christodoulou et al., 2016;

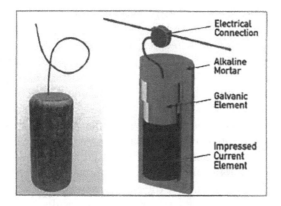

Figure 10.10 Two-stage anode (left) and schematic illustration of the anode (right). (Courtesy of Whitmore et al., 2019)

Glass et al., 2016; Dodds et al., 2017). Assessment to CP criteria, as per Section 9.9, is not typically undertaken.

The UK patent holders (Glass et al., 2007) in one of their first published papers (Glass et al., 2008) about the first generation system also describe performance assessment to the BV equation rather than the accepted and proven CP criteria of Section 9.9.

Thompson et al. (2012) provide some details on how the corrosion rate of a first generation HEP system is determined namely:

- The test first measures the current applied to a certain steel area and the user must calculate the steel area it is applied to.
- The instant off potential must then be recorded (reference potential measure between 0.1 and 0.5 s of the system being switched off).
- The difference between the instant off potential and the depolarised potential (between 4 and 24 hrs) is then used in the BV equation as ΔE.

It is beyond the scope of this book to explore the various concerns that apply to the corrosion risk (rate) assessment method of the hybrid treatment methods, the use of the BV equation, etc. suffice to say, for example, that as previous, Angst & Büchler (2015) describe various uncertainties in the results of corrosion rate measurements in galvanic corrosion cells, in relation to concrete, including errors within presently used standard formulae. Flitt et al. (2015), as previous, comment that the Wagner and Traud equation has been erroneously called the BV equation and that the use of the BV equation to estimate reinforcement corrosion rate as a means of corrosion risk assessment in the case of CP systems that do not meet protection criteria, is not correct as it eliminates important electrokinetic and corrosion kinetic information. Preston et al. (2017), comment that practical field experience has shown that using the BV approach is not straightforward, refer

Section 10.1.5 as previous. Elsener (2002) also, for example, advises that an error to a factor of ten can arise for corrosion rate estimations of chloride-induced (pitting) corrosion of steel reinforcement in concrete.

Glass et al. (2008) in one of the first published papers about the first generation hybrid treatment system describe laboratory trials applied to chloride contaminated reinforced concrete blocks and the achievement of in excess of 100 mV potential decay. However, for a first generation hybrid treated bridge in the same paper, emphasis is placed on corrosion rate estimation using the BV equation and no data is provided showing whether CP was achieved to accepted criteria in Standards, refer to Section 9.9.

Thompson et al. (2012) advise that 'the monitoring of a HEP system is similar to an ICCP system in that the current is recorded and a depolarisation test is conducted but the evaluation of the results is carried out in a different manner. ICCP systems are evaluated using the 100mV depolarisation criteria set out in the AS 2832.5. (Other criteria are used also). HEP, however, uses a much smaller current in the galvanic phase and therefore produces a much smaller potential shift and can't be evaluated in the same way'.

Christodoulou and Kilgour (2013) describe HCP treatment of the prestressed concrete substructure elements of the Kyle of Tongue Bridge in Sutherland, Northern Scotland, and the Tiwai Point Bridge in Ivercargill, Southland, New Zealand. For the Kyle of Tongue Bridge, the authors state that 'HCP was used to arrest existing corrosion activity to the prestressed concrete beams of the superstructure and extend their service life for a 30-year period'. For the Tiwai Point Bridge, the authors state that 'a trial HCP system was developed to provide corrosion prevention to the prestressed concrete piles within the tidal zone with a targeted service life of 50 years'. Performance assessment by Christodoulou and Kilgour (2013) of both systems has been made solely on the basis of corrosion rate estimation and utilisation of the BV equation. No data is provided to enable performance assessment to CP criteria, even for the impressed current phase (which was of 8 to 12 weeks duration) of the treatments.

Christodoulou et al. (2016) then describe the first generation hybrid treatment system as HCP for two existing reinforced concrete bridges as part of the State Highway 16 Motorway in Auckland, New Zealand. Systems performance assessment is again on the basis of corrosion current density calculations based on applied currents and applied potential shifts and insertion into the BV equation. In this paper some potential depolarisation (decay) results at a limited number of locations for one time during the impressed current phase only of the treatment are provided. Some depolarisations (decays) are more than the CP criteria requirement of 100 mV but the data is limited in quantum and location, the time during which the data was taken during the impressed current stage is not provided and no data is provided for the galvanic phase of the treatment.

Glass et al. (2016) describe the first generation hybrid treatment process as HCP and discuss 'long term' performance (i.e. dating back to 2006) on three reinforced concrete bridges. Emphasis is again placed on steel corrosion rate data, presumably by the same methods as previous, i.e. on the basis of corrosion current density calculations based on applied currents and applied potential shifts and insertion into the BV equation. There is yet again no data to enable systems assessment to CP criteria. However, in this paper, there is also some steel potential data (presumably instant off potentials) that show a trend to less negative (more positive) potentials with time of up to 9–10 years but it is not known from the paper where the reference electrodes that are providing such potential data are positioned, i.e. at corroding locations, in repair areas (passive reinforcement), etc. Glass et al. (2008) have proposed previously that such a trend in reinforcement electrode potentials is indicative of steel passivity.

Dodds et al. (2017) discuss the performance of the first generation HCP system installed to reinforced concrete substructure elements of six bridges, 1–8 years after treatment. Performance assessment in all instances is again by corrosion rate estimation and there is no data provided to enable assessment to CP criteria. Some steel on-potential data with time up to 8 years is provided and shows a trend towards less negative (more positive) values but again it is not known from the paper where the reference electrodes that are providing such potential data are positioned, i.e. at corroding locations, in repair areas, etc.

The performance of hybrid electrochemical treatment systems is therefore considered questionable. Application of cathodic current to steel in concrete is of course beneficial, unless it is excessive thereby leading to the risk of hydrogen embrittlement of prestressing steel or impacting of the bond for conventional reinforcement. However, the extent of cathodic current application needs to be sufficient to effect CP or else merely a reduction in corrosion rate (risk) is achieved. For reinforced concrete structures in saline environments that are genuinely suffering from chloride-induced pitting, as a structure owner, one would want to be confident that 'a reduction in corrosion rate (risk)', provision of 'corrosion protection' or provision of 'galvanic protection' is adequate. As a structure owner it may be more appropriate to have certainty and confidence that corrosion is being arrested by CP (to long-accepted protection criteria) so as to sustain (and extend) service life (cost-effectively).

Holloway et al. (2012) believe that there are some key gaps in the general understanding at the moment of galvanic anode and hybrid electrochemical treatments that need to be addressed including:

1. Suitable monitoring methods for determining the performance of galvanic anode systems, especially if they do not meet the 100mV decay criteria need to be assessed and accepted by the greater community. While corrosion rate monitoring may provide one option it is

important to note that this method can prove difficult to apply and requires a much more detailed understanding of the electrochemical response of steel in concrete to be applied proficiently.

2. Independent and robust long term studies on the actual performance of hybrid, sheet, and embedded galvanic anode systems. Many of the papers available in the literature have been authored or co-authored by representatives from the anode manufacturing companies. It is proposed that independent academic and industry-based studies will provide a more objective view on performance and their relative limitations. This knowledge can only assist in the successful implementation of such systems.

3. In addition, the mechanistic aspects, which underpin the efficacy of galvanic CP, and how the performance, and technology of hybrid anode systems perform in a mechanistic sense, remain unclear. This is not necessarily a negative issue, but one that will need to be addressed by future works in a scientific setting.

It should also be noted that in addition to the performance limitations already discussed, the performance limitation of galvanic discrete zinc anode electrochemical treatments, refer to Section 10.1.4, apply also to hybrid electrochemical treatments.

10.3 ELECTROCHEMICAL CHLORIDE EXTRACTION

Electrochemical chloride extraction is sometimes referred to as electrochemical chloride removal or desalination. Desalination is a method employed for the remediation of reinforced concrete suffering or at risk from chloride-induced corrosion. The set-up is similar to that which is used for impressed current CP, but the current density applied is up to a hundred times higher (i.e. 1–2 A/m^2 of concrete surface) and it is only applied for a short time (6–10 weeks). The principles of the desalination process are shown in Figure 10.11.

As chloride ions are discharged at the anode yielding a highly corrosive environment the anode is usually a mixed metal oxide (MMO) coated titanium mesh which can resist such attack. Steel mesh is generally not used as it would corrode. Typically, a MMO titanium mesh embedded in a paper pulp which is continuously kept wet is used as the anode/electrolyte system, refer to Figure 10.12 (as an example of paper pulp application in particular). Because a quantity of hydrochloric acid is transported into the anodic solution a solution of calcium hydroxide is often used. Other methods include a felt cloth blanket which is applicable for horizontal surfaces and an enclosed coffer tank system.

The current flowing from the cathode is carried by the chloride ions and hydroxyl ions towards the positively charged external electrode (anode)

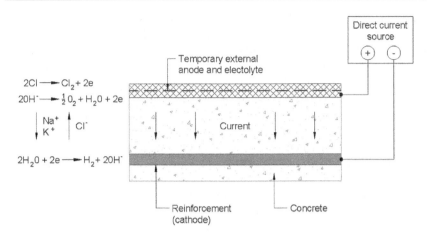

$$2Cl \longrightarrow Cl_2 + 2e$$
$$2OH^- \longrightarrow \tfrac{1}{2}O_2 + H_2O + 2e$$

Na$^+$ Cl$^-$
K$^+$

$$2H_2O + 2e \longrightarrow H_2 + 2OH^-$$

Temporary external
anode and electolyte

Direct current
source

(+)   (−)

Current

Reinforcement
(cathode)

Concrete

Figure 10.11 Principal reactions involved in chloride extraction. (Courtesy of Bertolini et al., 2004)

Figure 10.12 Application of electrochemical chloride extraction to a bridge structure. (Courtesy of Broomfield, 1997)

mesh where they are collected in the electrolyte reservoir. Initially, because there is a high concentration of chloride ions, much of the current is carried by them and the ratio of chloride to hydroxyl ions decreases. The desalination process ceases to be effective when the ratio of chloride to hydroxide has fallen to such a level that most of the current is being carried by hydroxyl ions and the rate of removal of chloride ions falls to a small value.

Potable water is the most commonly used electrolyte because additional ions which could compete as current carriers are kept to a minimum. In enclosed situations where the evolution of hypochlorous acid could pose a safety problem a saturated calcium hydroxide reaction is often used to neutralise the acid generation.

After a few weeks of treatment concrete cores or dust samples are analysed to test the extent to which the chloride content of the concrete has been reduced. When sufficient desalination has occurred, the DC power supply (transformer/rectifier) is switched off. If there is danger of future chloride ingress into the structure, a chloride protective coating has then to be applied to limit future ingress of the corrodent.

The advantages of desalination compared with impressed current CP are (Hudson, 1998):

a. No continuing maintenance or monitoring programme is necessary.
b. The very specific detailing that is carried out for CP as a permanent installation is not required.
c. There is no risk that the long term effectiveness of the system will be negated by, e.g. vandalism.
d. The performance criteria, i.e. reduction in chloride levels below the corrosion threshold, is simpler than that for CP.
e. There is no additional long term load imposed on the structure as a result of the process.
f. As the method is short term, it may be possible to treat structures with steels that would be susceptible to hydrogen embrittlement by temporarily supporting the structure.

The disadvantages are:

a. If there will be further contamination with chlorides, the future integrity of a structure will be dependent upon the performance of the protective coating applied and monitoring and maintenance may be necessary.
b. It may not be possible to remove chlorides from below the first level of reinforcement. In such cases there would be some uncertainty as to whether corrosion could be activated in the future. This could occur as a result of dissipation of hydroxyl ions over time from around the steel causing an increase in the ratio of chloride ions to hydroxyl ions above that necessary for corrosion.

c. The consequences of 'shadow spots' or 'silos' of chlorides left between rebars after treatment may need to be considered.
d. For severe pitting corrosion situations, the proportion of chlorides removed by this technique may not ensure repassivation of the reinforcement.
e. Prestressing steel can be severely damaged by hydrogen embrittlement due to the high levels of current applied, although as above some structures may be temporarily supported during treatment.
f. There are also concerns regarding the possibility of reduction in bond strength which is of a particular concern where structures have been constructed using smooth surface reinforcement rather than a mechanical interlock of ribbed steel.
g. No benefits will be attained by any isolated steel work.

Experience has shown that typically 50–90% of the chlorides may be completely removed from the concrete and good levels of passivation of the steel can be obtained if correctly applied. However, only partial success may be achieved on complicated elements.

Desalination can be applied in many of the situations where CP can also be applied. It has been applied to bridges, buildings, and carparks with variable success.

As for impressed current CP treatment of concrete structures, cracked, spalled, or delaminated concrete needs to be repaired (but there is no need to 'chase' corroded bars, breakout behind bars, use rebar coatings, etc.) prior to electrochemical chloride extraction treatment.

The possibility of initiating alkali aggregate reaction (AAR) as a side effect of electrochemical treatments arises from the production of hydroxyl ions at the rebar surface and the consequent attraction of alkali metal ions. In general, concrete containing alkali reactive aggregates will be susceptible to AAR if the alkali metal content expressed as Na_2O is more than about 2.5 kg/m^3 of concrete.

During desalination, although no alkali metal ions are added to the concrete. There is a tendency for alkali metal ions to migrate towards the reinforcement and this has been found to be capable of initiating AAR. In such cases the use of lithium compounds has been found to suppress the problem.

10.4 ELECTROCHEMICAL REALKALISATION

Electrochemical realkalisation is a very similar process to electrochemical chloride extraction except it is applicable for carbonated concrete structures. It is a shorter-term treatment, days instead of weeks. It has been used for buildings and bridges with concrete cover as low as 10 mm.

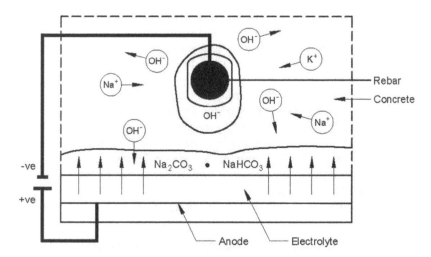

Figure 10.13 Electrochemical realkalisation. (Courtesy of Hudson 1998, p. 154)

Electrochemical realkalisation aims to restore the pH to a value greater than 11 and involves very much higher current densities than are used for impressed current CP. The principles of electrochemical realkalisation are shown in Figure 10.13.

Electrochemcal realkalisation is often carried out with an externally mounted anode immersed in an alkaline electrolyte normally a sodium carbonate solution. Three principal methods are used: cellulose fibre which can be spray applied with the electrolyte onto the concrete surface; a felt cloth blanket which is applicable for horizontal surfaces; and, an enclosed coffer tank system, refer to Figure 10.14.

Hydroxyl ions generated by the cathodic reaction restore the alkalinity of the environment of the reinforcing steel. Electroendosmosis plays an additional role in electrochemical realkalisation. Electroendosmosis is the process in which if an electric current passes through a porous material, then liquid within the pores has a tendency to move towards the negative electrode. The principle of the mechanism of electroendosmosis is shown in Figure 10.15.

In the capillary pores of concrete, the polar water close to the pore walls is negatively charged and tightly bound to the pore walls. Adjacent to the tightly bound inner layer, is a more loosely bound positively charged layer and under the influence of a potential field the free water moves towards the cathodic area. Typically, the size of the double layer might be up to half a micron if the liquid is distilled water while it is under a nanometre in concentrated salt solution. Consequently, electroendosmosis plays little role in desalination, but is made use of in the electrochemical realkalisation process as the anode is usually mounted in a tank of sodium carbonate solution attached to the side of the structure so that the alkaline solution is drawn

Figure 10.14 Electrochemical realkalisation in progress. (Courtesy of Broomfield, 1997, p157)

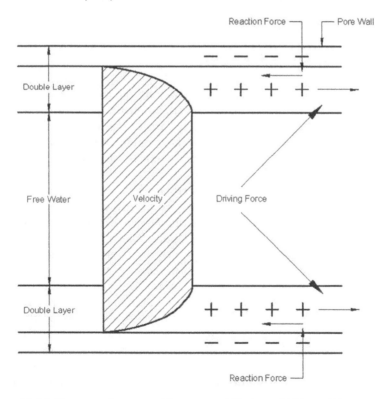

Figure 10.15 Electroendosmosis. (Courtesy of Hudson, 1998, p.155)

into the concrete by the electroendosmotic effect. A nominal 1 A/m^2 of concrete surface area current density is often used and this may require a drive voltage of 10–40 Volts.

Electrochemical realkalisation is often carried out in conjunction with other repairs which may involve patching. In this case special proprietary materials may be used that allow the transport of sodium carbonate and are resistant to solutions of up to pH 14.

As previous, the possibility of initiating AAR as a side effect of electrochemical treatments arises from the production of hydroxyl ions at the rebar surface and the consequent attraction of alkali metal ions. In general, concrete containing alkali reactive aggregates will be susceptible to AAR if the alkali metal content expressed as Na_2O is more than about 2.5 kg/m^3 of concrete.

It is rare for the final pH of realkalised concrete to reach high enough values to initiate AAR, but if the electrochemical realkalisation process is continued for an excessive time then there is a danger that sufficient sodium ions could be moved into adjacent uncarbonated concrete that its original high pH could be raised further to dangerous levels. It is therefore necessary to monitor the alkali metal ions (Na^+ and K^+) content of the concrete throughout the treatment to prevent any chance of AAR.

10.5   REPAIR AND PROTECTION OPTIONS – COSTS ASSESSMENT APPROACHES

The service life of a reinforced concrete structure can most certainly be maximised (and duly extended if required) with minimal expenditure (operating and/or capital). There is an appropriately varied range of protection, repair, and durability options available so that the maintenance phase in the operational service life of a structure can be optimised at minimal expenditure and whole of life costs (Chess & Green, 2020).

Whether the repair and protection methods that are being considered for a reinforced concrete structure are mechanical methods (e.g. patch repair, coatings, and penetrants, etc.), CP (e.g. impressed current and/or galvanic) or electrochemical treatment methods (e.g. galvanic, hybrid, chloride extraction or realkalisation) or combinations thereof, often a costs assessment is necessary as part of the decision-making process.

Life cycle cost models using investment analyses and discount rates are one commonly used method for analysis of repair and protection costs over time. These models are highly dependent on discount rates and that they assume that future funds will be provided for repair and protection. Discount rates between 0 and 10% are found in the literature leading to significantly different choices in materials. When discount rates are low, long term durable systems at initial construction are favoured, while at high discount rates, more frequent repairs are favoured as the current value of future costs are low.

Regular inspection and condition survey of concrete structures is essential, and it is desirable that this is more than a visual inspection. Preventive maintenance can then be performed before degradation becomes advanced and the remedial work will be less extensive and less expensive. The fundamental factors that govern concrete durability need to be well understood to enable practices that ensure satisfactory asset maintenance and enhanced service life performance (Standards Australia, 2018).

Dutch engineer Reinhold De Sitter first postulated the 'Law of Fives' in 1984, that in terms of the life of a concrete structure, concrete durability problems can be divided into four phases (Standards Australia, 2018):

- **Phase A:** design, construction, and concrete curing.
- **Phase B:** corrosion initiation processes are underway, but propagation of damage has not yet begun.
- **Phase C:** propagating deterioration has just begun.
- **Phase D:** propagation of corrosion is advanced, with extensive damage manifesting.

The 'Law of Fives' then states that **$1 extra spent in Phase A is equivalent to saving $5 of remedial expenditure in Phase B, or $25 remedial expenditure in Phase C, or $125 remedial expenditure in Phase D** (Standards Australia, 2018).

While the ratios may differ between structures it is now well understood that the compounding effects of untreated deterioration results in rapidly escalating costs over time.

The law demonstrates that quality design and construction, along with early intervention or preventive maintenance throughout an asset's life cycle, saves substantial money in the long term.

Collins and Blin (2019) explore whole of life (WOL) considerations and indicate that the approach for option selection can include a mix of quantitative (e.g. capital expenditure/CAPEX and operating expenditure/OPEX estimates) and qualitative measures (e.g. social, environmental, technical and reputational impacts that can be evaluated in a multicriteria type assessment (MCA)).

Risk considerations are also important, and these alone may not allow for the optimised prioritisation of interventions on reinforced concrete structures to achieve (or extend) service life, particularly in a funding-constrained environment. Collins and Blin (2019) advise that this is where intervention strategies can be driven by considerations of balancing life cycle risks and costs and they cite an example of an approach on a marine jetty could be as follows:

- A jetty has a replacement value of $X and a probability of failure based on its condition of 75% within the next five years; that is a risk cost of $0.75X.
- As the asset owner cannot afford to replace the asset, it is considering two options:
  1. The repair and protection of all crossbeams, which would reduce the probability of failure down to 10% (a reduction in risk costs of $0.6X) for a cost of $0.3X (a 1:2 ratio between the sum spent and the risk reduction).
  2. The repair and protection of selected beams (including all critical ones) only, leading to a risk reduction of $0.5X (a probability of failure of 25%) for a cost of $0.165X (a ratio cost to risk reduction close to 1:3).
- As the organisation may only be able to fund $0.2X of repairs, it could select option (2) and use the balance of funds to apply protective treatment or protection to the elements in the next risk category.

10.6 REPAIR AND PROTECTION OPTIONS – TECHNICAL ASSESSMENT APPROACHES

10.6.1 General

Structure owners not only need to utilise costs assessment methods to enable decision on repair and protection approaches but also technical assessment methods are often necessary to arrive at a considered and balanced decision. Each option has its pros and cons and a structure owner cannot make a considered decision unless all options (including 'do nothing') are presented.

Chess and Green (2020) advise that decisions on remediation procedures for deteriorated concrete structures or elements thereof should be based on knowledge of:

- The cause(s) of deterioration.
- The degree and extent of deterioration.
- The expected progress of deterioration with time.
- The effect of the deterioration on structural behaviour and serviceability.
- The money available.
- The level of disruption to operations allowable.

Knowledge of where deterioration is occurring, together with information regarding the causes of deterioration, is always necessary before deciding on the most appropriate, or combinations of appropriate, remediation procedures (Chess & Green, 2020).

Other factors that need consideration include, but are not limited to (Chess & Green, 2020):

- Structure type.
- Element types.
- Future service life.

Technical assessment methods to assist with decisions of the types of, or combinations of, repair and protection methods can include: multicriteria type assessments (MCA) as referred to above (Collins & Blin, 2019); multi-criteria decision analysis (MCDA) methods; 'spider web' methods; scenario analyses (including the risks associated with each); etc. The scenario analysis approach is the focus below and example analyses for marine wharf sub-structure elements are provided.

As an asset owner, if advice on repair, and protection options is being sought, where it is sought from is also often of some consideration. Chess and Green (2020) comment that it is strongly recommended that conduct of such, including presentation of remediation options, be undertaken by independent consultants and not consultants that are conflicted nor by contractors (who are obviously conflicted) or organisations that wear multiple hats (e.g. consultant/supplier/installer).

10.6.2 'Do nothing' option

As advised at Section 8.3, Chess and Green (2020) point out that whenever a structure is being investigated all the available options (including 'do nothing') should be presented to the owner/stakeholder for consideration.

Chess and Green (2020) further note that in terms of the 'do nothing' option, that this option has led to successful long service lives of various reinforced concrete structures (or for large sections of structures) where the structure owners have not needed to undertake any repair and protection works for their 50 years and more of service life achieved thus far.

10.6.3 Scenario analyses approach

One means of deciding the most appropriate remediation approach(s) for a reinforcement corrosion affected structure, for example, is to present a scenario analysis of concrete repair and protection options (including the risks associated with each) to a structure owner. An example scenario analysis for marine wharf substructure elements is provided below, Table 10.1 (Chess & Green, 2020).

Chess and Green (2020) also note that it is often relevant to present to a structure owner a scenario analysis of other remedial options that are available together with the reasons they are not considered appropriate. An

Table 10.1 Marine concrete substructure reinforced concrete elements – example scenario analysis – preferred options

Remedial type	Comments	Risks
Do nothing, make safe, and monitor	• Deflection of deck/beams/pile movements may be the monitoring means • Diligent and regular inspection required • Make safe as required • Local structural failure likely including at 'lime leach cracks' • Local strengthening (e.g. fibre-reinforced polymer (FRP)) may be possible after deflection • Reactive strategy	• Localised structural failure • Reduction in structural capacity and acceptable operation loads • Re-application of local strengthening (at intervals dependent on strengthening material type) • Highest risk option
Patch repair and cracks repair	• Proprietary systems • Conventional techniques • Concrete breakout of deteriorated concrete • Breakout and repairs will need to be staged to not affect structural integrity • Reinforcement to be treated • Polymer modification and the composition of proprietary cementitious mortars leads to high performance properties (wet and physical) and low penetrability • Surface coating may be necessary • Cracks to be widened by saw cutting and breakout • Crack movement to be determined to ensure structural bond • Special design requirements to stop ongoing localised corrosion • Polymeric crack injection materials not to be used • Formation of incipient anodes at patch and crack repair locations (but can be eliminated with rebar coatings) • Low capital cost • Maintenance regime with monitoring required • Reactive strategy	• Risk of structural failure since reactive strategy • Regular structural inspections (every 1–3 years) after initial patch repair • Patch repairs will need to be ongoing at regular intervals (e.g. every 5–8 years) during life of structure • Re-repair of patch repairs may be necessary over a future berth service life (>25 years) • Future service life may become limited and less than desired

CP (Impressed current)	• Can completely halt corrosion if correctly designed (regardless of chloride content of concrete, extent, and rate of reinforcement corrosion, etc.) • Zoning, sub-zoning, current control, etc necessary given different micro-exposure environs (tidal, splash, atmospheric), etc. • Likely anode types are mesh, ribbon, discrete, or a combination thereof • Multi-channel transformer/rectifier (T/R) unit or distributed T/R units (power supplies) • Remote monitoring or remote monitoring and control systems can be incorporated • Reinforcement electrical continuity required • Less concrete breakout • Permanent solution • Routine maintenance (connections, conduits, T/R unit modules, etc) • Replacement of some components (impressed current systems) every 25 years or so (e.g. T/R units, conduits, junction boxes, etc) • Annual monitoring and reporting • High initial capital cost • Likely lowest life cycle cost • Ongoing cost to maintain	• Uneven current distribution, uneven protection, acidification issues, etc. if not correctly designed • Poor component reliability if not correctly designed and specified • Lowest ongoing risk option
CP (Galvanic)	• Trial would be required given resistivity of back of berth lower beam and front beam concrete • Al-Zn-In metal spray system likely because of higher drive voltage and current output and ease of touch-up • May not completely halt corrosion but will reduce rate • Still requires structural monitoring but at less frequent than Do Nothing case • Reinforcement electrical continuity required • Permanent solution • Annual monitoring and reporting • Routine maintenance commitment • Anode touch-up repair (and likely replacement) required every 8–12 years • Anode replacement every 8–12 years (sprayed metal system) • Annual monitoring and reporting • Lower initial capital cost than impressed current CP but likely higher life cycle cost than impressed current CP	• May not halt corrosion but will reduce rate of corrosion • Structural risk reduced compared to Do Nothing

Table 10.2 Marine concrete substructure reinforced concrete elements – example scenario analysis – other remedial options

Remedial type	Reasons not considered
Surface penetrants/ coatings	• Sufficient chlorides already present to initiate and propagate corrosion at lower section front beam and lower section back beam locations • Not likely to slow the rate of corrosion (dependent on corrosion kinetics) of lower section front beam and lower section back beam areas • No need to apply to other substructure beam and soffit areas as chloride build-up within cover concrete is insignificant
Corrosion inhibitors	• Contentious performance issues (extent of migration; timely migration; locale measurement; longevity; stability) • Accelerate corrosion if insufficient concentration or only partial coverage of the steel reinforcement • Excessive applications can cause chemical attack of concrete • Insitu performance monitoring necessary to confirm effectiveness • Limited performance data most particularly on marine structures • Mechanisms not understood
'Hybrid' treatment	• Anode system that is operated in impressed current mode initially (1–3 weeks) and then operates in galvanic mode • This system is not in the 'Concrete CP Standard' (AS 2832.5-2008) • May not completely halt corrosion but may reduce corrosion rate, therefore may not provide cathodic protection • Reinforcement electrical continuity required • Difficult to install (cylindrical type) if congested reinforcement • Not likely to provide adequate cathodic current given resistivity of lower section front beam and lower section back beam concrete and congested reinforcement • Recent system, no long term performance data in a marine environment • Lack of track record • Anodes not easy to find and replace (every 15 years or so) since buried in cover concrete • Anode replacement every 15 years or so (but may be less dependent on anode consumption rates) • Trials are recommended given likely corrosion activity of lower section front beam and lower section back beam reinforcement together with the resistivity of the concrete • Trials would need to be undertaken over a number of years so as to also confirm anode output decrease and extent of reinforcement corrosion rate reduction

Galvanic current devices	• Proprietary, discrete anodes (cylindrical, strip, tubular) embedded within cover concrete • Are not included in AS 2832.5-2008 because do not provide CP per se but rather only some corrosion rate reduction • Not likely to provide adequate cathodic current given resistivity of lower section front beam and lower section back beam concrete and congested reinforcement • Difficult to install if congested reinforcement • Anodes not easy to find and replace (every 5-15 years or so) since buried in cover concrete • Anodes will attract chloride ions into concrete and may be source of localised corrosion risk to reinforcement when anodes consumed • Trials would need to be undertaken over a number of years so as to confirm anode output decrease and extent of reinforcement corrosion rate reduction
Electrochemical chloride extraction (desalination)	• Reinforcement electrical continuity required • Application constraints • Ongoing maintenance costs (i.e. chloride barrier coating) • Not all chlorides can be removed • May cause alkali silica reaction (ASR) of concrete • Routine maintenance (e.g. touch-up, etc. of chloride barrier coating) • Local concrete patch repairs • Re-application every 10–15 years of chloride barrier coating
Replacement	• Impact on operations • Berth (or sections of berth) would need to be closed • High cost

example scenario analysis of non-preferred remedial options for marine wharf substructure elements is provided at Table 10.2.

REFERENCES

Angst, U and Büchler, M (2015), 'On the applicability of the Stern–Geary relationship to determine instantaneous corrosion rates in macro-cell corrosion', *Materials and Corrosion*, 66, No.10, 1017–1028.

Allan, AG, Leng, DL, Davison, N (2000), '*Sacrificial Anode Systems for Repair of Reinforced Concrete*', Proc. Corrosion and Prevention 2000 Conference, Paper 116 Auckland, New Zealand, November.

Ball, JC and Whitmore, DW (2008), '*Galvanic Protection for Reinforced Concrete Bridge Structures: Case Studies and Performance Assessment*', Proc. Corrosion and Prevention 2008 Conference, Paper 049, Wellington, New Zealand, November.

Bertolini, L, Elsener, B, Pedeferri, P and Polder, R (2004), '*Corrosion of Steel in Concrete*', Wiley Weinheim.

Broomfield JP (1997), '*Corrosion of Steel in Concrete*', Spon, London, UK.

Broomfield, JP (2017), 'Up-to-date overview of repair & protection aspects for reinforced concrete', in *Reinforced concrete corrosion, protection, repair and durability*, Eds W K Green, F G Collins and M A Forsyth, ISBN 978-0-646-97456-9, Australasian Corrosion Association Inc, Melbourne, Australia.

BSI (2016), BS EN ISO 12696, '*Cathodic Protection of Steel in Concrete*', British Standards Institute, London.

Chess, PM (2019), '*Cathodic Protection for Reinforced Concrete Structures*', CRC Press, Boca Raton, FL, USA.

Chess, P. and Green, W. (2020), '*Durability of Reinforced Concrete Structures*', CRC Press, Boca Raton, FL, USA.

Christodoulou, C and Kilgour, R (2013), '*The World's First Hybrid Corrosion Protection Systems for Prestressed Concrete Bridges*', Proc. Corrosion & Prevention 2013 Conf., Australasian Corrosion Association Inc., Brisbane, Australia, November, Paper 076.

Christodoulou, C, Corbett, P and Coxhill, N (2016), '*Service Life Extension of State Highway Bridges – New Zealand's First Hybrid Corrosion Protection Application*', Proc. Corrosion & Prevention 2016 Conf., Australasian Corrosion Association Inc., Auckland, New Zealand, November, Paper 65.

Collins, F. and Blin, F. (2019), '*Ageing of Infrastructure*', CRC Press, Boca Raton, FL, USA.

Concrete Society (2011), '*Cathodic protection of steel in concrete*', Technical Report 73, Surrey, UK.

Dodds, W, Christodoulou, C and Goodier, C (2017), 'Hybrid corrosion protection – An independent appraisal', *Proceedings of the Institution of Civil Engineers-Construction Materials*, 171 (4), 149–160.

Elsener, B (2002), 'Macrocell corrosion of steel in concrete – Implications for corrosion monitoring', *Cement & Concrete Composites*, 24, 65–72.

Flitt, H, Will, G and Green, W (2015), '*Electrokinetic Review and Appraisal of Steel Reinforcement Corrosion in Concrete*', Proc. Corrosion & Prevention 2015 Conf., Australasian Corrosion Association Inc., Adelaide, November, Paper 16.

Giorgini, R and Papworth, F (2011), '*Galvanic Cathodic Protection System Complying with Code Based Protection Criteria*', Concrete Institute of Australia Biannual Conference, Perth, Western Australia, October.

Glass, GK, Roberts, AC and Davison, N (2007), 'Treatment Process for Concrete', UK Patent GB2426008B, November.

Glass, GK, Roberts, AC and Davidson, N (2008), 'Hybrid corrosion protection of chloride-contaminated concrete', *Proceedings of Institution of Civil Engineers, Construction Materials*, 161, 4, 163–172.

Glass, GK, Davidson, N and Roberts, AC (2016), '*Hybrid Corrosion Protection of Reinforced Concrete Structures*', Proc. Corrosion & Prevention 2016 Conf., Australasian Corrosion Association Inc., Auckland, New Zealand, November, Paper 164.

Green, W, Dockrill, B and Eliasson, B (2013), '*Concrete repair and protection – overlooked issues*', Proc. Corrosion and Prevention 2013 Conf., Australasian Corrosion Association Inc., November, Paper 020.

Holloway, L, Green, W, Karajayli, P and Birbilis, N (2012), '*A review of sacrificial and hybrid cathodic protection systems for the mitigation of concrete reinforcement corrosion*', Proc. Corrosion & Prevention 2012 Conf., Australasian Corrosion Association Inc., November, Paper 062.

Holmes, S, Christodoulou, C and Glass GK (2013), '*Monitoring the Passivity of Steel Subject to Galvanic Protection*', Proc. Corrosion & Prevention 2013 Conf., Australasian Corrosion Association Inc. November, Brisbane, Australia.

Hudson, D (1998), 'Current developments and related techniques' in '*Cathodic Protection of Steel in Concrete*' Ed Paul Chess, E & FN Spon, London, UK.

NACE (2005), 'Sacrificial Cathodic Protection of Reinforced Concrete Elements – A State of the Art Report', Publication 01105, Task Group 047, Houston, USA.

NACE (2016), 'Sacrificial Cathodic Protection of Reinforcing Steel in Atmospherically Exposed Concrete Structures', Standard Practice 0216, NACE International, Houston, USA.

Page, CL and Treadaway, KWJ (1982), 'Aspects of the electrochemistry of steel in concrete', *Nature*, 297, 109–115.

Preston, J, Wyatt, BS, and Spring, IT (2017), 'Cathodic protection of reinforced concrete: Is there anything still to learn?', in *Reinforced Concrete Corrosion, Protection, Repair, and Durability*, (Eds) W K Green, F G Collins, and M A Forsyth, ISBN 978-0-646-97456-9, Australasian Corrosion Association Inc., Melbourne, Australia.

Sergi, G and Page, C L (1999), '*Sacrificial anodes for cathodic protection of reinforcing steel around patch repairs applied to chloride-contaminated concrete*', Proc. Eurocorr '99, European Corrosion Congress, Aachen, Germany.

Sergi, G (2009), '*Ten year results of galvanic sacrificial anodes in steel reinforced concrete*', Proc. Eurocorr '09, Nice, France.

Savcor Products Australia (2020), '*PatchGuard Ultra*', Technical Data Sheet.

Standards Australia (2008), 'AS 2832.5 Cathodic Protection of Metals Part 5: Steel in Concrete Structures', Sydney, Australia.

Standards Australia, SA HB 84: 2018, 'Guide to concrete repair and protection', Sydney, Australia.

Thompson, L, Godson, I and Heath, J (2012), '*Introduction to Design Techniques Used for Hybrid Corrosion Protection Systems*', Proc. Corrosion & Prevention 2012 Conf., Melbourne, Australia, November, Paper 148.

Whitmore, D, Simpson, D, Beaudette, M and Sergi, G (2019), 'Two-stage, self-powered, modular electrochemical treatment system for reinforced concrete structures', *Corrosion & Materials*, August, 54–58.

Chapter 11

Preventative measures

11.1 INTRODUCTION

Very many durability problems can be avoided at the design stage if the environmental conditions are appropriately taken into account and appropriate preventative measures applied at the construction stage. This chapter is therefore concerned with the prevention of corrosion by measures that can be taken before or during construction rather than protection from corrosion that has already initiated.

Concrete technology options present themselves including the availability of different binder types, aggregates, water content and admixtures in terms of mix design (mix selection) durability. The simple 'five Cs (pentagon of Cs)' of concrete construction durability are presented, namely: Cover; Constituents; Consistency; Compaction; and Cure.

Coating, penetrant, overlay, and membrane treatments to concrete are possible preventative measures. Coated and alternate reinforcement options are available durability measures. Permanent corrosion monitoring and integrated health monitoring (IHM™) can be incorporated to increase confidence of preventative measures achievement.

11.2 CONCRETE TECHNOLOGY ASPECTS

11.2.1 General

It is beyond the scope of this book (and the competences of the authors) to detail concrete technology factors in terms of preventative measures to contribute to concrete durability achievement. As there are few concrete technologist's that reside nowadays within principal organisations, within consultancies, or within contracting organisations, concrete technology competences are more often than not with the concrete producers and

construction material supply organisations. This is not necessarily a negative but more a recognition that only some concrete technology considerations in terms of preventative measures can be conveyed here (and that are
in addition to the concrete technology considerations already conveyed in
Chapter 1).

It is considered fair to say that many of the problems that arise later in the
life of a reinforced concrete structure would have been avoided if the appropriate precautions had been taken at the design and construction stage, in
particular, for example, it is a good idea to ensure that:

- The appropriate concrete was chosen.
- The concrete was well constituted.
- The concrete had been properly placed around the reinforcement.
- The concrete was properly cured.
- Coatings were applied as necessary.

11.2.2 Mix design/mix selection

Concrete mix design nowadays is almost solely vested with the concrete
producers or specialist construction material supply organisations. Gone are
the days it would seem when concrete mix design was able to be undertaken
in the laboratory by people within principal or consultancy organisations.

At Section 1.4.1, mix design is considered to involve the specification of
the relative amounts of cement (binder), coarse aggregate, fine aggregate,
and water, and has four major objectives which apply to nearly all structures. The wet freshly mixed concrete must be sufficiently workable that it
can be placed in position, and this often involves the ability to penetrate the
small gaps between reinforcing bars. The cured concrete must have a
required strength. The cured concrete must crack where it is permissible and
for the cracks to be of a surface width sufficient to not compromise durability. The cured concrete must be sufficiently impenetrable (impermeable) to
the ingress of aggressive agents that its durability is assured. These qualities
of the concrete mix, in their different ways, are affected by the concrete mix
design and also by the conditions under which it is placed in position, by the
curing conditions, and by additions that may be incorporated into the mix
to modify its properties.

Neville (2011) believes that *mix selection* rather than *mix design* is a more
appropriate descriptor for the selection of concrete mix ingredients. In the
British usage, the selection of the mix ingredients and their proportions is
referred to as *mix design*. This term, according to Neville (2011), although
common, has the disadvantage of implying that the selection is a part of the
structural design process. Neville (2011) then states that this is not correct
because the structural design is concerned with the required performance of

concrete, and not with detailed proportioning of materials that will ensure that performance.

Neville (2011) indicates that although the structural design is not normally concerned with mix selection, the design imposes two criteria for this selection: strength of concrete and its durability. He then considers it important to add an implied requirement to the effect that the workability *must* be appropriate for the placing conditions and that it applies not only to, say, slump at the time of discharge from the mixer but also to a limitation on slump loss up to the time of placing of concrete.

In addition, Neville (2011) advises that the selection of mix proportions has to take into account the method of transporting the concrete, especially if pumping is envisaged. Other important criteria he considers to be are: setting time, extent of bleeding, and ease of finishing; and that these three are interlinked. Considerable difficulties can arise, he advises, if these criteria are not properly taken into account during the selection of the mix proportions or when adjusting these proportions.

The selection of mix proportions should thus, simply, be the process of choosing suitable ingredients of concrete and determining their relative quantities with the object of producing as economically as possible concrete of certain minimum properties, notably strength, durability, and a required consistency (Neville, 2011).

11.2.3 Mix selection process

Some basic factors which have to be considered in the mix selection process are represented schematically at Figure 11.1.

It can be seen, that mix selection requires both a knowledge of the properties of concrete and experimental data or experience (Neville, 2011).

11.2.4 Binder types

A particularly effective preventative measure concrete mix selection wise, is the improvement in durability associated with the so-called 'Supplementary Cementitious Materials', or SCMs, and the consequent use of blended cements. As indicated at Section 1.4, the three most common SCMs are:

- Fly ash, complying with for example AS 3582.1 (Standards Australia, 2016a).
- Ground granulated iron blast furnace slag, complying with for example AS 3582.2 (Standards Australia, 2016b).
- Amorphous silica, complying with for example AS 3582.3 (Standards Australia, 2016c).

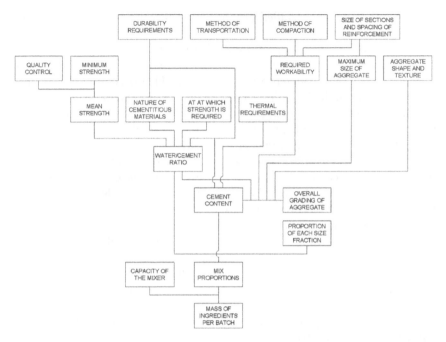

Figure 11.1 Basic factors in the mix selection process. (Courtesy of Neville, 2011, p.730)

Chloride durability is increased markedly with the use of slag and fly ash blended cement based concretes, triple blend cement based concretes and even quaternary blend cement based concretes.

Alkali aggregate reaction (AAR) deterioration can be managed with the use of blended cements.

One of the means to reduce the risk of early age thermal contraction cracking to manageable levels, is the use of blended cement based concretes, most particularly high (60–70%) slag.

Chemical and biological deterioration of concrete is reduced by the use of blended cement based concretes.

11.2.5 Water/cement (water/binder) ratio

Obviously, the more water that is added to the concrete mix, the easier it is to place or pour, but Section 2.3 Figure 2.10, reproduced here as Figure 11.2, showed the variation in permeability of a cement with the water/cement (w/c) ratio.

Since the penetrability (permeability) of the hardened cement (binder) paste is a controlling factor of the ability of the aggressive agents to reach reinforcing steel, a balance has to be struck between a mix which is

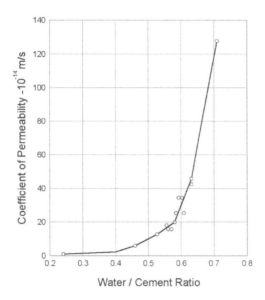

Figure 11.2 The relationship between water permeability and water/cement ratio for a cement paste. (Courtesy of Neville, 1975, p.386)

placeable and one which is sufficiently impenetrable to confer the necessary durability.

11.2.6 Concrete strength

The compressive strength is probably the most commonly measured property of a cured concrete. It may not in fact be the most important characteristic of the concrete as its durability or penetrability may play a greater role in the determination of the concrete's utility. However, the compressive strength gives a general indication of the properties of the cement paste and methods for its determination are well developed e.g. Australian Standard AS 1012.9 (Standards Australia, 2014). The strength of the resultant concrete depends upon the mix proportions.

When concrete is fully compacted, its strength is taken to be inversely proportional to the w/c ratio. This relationship was preceded by the rule established by Abrams in 1919 where he found strength to be equal to (Neville, 2011):

$$f_c = K_1 / K_2^{w/c} \tag{11.1}$$

Where K_1 and K_2 are empirical constants. The general form of the strength versus w/c ratio curve is shown at Figure 11.3.

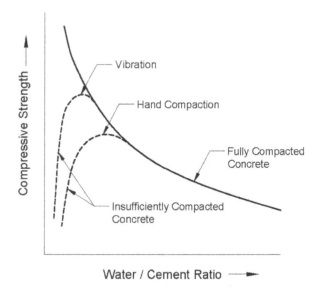

Figure 11.3 The relationship between strength and the water/cement ratio for a concrete. (Courtesy of Neville, 2011, p.272)

Abrams' rule, although established independently, is similar to a general rule formulated by Feret in 1896 where the strength of concrete is related to the volumes of water and cement (Neville, 2011):

$$S = K(c/c + e + a)^2 \qquad\qquad (11.2)$$

In this equation S is the strength of the concrete, c the volume of the cement, e the volume of the water and a, the volume of the air in the concrete. This rule is based upon the amount of void space left in the concrete. The strength of the resultant concrete is therefore markedly affected by the degree of compaction, that is, the extent to which any entrained air is removed.

The major factors controlling concrete strength are the w/c (w/b) ratio and cement content, the aggregate type, the cement type, the compaction, and the curing conditions. The compressive strength as well as the durability decreases as the w/c (w/b) ratio increases and the compressive strength is often a good indicator of whether an appropriate w/c (w/b) ratio has been used.

As can be seen from Equation 11.2, increasing the cement content increases strength and generally, the higher the compressive strength, the greater is the durability and the impenetrability.

Equation 11.2 also shows that an increase in void content due to insufficient compaction, that is 'a', results in a decrease in compressive strength.

Because the voids (capillary pores) also provide easy passage for aggressive agents through the cured concrete they also affect the durability. Consequently, adequate vibration after placing is a vital requisite for the preparation of good concrete. Figure 11.3 also illustrates the vital importance of good compaction, either by manual action, or more likely by vibration, if the concrete is to develop the strength and impenetrability of which it is capable.

11.3 CONSTRUCTION CONSIDERATIONS

11.3.1 General

The Concrete Institute of Australia has developed the Durability Series, a set of Recommended Practice documents, that provide deemed to satisfy requirements applicable to all concrete structure types based on standard input parameters for design life, reliability, and exposure.

Recommended Practice Z7/04 (Concrete Institute of Australia, 2014) of the series titled 'Good Practice Through Design, Concrete Supply, and Construction' has applicability to more general concrete design and construction as well as concrete requiring specifically higher levels of durability. Z7/04 provides specific detail covering areas such as: the impact of specifications and the contract process; impacts of design on construction; more detailed view of the materials used in construction; material quality control processes; construction process and supervision; as well as some detailing issues in common structural elements; that may present potential durability issues to the designer and constructor. In addition to this an appendix section is included on reinforcement spacers and chairs as this is an area that has demonstrated to cause weakness in durable construction and is rarely adequately specified.

Recommended Practice Z7/04 (2014) of the Concrete Institute of Australia Durability Series then further proposes that the designer and durability planner must understand not only the intended design but must understand the material properties and consider how these properties can be delivered during the construction process. They consider that there are many elements to this delivery process that impact on the final structure's durability and Document Z7/04 provides information that helps highlight the more critical areas of concern from design detailing through material supply to construction of the structure for all concrete construction stakeholders.

Recommended Practice Z7/04 (2014) also informs all parties involved in the design and construction about the benefits of durability planning and subsequent control of implementation so they can deliver the expected level of maintenance and life of the structure to the asset owner.

Further discussion of Durability Planning is provided at Chapter 12.

11.3.2 The 5 Cs/Pentagon of Cs

A phrase of the '5 Cs' or the 'Pentagon of Cs' has been coined to summarise some important simple construction considerations in terms of preventative measures for concrete durability, namely:

- **Cover:** the reinforcement in concrete must be correctly located if the design durability is to be achieved. Although considerable sums are often spent on durability design a common cause of premature reinforcement corrosion is low cover.

  Modelling using the approaches of Sections 5.4 and 5.5 calculates minimum cover required to achieve a stated design level at a stated reliability. Minimum covers often do not include a negative tolerance. To achieve minimum covers an appropriate construction allowance must be added to give a target cover for construction (Concrete Institute of Australia, 2014).

  The concept of minimum cover is generally taken as an absolute value but it is more realistic to consider cover as a probability distribution. Hence, minimum cover could be better specified as the characteristic cover, which must be used in the durability design (Concrete Institute of Australia, 2014).

  In an ideal world the entire surface of an element would be tested for cover compliance but like most quality assurance procedures this is not practical and a statistically based acceptance method has to be implemented (Concrete Institute of Australia, 2014).

  Minimum covers often may need to be increased, e.g. where concrete is cast against ground or where special finishes are applied to compensate for deleterious chemical actions or fracturing of the concrete (Concrete Institute of Australia, 2014).

  Based on design life being proportional to the square of the cover in many circumstances, increasing the cover could more than double the maintenance free service life of reinforced concrete structures. This highlights the importance of the correct choice and control of concrete cover to the long term durability of concrete structures (Concrete Institute of Australia, 2014).

  However, no quality control guidelines or compliance criteria for assuring adequate cover are given in standards or codes. Common practice is the checking of the gap between steel reinforcement and the formwork during a pre-pour inspection. The adequacy of such an approach to quality control of concrete cover can be gauged by the frequency of revelations of the lack of cover as the main cause of the premature deterioration due to steel corrosion of many concrete structures. In most cases, the achieved cover is significantly less than that specified indicating the problem to be caused by the lack of adequate quality control procedures (Concrete Institute of Australia, 2014).

Recommended Practice Z7/04 (2014) of the Concrete Institute of Australia Durability Series provides some: general considerations for improving the adequacy and consistency of concrete cover; recommendations to improve the compliance of reinforcement cover; and, some post pour cover quality assurance checks.

- **Constituents:** spend time and effort on optimising the constituents of the concrete. There are a wide array of materials and design options available to the concrete mix selector that are not discussed in Standards.

  As indicated previously, cement type has a considerable influence on mix design, transport, and constructability. The use of materials, such as fly ash, slag, and silica fume has long been known to improve concrete durability under certain conditions.

  The properties of the aggregate used in concrete can have a significant effect on its durability. The volumetric stability of the aggregate can influence the drying shrinkage and creep of the concrete. Poor shape can increase water requirements, with a subsequent increase in drying shrinkage and penetrability.

  Chemical admixtures are important to modify a number of properties of concrete (e.g. setting characteristics, water demand, air entrainment and chemistry) and reduce its cost.

  'Water must be clean, fresh and free from any dirt, unwanted chemicals or rubbish that may affect concrete' (Concrete Institute of Australia, 2014).

- **Consistency:** concrete should be ordered from a supplier who has been evaluated and selected on its capacity to produce concrete in accordance with the contract documents and deliver the concrete at a rate consistent with the project schedule.

  The concrete should be mixed in a properly designed mixer, and mixing should continue until there is a uniform distribution of the materials.

  Concrete should be transported from the place of mixing to its final position as quickly as is practicable and by means that will prevent segregation or loss of any part of the mix materials.

  Particular care should be taken on hot, dry weather, or windy conditions to prevent evaporation or drying associated slump regression.

  It is important for durability that concrete is delivered in a sufficiently workable state to ensure that it can be placed and that it is placed within a time frame that will ensure compaction is possible, compaction does not affect already developing material hydration bonds significantly and that vibration at placement interfaces is possible to ensure that cold joints do not form (Concrete Institute of Australia, 2014).

- **Compaction:** concrete should be thoroughly compacted by vibration or other suitable means and thoroughly worked around reinforcement and fixtures so as to form a solid mass with as minimal voids as practicably possible. Particular care should be taken to ensure that there is sufficient equipment with appropriate capacity to compact the concrete. The capacity will depend on the geometry of the concrete element, access to the element, restrictions imposed by reinforcement, the workability of the concrete, placement rate, and rate of supply.

Even if the mix is well designed, and the concrete properly cured, low penetrability will be achieved only if the concrete is properly compacted. Porous concrete resulting from unsatisfactory compaction can permit penetration of aggressive agents to the reinforcement, with consequent corrosion problems.

Proper compaction is essential to expel air entrapped in the concrete during placement, to consolidate the concrete without causing segregation, to reduce the risk of settlement cracking and to ensure good bond between concrete layers and between concrete and reinforcement.

Internal vibrators should not be used to move the concrete laterally. Care should be taken to ensure the reinforcement is not displaced by prolonged contact with such vibrators.

Thorough compaction is particularly important at edges and corners, where the concrete is exposed on more than one face. When concrete is placed in a number of layers, the internal vibrators should penetrate into the layer below that being placed.

When concrete is placed in high lifts, the discharge stream should avoid displacing the reinforcement. Where necessary, the concrete should be deposited in its final position through chutes or tremie pipes.

Particular care is required in areas of congested reinforcement, where placing, and compaction is difficult, as such areas can exhibit segregation or voids if not carefully compacted. Where plastic shrinkage or settlement cracks occur, these can often be closed by re-vibrating the concrete (Concrete Institute of Australia, 2014).

- **Curing:** curing of well constituted and compacted concrete is then an essential fifth step.

    There are three primary aspects to curing (Concrete Institute of Australia 2014), i.e.:
    1. Between screeding and finishing to prevent plastic shrinkage cracking and top down setting.
    2. After finishing to prevent interference with hydration due to water loss from the hardened concrete.
    3. Thermal curing to reduce temperature differentials.

    Curing should be maintained in place for at least the period where the concrete has attained the properties that are assumed in the design including durability properties of the layer of concrete protecting the reinforcement (Concrete Institute of Australia, 2014).

- **Curing After Placement**
  An effective means of minimising loss of water by evaporation is to spray exposed surfaces with an evaporation retardant such as aliphatic alcohol immediately after screeding and as required until the concrete is ready for finishing operations. If this is not achieved successfully, finishing procedures may close the top of plastic cracks such that they only become apparent sometime later (Concrete Institute of Australia, 2014).

- **Curing After Hardening**
  Some concrete members are cured by leaving the formwork in place, particularly on vertical surfaces, and the underside of horizontal members. After stripping the forms, curing can be continued by the application of an impermeable covering to minimise moisture loss. However, this may still not provide the same level of performance as water curing.

  Curing compounds (external and internal) may be the only practical method of curing some surfaces where early removal of formwork is required. Although curing compounds that meet, for example, the requirements of 90% water retention at 72 hours when tested in accordance with AS 3799 (Standards Australia, 1998) improve concrete quality significantly relative to no curing they still do not lead to the same performance of water curing.

  Curing of concrete takes place after all finishing processes are complete. Suitable methods of curing (in descending order of effectiveness) are (Concrete Institute of Australia, 2014):
  - Maintaining water saturation of concrete.
  - Covering the exposed surfaces of concrete with wet absorptive material and wrapping this with sealed plastic.
  - Covering with sealed plastic film.
  - Retaining formwork in place.
  - Applying a curing compound.

- **Thermal Curing**
  Thermal curing refers to control of the maximum temperature or temperature differential by curing methods. This might comprise use of insulation or cooling pipes and is discussed in Recommended Practice Z7/06 (Concrete Institute of Australia, 2017) of the Concrete Institute of Australia Durability Series and has been mentioned briefly previously at Section 2.2.4.

11.4 COATINGS AND PENETRANTS

11.4.1 General

A wide range of coatings and sealants may be applied to concrete surfaces in an attempt to improve the abrasion resistance, the corrosion resistance, the aesthetic appearance, or for a variety of other reasons. The material in this section will be primarily concerned with the various surface treatments that are applied with the objective of improving the corrosion resistance of the reinforcement in the concrete.

Such surface treatments are designed to decrease the rate at which chloride ions, carbon dioxide, and oxygen can enter the concrete. Surface treatments can in general be divided into four classes, refer also Figure 11.4:

- **Organic coatings:** These may be polymer emulsions that can be applied to the surface of the concrete and which key into the surface and produce films of greater or lesser thickness or they may be solvent borne coatings. These are organic coatings that polymerise on the surface of the concrete to produce heavy duty coatings.
- **Penetrants:** These may be solvent borne small molecules which are absorbed on the surface of the pores in the concrete and which because of their hydrophobic nature reduce the movement of water through the concrete. This in turn restricts the transport of aggressive species that might be carried by the water.

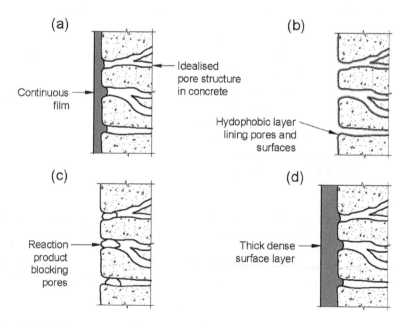

Figure 11.4 Surface treatments for concrete a) organic coating, b) hydrophobic penetrant, c) pore blocker, d) cementitious overlay. (Courtesy of Bertolini et al., 2013, p.245)

- **Pore blocking treatments:** Materials that penetrate the pores, react with the cement material and block the pores.
- **Cementitious overlays:** These are layers of mortar specially formulated to have a reduced penetrability in particular to chlorides and carbon dioxide.

Sheet membranes may also be considered a class of protective surface treatment of concrete. These may be applied to provide the durability required in harsh chemical environments. These often consist of thermoplastic sheets that have to be welded in situ or even when it is necessary to resist extreme chemical attack, acid resistant ceramic tiles.

11.4.2 Organic coatings

Water borne coatings are lattices of organic polymers in water generally described in AS/NZS 4548 'Guide to long-life coatings for concrete and masonry' (Standards Australia, 1999), for example. They are generally applied by a roller or by spray gun to surfaces that have been sealed for the following coats. The sealer is the first coat applied to a new concrete surface and is often a low solids content polymer solution that penetrates the pores of the concrete and provide moisture barriers in them. The second coat then often consists of a high solids emulsion of a similar polymer. These coatings are required to have sufficient elasticity to be able to bridge cracks and provide an adequate resistance to water permeability. They also act as anti-carbonation coatings and chloride ion barriers. Tests for all these properties are included in AS/NZS 4548.5 (Standards Australia, 1999), for example.

Solvent borne coatings are either reacting polymeric coatings such as epoxy or polyurethane resins or elastomers such as chlorinated rubber dissolved in an organic solvent. They are applied, usually over a sealant by spray gun or less often by roller to a thickness of 100 – 400 microns. Unlike the water borne coatings they tend not to penetrate the concrete but to provide a durable wear resistant coating on the surface.

11.4.3 Penetrants

Many of the hydrophobic penetrants are based upon silicon chemistry, the silanes which contain one silicon atom and one or more alkoxy groups and the siloxanes which are short chains of silicon atoms with attached alkyl groups (Polder et al., 1996), refer also Figure 11.5 (Bertolini et al., 2013).

These materials are often applied dissolved in an organic solvent, they enter the pores of the concrete and react with the hydrated silicates of the concrete producing alcohol and remaining bonded to the pore wall exposing alkyl groups (CH_3-) which are water repellent as is illustrated in Figure 11.6 (Polder et al., 1996) and Figure 11.7. (Bertolini et al., 2013)

The effect of hydrophobic treatment in reducing the transport of chloride ion into concrete is illustrated in Figure 11.8 (Polder et al., 1996), for example.

Figure 11.5 Silane and siloxane molecules. (Courtesy of Bertolini et al., 2013, p.252)

Figure 11.6 Chemical bond between the hydrophobic agent and concrete. (Courtesy of Polder et al., 1996, p.548)

Figure 11.7 Reaction of (methyl-methoxy) silane with a concrete substrate. (Courtesy of Bertolini et al., 2013, p.252)

More recently these materials have become available as thixotropic creams or gels so that they can be applied without spraying.

11.4.4 Pore blocking treatments

Pore blocking involves treating the surface with a material that penetrates the pores of the cement paste and then reacts with the calcium hydroxide within the pores to produce solid compounds that block the pores. Sodium silicate (water glass) was used originally, it reacted as:

$$Na_2SiO_3 + Ca(OH)_2 + CO_2 \rightarrow C-S-H + Na_2CO_3 \qquad (11.3)$$

Bertolini et al. (2013) notes that their penetration into concrete is superficial, except in the most porous concrete. Also, given a reaction with CO_2 is often involved, the alkali carbonates that are formed cause efflorescence. In

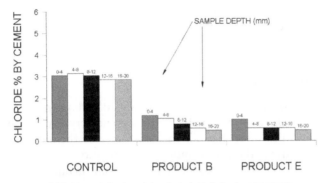

a) Chloride penetration in control and hydophobic Portland cement concrete
after 12 months weekly ponding

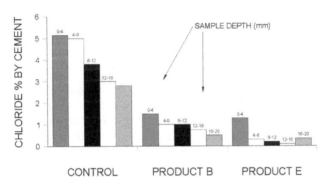

b) Chloride penetration in control and hydophobic blastfurnace slag cement
concrete after 12 months weekly ponding

Figure 11.8 **Effect of hydrophobic treatment on chloride penetration of laboratory concrete specimens exposed to wetting and drying cycles of salt solution (to simulate deicing salt application) for 52 weeks. (Courtesy of Polder et al., 1996, p.551)**

addition, Bertolini et al. (2013) comment that their effectiveness with regard to reducing water absorption is relatively poor because essentially they form C-S-H gel which is hydrophilic. On the other hand, pores are also blocked for water vapour present inside the concrete, so the risk of damage due to freezing may increase, in particular in porous, mechanically weak, concrete.

Aqueous solutions of metal-hexaflourosilicates can be applied to concrete surfaces, which form solid CaF_2, metal fluorides and silica (gel):

$$MgSiF_6 + Ca(OH)_2 \rightarrow CaF_2 + MgF_2 + SiO_2 + H_2O \qquad (11.4)$$

all of which contribute to filling up the pores (Bertolini et al., 2013).

Treatment with gaseous silicon fluoride SiF_6 that transforms parts of the concrete surface layer into CaF_2 has been applied in the past to concrete sewer pipes to increase their resistance to acid attack (Bertolini et al., 2013).

No test results in terms of the carbon dioxide or chloride penetration resistance of the inorganic pore blockers mentioned above are available (Bertolini et al., 2013).

11.4.5 Cementitious overlays

These are mortars usually fairly heavily modified with polymer lattices, such as acrylics, styrene butadiene rubbers, or styrene acrylic copolymers. They often contain silica fume that fills the gaps between the fine aggregate particles and so reduces the permeability to carbon dioxide. The materials are usually applied to thicknesses of 2 to 10 mm. One advantage of these overlays is that because they are more permeable to the small water molecule than to the larger carbon dioxide molecule, they allow the concrete to 'breathe'.

Two component, thixotropic, cementitious modified polymer coatings that are applied mms thick have been utilised for extended protection of reinforced concrete structures in saline environments (Lloyd, 2017).

11.4.6 Sheet membranes

Polymeric membranes with thicknesses up to 5 mm are used for the linings of chemical tanks or the facing (on the soil side) of underground structures potentially exposed to very aggressive groundwaters. These are in general not bonded to the concrete but may have a ribbed or even keyed surface against which concrete is poured so that they are held mechanically to it.

11.5 COATED AND ALTERNATE REINFORCEMENT

11.5.1 General

Although 'good' concrete (blended cement, low water/binder ratio, adequate binder content), well constituted, compacted and cured with an adequate cover is capable of providing sufficient durability for the design life of most structures, circumstances arise when additional protection is necessary for the reinforcement. A structure which has to be light may inevitably have inadequate cover. Connections may have to be made between plain reinforcing steel and decorative features exposed to the atmosphere. Cracking may be anticipated as the result of expected impacts. The structure may be exposed to high chloride or carbonating environments. In these circumstances when it is not possible to increase the quality of the concrete or the thickness of the cover additional protection in the form of a surface coating may be applied to the reinforcement. Three systems have been used with different levels of success to obtain enhanced durability under more or less extreme applications:

- Galvanising (coating with zinc).
- Painting (usually an epoxy coating).
- Stainless steel (protection by means of the resistant oxide film).

Metallic clad reinforcement is also available as is fibre-reinforced polymer (FRP) reinforcement.

11.5.2 Galvanised reinforcement

Galvanised reinforcement is readily available and has been used since the 1930s in structures around the world. Such structures in Australia include Parliament House in Canberra, the High Court in Canberra and the National Gallery of Australia in Canberra within the Australian Capital Territory.

Zinc is an amphoteric metal that is inherently corrosion resistant between pH 5.5 and 12.5, refer Figure 11.9.

When freshly galvanised reinforcing bar is embedded into wet concrete, generally with a pH about 13, a tightly adhered layer of calcium hydroxyzincate salts forms on the bar surface which inhibits further attack on the coating due to its passivating effect. This reaction consumes about 5–10 μm of the original zinc coating at the surface, refer Figure 11.10.

This passivating film layer isolates the zinc coating from the surrounding cement-rich matrix and once the concrete has hardened the reaction effectively ceases. The calcium hydroxyzincate layer so formed is quite stable and remains intact on the bar surface as long as the passivating conditions at the bar surface are maintained (Yeomans, 2018a, b).

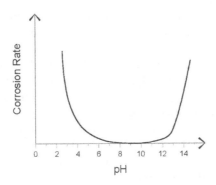

Figure 11.9 Effect of pH on the corrosion rate of pure zinc. (Courtesy of Bertolini et al., 2013, p.278)

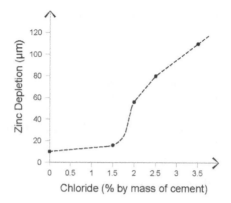

Figure 11.10 Depletion of the zinc coating of galvanised bars after 2.5 years of exposure in different chloride contaminated concretes (the initial thickness was 160 μm). (Courtesy of Bertolini et al., 2013, p.279)

Bertolini et al. (2013) note that the passivation of zinc depends on the pH of the pore solution. In contact with alkaline solutions, as long as the pH remains below 13.3, zinc can passivate due to the formation of calcium hydroxyzincate. Figure 11.9 shows the typical corrosion rate of zinc as a function of pH; however, even at pH values higher than 12, in the presence of calcium ions, such as in concrete pore solution, zinc can be passive, and has a very low corrosion rate. In saturated calcium hydroxide solution, it was found that for pH values up to about 12.8 a compact layer of zinc corrosion product forms, which will protect the steel even if the pH changes in a subsequent phase. For pH values between 12.8 and 13.3, larger crystals form that can still passivate the bar. Finally, above 13.3, coarse corrosion products of zinc form that cannot prevent corrosion of the underlying steel (Bertolini et al., 2013).

Bertolini et al. (2013) note further that the behaviour of galvanised steel may be influenced by the composition of the concrete, and especially by the cement type and its alkali content. In practice, however, the pH of the pore solution in concrete usually is below 13.3 during the first hours after mixing, due to the presence of sulphate ions from the gypsum added to the Portland cement as a set regulator. A protective calcium hydroxyzincate passive layer thus can be formed on galvanised bars.

The passive film/layer of galvanised rebars is stable even in a mildly acidic environment, so that the zinc coating remains passive even when the concrete is carbonated. The corrosion rate of galvanised steel in carbonated concrete is significantly lower than that of carbon steel and it remains negligible even if a low content of chloride is present (Bertolini et al., 2013).

In chloride contaminated concrete, galvanised steel may be affected by pitting corrosion, although the chloride threshold is somewhat higher than conventional steel (Bertolini et al., 2013). There is no universal agreement on the chloride threshold for galvanised steel reinforcement but a brief review of the literature indicates the following:

- Broomfield (2004) found that galvanised steel reinforcement resisted chloride levels in concrete at least 2.5 times higher than conventional steel reinforcement and delayed time to corrosion initiation by 4 to 5 times.
- Yeomans (2004) completed a review of laboratory and field data and determined that the evidence suggests the chloride threshold is several times higher than that of carbon steel reinforcement (2 to 4 times and up to ten times).
- Maldonado (2009) looked at the chloride threshold of galvanised steel reinforcement in concretes exposed to coastal conditions in the Caribbean. It was found that the galvanised reinforcement chloride threshold in these conditions was 2.6 to 3 times higher than carbon steel reinforcement.
- Darwin et al. (2007) found that galvanised steel reinforcement had a chloride threshold ranging from 0.04 to 0.17% by wt. concrete, with the average being around twice that of conventional steel reinforcement.
- Bertolini et al. (2013), refer Figure 11.10, has suggested an approximate critical chloride level in the range of 1–1.5% by mass of cement (~0.14–0.21 by wt. concrete) for galvanised reinforcement.
- Additionally, Yeomans (2018a, b) completed a review and determined that the evidence suggests the chloride threshold of galvanised steel reinforcement is 2 to 6 times higher than that of carbon steel reinforcement.

Bertolini et al. (2013) then comment that even if pitting corrosion has initiated, the corrosion rate tends to be lower for galvanised steel, since the zinc coating that surrounds the pits is a poor cathode and thus it reduces the effectiveness of the autocatalytic mechanism that takes place inside pits on bare steel (refer Section 4.7, for the latter).

11.5.3 Epoxy coated reinforcement

Epoxy coated rebar is used in the USA to provide additional protection in chloride aggressive environments. It is usually prepared by a powder coating method. The cleaned bars are pre-heated and then sprayed with a fine powder mixture of epoxy pre-polymer (a diglycidyl ether of bisphenol A) and a curing agent (amine or acid anhydride) and passed through a heated chamber where the curing reaction takes place to produce a tough polymer that is largely impermeable to chloride ions. The coating should be sufficiently thick to act as a barrier to the ingress of chloride ions but not so thick as to inhibit the curing process.

A prime requirement of the coating is that it is sufficiently flexible to withstand the bending of the bar without cracking and that bare metal such as cut ends is adequately coated. ASTM A775/A775M-07b (2014) covers the manufacture and performance of the bars. This standard requires that the thickness after curing should be 175–300 µm and should be sufficiently flexible to be bent around a mandrel (the diameter of which is specified for each class of bar) without cracking.

An intact epoxy coating delays the initiation of chloride corrosion but the major problem with this system is that any defects of the coating process or, more likely, damage caused by handling, or installation can shorten the life considerably and lead to serious failures. The Appendix to ASTM A775/A775M-07b (2014) describes in some detail the handling and storage procedures that should be adopted when using this material.

11.5.4 Stainless steel reinforcement

Although stainless steel is often avoided by design engineers because of the perceived cost disadvantages, an actual total lifetime costing may indicate that in particularly corrosive environments or when the cost occasioned by the necessity of closing the structure in order to effect repairs is large, the use of stainless steel may be the economic solution to the design. Selective use of stainless steel in more-corrosive locations, critical parts, outer section of the structure (skin reinforcement) is also possible.

The commonly used stainless steels described in the UK Concrete Society Technical Report No 51 (1998) are:

- Type 304 18.5 Cr 9.0 Ni 0.06 C
- Type 316 17 Cr 12 Ni 2.25 Mo 0.06 C
- Type 329 26 Cr 4.5 Ni 1.5 Mo 0.08 C

Type 304 is the most common austenitic stainless steel which is used in a wide variety of applications. Type 316 is the molybdenum modified version of type 304 and is more resistant to pitting attack by chloride ions. Type 329 is more commonly known as a 'duplex' stainless steel because it has a two phase (ferritic and austenitic) structure and is more resistant again to pitting attack by chloride ions.

In terms of types and grades of stainless steel rebars that are used in standards, Sussex (2017) makes two initial points about standards. When talking to materials engineers, grade usually means chemical composition. For design engineers, grade means a specific design strength which, mainly for reasons of continuity, has been aligned around the grade strengths of carbon steel bar. This possible confusion is generally avoided by referring to the different compositions as type or by providing the material designation.

Secondly Sussex (2017) makes the point that while usually the switch from Euronorm to ASTM/SAE designations is routine, a comparison of the composition limits shows that the BS 6744 (2016) is aiming for greater strength from a nominal material, e.g. it lists a high nitrogen (0.12–0.22%) version of 304 compared to the maximum of 0.10% in ASTM A955 (2014b) for 304 and the unspecified norm of about 0.05%. The UK standard also requires the 2.5–3% version of 316/316L whereas ASTM A995 (2014b) allows molybdenum to be as low as 2.00%. However, while both standards list alloys commonly used for reinforcement, they both allow other alloys subject to verification of corrosion resistance – and both provide normative block test methods which are protracted but realistic (Sussex, 2017).

The common use table in BS 6477 (2016) has six austenitic alloys ranging from '304LN' to a 6% Mo super austenitic and four duplex alloys ranging from utility 2101 to super duplex that are available as stainless steel rebars for concrete. These duplex alloys all require tests to confirm the absence of deleterious phases. There are no ferritic grades nor FeCrMn austenitics available (Sussex, 2017).

The common use table in ASTM A955/A955M (2018) has an austenitic alloy (XM-28) with more than 11% manganese and low levels of nickel. Because manganese is less effective in maintaining the ductile austenitic structure, the chromium content is lower than for 304 but the corrosion resistance is slightly enhanced by high nitrogen levels. Their inclusion appears to be due to requests from State Transportation bodies in the United States who had commissioned test work and were concerned about material costs during the period of high nickel prices (Sussex, 2017). The list of materials in common use also shows austenitics 304, 316L and 316LN and duplex (2205 and 2304). There are no ferritic alloys listed although there has been corrosion test work on 12% chromium alloys (Sussex, 2017).

As a major stainless steel producer, India (Sussex, 2017) has a stainless steel reinforcement standard which lists austenitics (304, 304 LN, and 316Mo2.5–3.0%) plus duplexes LDX2101, 2304, and 2205 as well as a 11–13.5% chromium ferritic (410 L) (IS 16651, 2017). There are four strength grades based on 0.2% proof stress, i.e. 500 MPa, 550 MPa, 600 MPa, and 650 MPa (IS 16651, 2017).

Other, non-standard but authoritative documents include the *fib* Bulletin 49 (2009) which has a thorough review of stainless steel reinforcement including details of international standards at the date of publication and summaries of the data and discussion of 14 different investigations of chloride resistance.

In terms of the corrosion resistance of stainless steels in concrete, all types of stainless steel reinforcement are passive in carbonated concrete (Bertolini et al., 2013).

In chloride contaminated concrete, stainless steel reinforcement can suffer pitting corrosion like carbon steel reinforcement. For stainless steels, like for carbon steel, the susceptibility to pitting attack can be expressed in terms of both the pitting potential (i.e. for a given chloride content, the potential value above which pitting can initiate) or the critical chloride content (i.e.

for a given potential, the threshold level of chloride content for the onset of pitting). These two parameters are interrelated and depend on the chemical composition and microstructure of the steel, on the surface condition of the bars and on the properties of the concrete. Because of the higher stability of the passive film of stainless steel compared with carbon steel, their resistance to pitting corrosion is much higher. This is due to the formation of chromium and nickel oxyhydroxides in contact with the alkaline pore solution of concrete (Bertolini et al., 2013).

As a result, stainless steel bars have a much higher chloride threshold level compared to carbon steel bars. The chloride threshold for stainless steel reinforcement may be 5–10 times higher than that of conventional carbon steel reinforcement dependent on stainless steel type (Concrete Institute of Australia, 2015).

However, Bertolini et al. (2013) notes that even small variations in the chemical composition, thermomechanical treatment or the surface condition may significantly affect the corrosion resistance of stainless steel bars in chloride bearing concrete. Therefore, the chloride threshold should be measured for any specific type of stainless steel and its variability should also be evaluated. Unfortunately, the evaluation of chloride threshold for steel in concrete is rather difficult and there are no standardised or generally accepted methods for its evaluation (Bertolini et al., 2013), refer also to Section 4.7.6.

In order to give an indicative picture of the order of magnitude of the chloride content at which some stainless steel bar types can resist corrosion initiation, Figure 11.11 depicts fields of applicability in chloride contaminated concrete exposed to temperatures of 20°C or 40°C (Bertolini et al., 2013).

Bertolini et al. (2013) points out that these values are indicative only, since the critical chloride content depends on the potential of the steel, and thus it can vary when oxygen access to the reinforcement is restricted as well as when macrocells or stray currents are present. For instance, they indicate that the domains of applicability are enlarged when the free corrosion potential is reduced, such as in saturated concrete. Furthermore, the values of the critical chloride level for stainless steel rebars with surface finishing other than pickling can be lower.

Bertolini et al. (2013) further comment that since the pH for noncarbonated concrete is around 13, while in carbonated concrete it is near 9, the right-hand side of the graphs of Figure 11.11 is representative of alkaline concrete and the left-hand side of carbonated concrete. In alkaline concrete, austenitic 304L (1.4307) can safely be used in concrete up to 5% chloride by mass of cement (i.e. ~0.7% chloride by mass of concrete), and 316L (1.4404) and 2205 (1.4462) duplex stainless steel even higher than 5%, that is, for chloride contents that are rarely ever reached in the vicinity of the steel surface.

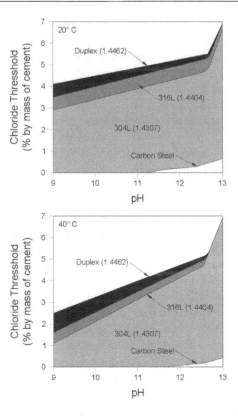

Figure 11.11 Schematic representation of fields of applicability of different stainless steel bars (pickled) in chloride bearing environments for 20°C and 40°C. The threshold levels are indicative only. They can decrease if oxides produced at high temperature, for example, welding or during manufacturing, are not completely removed, or the potential due to anodic polarisation increases (e.g., due to stray current) or the concrete is heavily cracked. Conversely, they can increase when there is a lack of oxygen or cathodic polarisation. (Courtesy of Bertolini et al., 2013, p.272)

In the presence of a welding scale (heat tint, heat affected zone/HAZ) on the surface of the reinforcement the critical chloride content is lowered perhaps to approximately 3.5% by mass of cement (i.e. ~0.5% chloride by mass of concrete for 304L) and the same reduction takes place if the surface is covered by the black scale formed at high temperature during thermomechanical treatments (Bertolini et al., 2013). Pickling of rebars is then more efficient in pushing up the critical chloride content than say sand blasting, which does completely free the surface from the oxide scale (Bertolini et al., 2013).

In carbonated concrete, or in the case where the concrete is extensively cracked, the critical chloride contents are remarkably lower. The more highly alloyed stainless steels should be preferred in these more aggressive conditions (Bertolini et al., 2013).

The critical chloride content also decreases as the temperature increases; for instance Figure 11.11 shows the expected variations between 20°C and 40°C. Thus, the more highly alloyed stainless steel bars should be preferred in hot climates (Bertolini et al., 2013).

Often, the use of stainless steel reinforcement is limited to the outer part (skin reinforcement) or to its most critical parts for economic reasons. Furthermore, when stainless steel bars are used in the rehabilitation of corroding structures, they are usually connected to the original carbon steel rebars (Bertolini et al., 2013).

In terms of examples of selective use of stainless steel reinforcement in structures, the concrete of (the 100 year design life) McGee Bridge in Tasmania, Australia (Figure 11.12) is mainly reinforced with carbon steel to save costs but in the tidal zone it uses 316 stainless steel for the near surface reinforcement. The submerged reinforcement in the piles is carbon steel because the saturated concrete has very low oxygen content. In the superstructure, low penetrability concrete controls chloride access (Sussex, 2017).

The Gateway Bridge (300 year design life) in Brisbane, Queensland, Australia generally has pickled LDX 2101 stainless steel reinforcement only near the surface where saltwater contact is possible but the deeper bars have more than sufficient cover to the carbon steel. The only exception is the ship fenders around the base of the piers (Figure 11.13) because they are in the intertidal zone with the possibility of surface concentration of chlorides – and impact damage from ships – so all their reinforcement is pickled LDX 2101 stainless steel (Sussex, 2017).

Figure 11.12 McGee Bridge (100-year design life) over tidal river, Tasmania, Australia. (Courtesy of Sussex, 2017)

Figure 11.13 Gateway Bridge (300-year design life), Brisbane, Queensland, Australia, stainless steel rebar for intertidal zone elements. (Courtesy of Sussex, 2017)

Connal and Berndt (2009) describe in more detail the durability approaches adopted for the 300 year design life Second Gateway Bridge in Brisbane, Queensland, Australia. The project scope and technical requirements (PSTR) for the bridge specified that durability assurance be applied diligently and continuously throughout the process of design, construction and throughout the maintenance period, and that the Second Gateway Bridge have a design life of 300 years, with some replaceable sub-items having design lives ranging from 20 years (wearing course) to 100 years (bearings). Design life was defined as the period assumed in design for which the structure or structural element is required to perform its intended purpose without replacement or major structural repairs. The philosophy adopted in meeting the extended design life was based on 'building in' the required durability at the outset, where feasible, and minimising the need to take measures later in the life of the bridge to achieve a 300 year service life. Integral with this philosophy was the appropriate selection of high quality materials chosen to address the particular durability issues that are posed by the range of exposure conditions.

The reinforced concrete pile caps of the bridge were in a tidal/splash exposure zone whereby the Brisbane River was determined to have a chloride concentration up to 18,000 ppm which is similar to that of seawater. The concrete mix proposed for the pile caps was a 50 MPa grade ternary blend consisting of 30% fly ash and 21% blast furnace slag. The total cementitious content was 560 kg/m^3 and the maximum water/cementitious material ratio was 0.32. The proposed pile cap design was to have 150 mm minimum cover to black steel reinforcement and 75 mm minimum cover to stainless steel reinforcement. A period of 280 years was then selected as the required time to corrosion initiation, following which corrosion, and spalling may take place over a subsequent 20 year period, resulting in a 300 year service life before major repairs (Connal & Berndt, 2009)

Predicted chloride ingress profiles (deterministic chloride diffusion modelling) indicated that 150 mm cover to carbon steel and 75 mm cover to LDX

2101 or 316LN stainless were likely to provide adequate protection and prevent initiation of corrosion within 280 years. The LDX 2101 had cost savings compared with 316LN and was therefore the favoured stainless steel (Connal & Berndt, 2009).

A probabilistic approach to modelling of chloride ingress was also adopted and this determined that the target reliability index ($\beta = 1.28$) is met over 300 years for 150 mm cover to black steel reinforcement steel with S50 ternary blend concrete even if the coefficient of variation for depth of cover is 20% (however, it was recommended that tight construction quality control be implemented to reduce the coefficient of variation). The reliability index calculations also predicted that the S50 ternary blend concrete with 75 mm cover to pickled LDX 2101 stainless steel would achieve the required life and that it was also noted that the predicted corrosion rate of stainless steel will be significantly lower than that for black steel and this further enhances design for durability (Connal & Berndt, 2009).

Selective use of stainless steel reinforcement begs the question as to whether there is a risk of galvanic corrosion of carbon steel reinforcement induced by coupling with stainless steel reinforcement. Experimental studies clearly show that the use of stainless steel in conjunction with carbon steel does not increase the risk of corrosion of the carbon steel. When both carbon steel and stainless steel rebars are passive and embedded in aerated concrete, macrocell action does not produce appreciable effects, since the two types of steel have almost the same corrosion potential. Indeed, in this environment, carbon steel is even slightly nobler than stainless steel. In any case, both carbon steel and stainless steel remain passive even after connection (Bertolini et al., 2013).

Only when the carbon steel corrodes does the macrocell current become significant. However, stainless steel is a poor cathode. Figure 11.14 shows that the consequences of coupling corroding carbon steel reinforcement with stainless steel reinforcement are generally modest, and they are negligible with respect to those coupling with passive carbon steel that always surround the corroding area. Consequently, the increase in corrosion rate on carbon steel reinforcement embedded in chloride contaminated concrete due to galvanic coupling with stainless steel reinforcement is significantly lower than the increase brought about by coupling with passive carbon steel reinforcement (Bertolini et al., 2013).

Bertolini et al (2103) confirm this behaviour by examination of the cathodic polarisation curves in saturated calcium hydroxide ($Ca(OH)_2$) solution (pH 12.6) of 316L stainless steel compared with carbon steel, refer Figure 11.15. Stainless steel is seen to be a less efficient cathode and a higher overvoltage for the cathodic reaction of oxygen reduction is necessary for stainless steel with respect to carbon steel.

Bertolini et al. (2013) note, however, that stainless steel with welding oxide (heat tint, HAZ) or with the black scale formed at high temperature is a better cathode and increased galvanic corrosion can occur in the presence

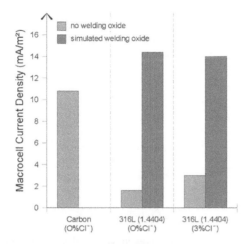

Figure 11.14 Macrocell current density exchanged between a corroding bar of carbon steel in 3% chloride contaminated concrete and a (parallel) passive bar of: carbon steel in chloride-free concrete, 316L stainless steel in chloride-free concrete or in 3% chloride contaminated concrete (20°C, 95% RH). Also results of stainless steel bars with oxide scale produced at 700°C (simulating welding scale) are reported. (Courtesy of Bertolini et al., 2013, p.274)

of these types of scale in stainless steel bars. However, in evaluating the effect of galvanic coupling, at least in the case of welds, it has to be considered that the area covered by the scale will normally be small compared to the total rebar area. The presence of these types of scale increases the macrocell current density generated by stainless steels, to the same order of magnitude or even greater than that produced by coupling with carbon steel (Figure 11.14), as a consequence of a change in the cathodic behaviour (Figure 11.15).

Sussex (2017) notes that cutting, bending, welding, cleaning, and packing of stainless steel reinforcement should ideally be carried out off site in controlled conditions. The stainless steel must be clean and dry before wrapping and transport requires non-ferrous materials including the truck tray and tie down cables. If the stainless steel is not to be used immediately, it should be stored in dry conditions, preferably out of the sun and protected from grime and damage until needed.

He goes on further to advise that there is a regrettable tendency for construction sites to consider stainless steel as invulnerable and, unless there is strong supervision, allow it to get dirty or greasy, walked on, sprayed with carbon steel grindings (Figure 11.16), smeared with carbon steel by using the same tools as for galvanised steel, splashed with acids used to remove excess mortar or concrete, overheated during cutting or grinding ... and the list goes on. Stainless steel is an expensive commodity with a carefully prepared protective surface which is very robust but can be degraded by ignorance whether wilful or through lack of training (Sussex, 2017).

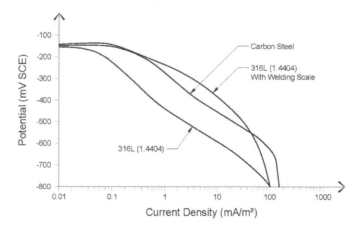

Figure 11.15 Cathodic polarisation curves in saturated Ca(OH)$_2$ solution (pH 12.6) of 316L (1.4404) stainless steel compared with curve of carbon steel. Results of stainless steel with simulated welding scale are also shown. (Courtesy of Bertolini et al., 2013, p.275)

Figure 11.16 Carbon steel grindings on site will degrade corrosion resistance. (Courtesy of Sussex, 2017)

The care and maintenance of stainless steel on site starts with incoming inspection and Figure 11.17 shows the requirements of the Ontario Ministry of Transportation, for example, to verify that the bar has been correctly pickled (Sussex, 2017). They regard A and B as acceptable but C has iron tint probably from rinse water and D appears to have been overpickled (Sussex, 2017).

Overpickling roughens the surface and has been associated with low chloride threshold (critical chloride level) results in some early lean duplex measurements (Sussex, 2017).

Figure 11.17 Site inspection of pickled rebar with iron tint (C) and overpickling (D). (Courtesy of Sussex, 2017)

Sussex (2017) advises any on site welding or grinding will have degraded the surface and to obtain the expected corrosion performance, these areas must be repickled. This will remove:

- Potential carbon steel contamination.
- Iron rich unprotective heat tint.
- Chromium depleted layer underneath the heat tint.
- In areas that have simply been ground, dissolve any exposed manganese sulphide inclusions that are the inevitable part of all steels including stainless steels (Standard bar stock tends to have higher sulphur content than sheet or plate although still well within standard material specifications).

Each of these four features can and will act as corrosion initiating points if exposed to the chlorides somewhat below the chloride threshold level, i.e. the stainless steel will locally act as if it is a lesser alloy (Sussex, 2017).

11.5.5 Metallic clad reinforcement

Stainless steel clad rebar (SCR) with a carbon steel core is commercially available. SCR types include 316L stainless steel, 316LN stainless steel, and 2205, and 2101 duplex grades. Investigation of the corrosion resistance of SCR dates back to at least the late 1990s/early 2000s (e.g. Hurley et al., 2001).

Copper-clad reinforcing was discussed as a corrosion resistant reinforcing bar as early as 1979. To date, it is understood that no significant structures have been constructed using these materials; however, several long term laboratory tests have been conducted showing that copper-clad bars may be a viable option for corrosion protection (McDonald et al., 1996).

Hot dipped aluminium as a coating for reinforcing bars has been investigated (Saremi et al., 2003) but the in-field success and performance is not known.

11.5.6 Non-metallic reinforcement

Non-metallic reinforcing bars, such as fibre-reinforced polymer (FRP) bars, which are manufactured from composite materials consisting of fibres embedded in a polymer matrix (resin) are commercially available. Glass (GFRP), aramid (AFRP) and carbon (CFRP) are the three most commonly used fibres (Cement & Concrete Association of New Zealand, 2009).

The advantages and disadvantages of FRP reinforcement are listed in American Concrete Institute (ACI) 440.1R-06 (2006) which is reproduced as Table 11.1 below.

FRP bars have been used in concrete close to sensitive electronic equipment (such as hospital MRI rooms), in sea walls, foundations, chemical

Table 11.1 Advantages and disadvantages of FRP reinforcement

Advantages of FRP reinforcement	*Disadvantages of FRP reinforcement*
High longitudinal tensile strength (varies with sign and direction of loading relative to fibres)	No yielding before brittle rupture
Corrosion resistance (not dependent on a coating)	Low transverse strength (varies with sign and direction of loading relative to fibres)
Nonmagnetic	Low modulus of elasticity (varies with type of reinforcing fibre)
High fatigue endurance (varies with type of reinforcing fibre)	Susceptibility of damage to polymeric resins and fibres under ultraviolet radiation exposure
Lightweight (about 1/5th to ¼ the density of steel)	Low durability of glass fibres in a moist environment
Low thermal and electric conductivity (for glass and aramid fibres)	Low durability of some glass and aramid fibres in an alkaline environment
	High coefficient of thermal expansion perpendicular to the fibres, relative to concrete
	May be susceptible to fire depending on matrix type and concrete cover thickness

Source: ACI (2006)

plants, reinforced shotcrete lined tunnels and masonry walls in corrosive environments (Cement & Concrete Association of New Zealand, 2009).

Although the use of FRP reinforcing bars has been steady since the mid-1980s, their wider adoption has been limited, most probably due to the absence of standard design codes. There have also been some concerns about the lack of ductility and fire resistance of the bars (Cement & Concrete Association of New Zealand, 2009).

However, these issues have been addressed by ACI Committee 440 (2006), whose mission it is too develop and report information on FRP reinforcement of concrete. As a result, FRP bars could become an increasingly common method of enhancing durability of concrete structures (Cement & Concrete Association of New Zealand, 2009).

11.6 PERMANENT CORROSION MONITORING

Permanent corrosion monitoring is an aspect of the integrated health monitoring (IHM™) of structures including reinforced concrete structures. It has predominantly taken the form of embeddable reference electrodes to measure the electrode potential of reinforcement as well as to a lesser extent, probes to measure the corrosion rate of reinforcement by the polarisation resistance (R_p) method.

Schiessel & Raupach (1992) first proposed a graduated series of probes, in a ladder arrangement ('Schiessel's Ladder Probe') to measure the ingress of chlorides or carbonation into cover concrete. The 'ladder steps' each comprise a carbon steel element and a galvanic corrosion current is measured between each steel element and noble cathode (stainless steel or platinised titanium) element located nearby. A reference electrode may also be incorporated to enable measurement of steel element potentials.

The 'Schiessel Ladder', by its distribution of steel elements at different depths of cover from the concrete surface allows the progress of chloride ingress or carbonation to be monitored. Other macrocell or galvanic current probes have since been developed and an example of some is shown at Figure 11.18.

These probes are incorporated into the new build of structures at an inclination to the surface down to rebar depth or at different heights within the cover concrete. As the chloride (or carbonation) front advances, each carbon steel element activates and the current flow from the anodic steel element and the cathodic element (stainless steel or mixed metal oxide titanium) can be monitored along with the potential. Periodic measurements are necessary in order to track the shifting chloride or carbonation front thereby providing valuable advance warning of the onset of corrosion across the structure. Obviously, these devices would need to be used in sufficient numbers at strategic locations across the structure in order to provide appropriate information for long term maintenance planning.

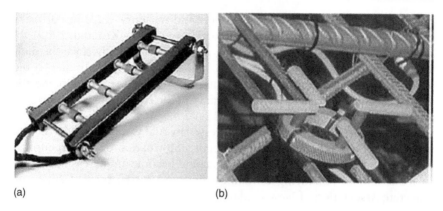

(a)   (b)

Figure 11.18 Examples of embeddable permanent macrocell or galvanic current probes

The corrosion rate of the carbon steel elements as well as the reinforcement can also be measured on some probes by the R_p method. Concrete resistivity measurements are also possible as is remote monitoring, if necessary.

If the macrocell, galvanic current, or ladder probes are properly interrogated and with appropriate data interpretation, notice can be given to a structure owner that corrosion is about to be initiated at the reinforcement. The extent of this notice should be sufficient to enable appropriate preventative corrosion control measures such as coating/penetrant treatment, cathodic prevention, or cathodic protection to be implemented.

Long term monitoring also means that the progression of changes can be determined. The growth of anodic areas, potential changes, changes in corrosion rates are all more helpful in predicting long term durability of a structure rather than a 'snap shot' approach of a condition survey, for example.

REFERENCES

ACI (2006), *Guide for the Design and Construction of Structural Concrete Reinforced with FRP Bars*, ACI 440.1R-06, American Concrete Institute, Farmington, MI, USA.

ASTM (2014), *Standard Specification for Epoxy-Coated Steel Reinforcing Bars*, A775/A775M-07b, American Society of Testing Materials, West Conshohocken, Pennsylvania, USA.

ASTM (2018), *Standard Specification for Deformed and Plain Stainless Steel Bars for Concrete Reinforcement*, A955/A955M, American Society of Testing Materials, West Conshohocken, Pennsylvania, USA.

Bertolini, L, Elsener, B, Pedeferri, P, Redaelli, E and Polder, R (2013), *Corrosion of Steel in Concrete Prevention, Diagnosis, Repair*, Second Edition, Wiley-VCH Verlag GmbG & Co. KGaA, Weinheim, Germany.

Broomfield, J P (2004), 'Chapter 9 – Galvanized steel reinforcement in concrete: A consultant's perspective', in *Galvanized Steel Reinforcement in Concrete*, (Ed) S R Yeomans, Elsevier Science, 271–272.

BSI 6477 (2016), *Stainless Steel Bars – Reinforcement of Concrete – Requirements and Test Methods*, British Standards Institute, London, UK.

Cement & Concrete Association of New Zealand (2009), 'FRP Reinforcement for Durable Concrete', *Concrete Newsletter*, 53, 2, July.

Concrete Institute of Australia (2014), *Z7/04 Good Practice Through Design, Concrete Supply and Construction*, Recommended Practice, Concrete Durability Series, Sydney, Australia.

Concrete Institute of Australia, (2015), *Z7/07 Performance Tests to Assess Concrete Durability*, Recommended Practice, Concrete Durability Series, Sydney, Australia.

Concrete Institute of Australia (2017), *Z7/06 Concrete Cracking and Crack Control*, Recommended Practice, Concrete Durability Series, Sydney, Australia.

Concrete Society (1998), '*Guidance on the use of stainless steel reinforcement*', Technical Report No 51, Camberley, UK.

Connal, J and Berndt, M (2009), '*Sustainable Bridges – 300 Year Design Life for the Second Gateway Bridge, Brisbane*', *Proc. 7th Austroads Bridge Conference*, Auckland, New Zealand.

Darwin, D, Browning, J, O'Reilly, M and Xing, L (2007), '*Critical Chloride Corrosion Threshold for Galvanized Reinforcing Bars*', The University of Kansas Center for Research, Inc., USA.

fib (2009), '*Corrosion Protection of Reinforcing Steels*', Bulletin 49, International Federation of Structural Concrete, Launsanne, Switzerland, February.

Hurley, MF, Scully, JR and Clemena, G (2001), '*Selected Issues in the Corrosion Resistance of Stainless Steel Clad Rebar*', Proc Corrosion 2001 Conference, NACE International, Paper 01646.

IS 16651 (2017), 'High Strength Deformed Stainless Steel Bars and Wires for Concrete Reinforcement – Specification', Bureau of Indian Standards, New Delhi, July.

Lloyd, C (2017), 'Concrete and chlorides: Extending the durability of reinforced concrete', Concrete, April, 29–31.

Maldonado, L (2009), 'Chloride threshold for corrosion of galvanized reinforcement in concrete exposed in the mexican Caribbean', *Materials and Corrosion*, 60, 7, 536–539.

McDonald, D B, Virmani, YP and Pfeifer, DF (1996), 'Testing the Performance of Copper-Clad Reinforcing Bars', Concrete International, November.

Neville, A M (1975), *Properties of Concrete*, Second Edition, Pitman International, London, UK.

Neville, A M (2011), *Properties of Concrete*. Fifth Edition, Pearson Education Limited, Essex, England.

Polder, R B, Borsje, H and de Vries, H (1996), 'Hydrophobic treatment of concrete against chloride penetration', in *Corrosion of Reinforcement in Concrete*, (Eds) C L Page, P B Bamforth and J W Figg, The Royal Society of Chemistry, Cambridge, UK, 546–555.

Saremi, M, Baharvandi, H R and Tafaghoudi, A (2003), '*Aluminium as a Protective Coating on Steel Rebar in Concrete*', *Proc. Eurocorr Conference*, Budapest, Hungary, 28 September–2 October.

Schiessel, P and Raupach, M (1992), 'Monitoring System for the Corrosion Risk of Steel in Concrete', Concrete International, July, 52.

Standards Australia (1998), 'AS 3799 Liquid membrane-forming curing compounds for concrete', Sydney, Australia.

Standards Australia (1999, Reconfirmed 2013), 'AS/NZS 4548.5 Guide to long-life coatings for concrete and masonry – Part 5: Guidelines to methods of test', Sydney, Australia.

Standards Australia (2014), 'AS 1012.9 Methods of testing concrete Compressive strength tests - Concrete, mortar and grout specimens', Sydney, Australia.

Standards Australia (2016a), 'AS 3582.1 Supplementary cementitious materials for use with Portland and blended cement – Fly ash', Sydney, Australia.

Standards Australia (2016b), 'AS 3582.2 Supplementary cementitious materials for use with Portland and blended cement – Slag – Ground granulated iron blast-furnace slag', Sydney, Australia.

Standards Australia (2016c), 'AS 3582.3 Supplementary cementitious materials for use with Portland and blended cement – Amorphous silica', Sydney, Australia.

Sussex, G A (2017), 'Stainless steel reinforcement performance in concrete', in *Reinforced Concrete Corrosion, Protection, Repair & Durability*, (Eds) W K Green, F G Collins & M A Forsyth, ISBN: 978-0-646-97456-9, Australasian Corrosion Association Inc. Melbourne, Australia.

Yeomans, S R (2004), 'Chapter 1 – Galvanized steel in concrete: An overview', in *Galvanized Steel Reinforcement in Concrete*, (Ed) S R Yeomans, Elsevier Science, 1–7.

Yeomans, S R (2018a), '*Galvanized Steel Reinforcement*', Proc. 2018 fib Congress 2018, Paper 38, Melbourne, Australia, 7–11 October.

Yeomans, S R (2018b), *Galvanized Reinforcement in Concrete Structures Questions and Answers*, Edition 2.0, Melbourne, Australia.

Chapter 12

Durability planning aspects

12.1 SIGNIFICANCE OF DURABILITY

All engineering materials deteriorate with time, at rates dependent upon the type of material, the severity of the environment, and the deterioration mechanisms involved.

In engineering terms, the objective is to select the most cost-effective combination of materials to achieve the required design life. In doing so, it is critical to realise that the nature and rate of deterioration of materials is a function of their environment.

Accordingly, the environment is a 'load' on a material as a force is a 'load' on a structural component. It is the combination of the structural load and environment load in synergy which determines the performance of the structural component.

Durability planning is a system which formalises the process of achieving durability through appropriate design, construction, and maintenance.

The advantages of durability planning are many, including, but not limited to:

1. Increased likelihood of achievement of design life of structures and buildings.
2. Reduced life cycle costs.
3. Reduced maintenance and repair costs.
4. Establishment of predictable maintenance actions and costs.
5. Initiation of accountable maintenance management.
6. Minimisation of down-time in operations.
7. Prediction of materials performance in their service environments.
8. Establishment of a continuous link (through-train) in durability objectives between design, construction, and maintenance.

12.2 DURABILITY PHILOSOPHY

The protective measures to be adopted for structures/elements within a project depend on the risk of deterioration, the cost of preventative measures, the feasibility and cost of remedial actions, and ongoing preventative maintenance. It is typical that these need to be balanced to arrive at the best whole-of-life cost and optimised value for money.

A holistic project-through durability philosophy is depicted in the flow chart shown at Figure 12.1.

12.3 PHASES IN THE LIFE OF A STRUCTURE

The service life of a structure can most certainly be maximised with minimal expenditure (operating and/or capital) (Chess & Green, 2020).

The phases in the life of a reinforced concrete structure are summarised at Figure 12.2.

Definitions of some of the terminology used in Figure 12.2 are thus considered warranted, as follows:

- **Birth Certificate:**
  A document, report, or technical file (depending on the size and complexity of the structure concerned) containing engineering information formally defining the form and condition of the structure immediately after construction (Concrete Institute of Australia, 2014).

  The birth certificate should provide specific details on parameters important to the durability and service life of the structure concerned (e.g. cover to reinforcement, concrete penetrability, environmental conditions, quality of workmanship achieved, etc.) and the basis upon which future through-life performance should be recorded (Concrete Institute of Australia, 2014).

  The framework laid down in the birth certificate should provide a means of comparing actual behaviour/performance with that anticipated at the time of design of the structure (Concrete Institute of Australia, 2014).

  The birth certificate should facilitate ongoing (through-life) evaluation of the service life which is likely to be achieved by the structure.
- **Condition Assessment:**
  A process of reviewing information gathered about the current condition of a structure or its components, its service environment, and general circumstances, whereby its adequacy for future service may be

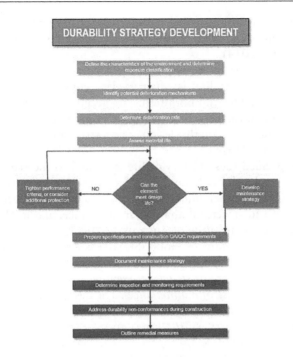

Figure 12.1 Project-through durability philosophy

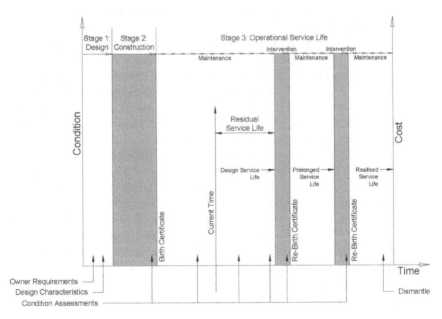

Figure 12.2 Phases in the life of a reinforced concrete structure. (Courtesy of *fib* 65, 2012a)

established against specified performance requirements for a defined set of loadings and/or environmental circumstances (Concrete Institute of Australia, 2014).

- **Design Life or Design Service Life (specified):**
  The term 'design life' is often used to convey the same intent as 'design service life' and both terms are acceptable to convey the same intent, namely, the period in which the required performance shall be achieved and as used in the design of new structures (Concrete Institute of Australia, 2014).

  The specified (design) service life is related to the required service life as given by the stakeholders (i.e. owners, users, contractors, society) and to the other implications of service criteria agreement (e.g. with regard to structural analysis, maintenance, and quality management) (Concrete Institute of Australia, 2014).

  Australian Standard AS 3600 (Standards Australia, 2018) for concrete structures design, for example, defines design life as the *'Period for which a structure or structural member is to remain fit for use for its designed purpose of maintenance'*. The design life in AS 3600:2018 applicable to plain, reinforced, and prestressed concrete structures and members is 50 years ± 20%. It is then noted that *'More stringent requirements would be appropriate for structures with a design life in excess of 50 years (e.g. monumental structures), while some relaxation of the requirements may be acceptable for structures with a design life of less than 50 years (e.g. temporary structures)'* (Concrete Institute of Australia, 2014).

  Australian Standard AS 5100.5 (Standards Australia, 2017) for concrete bridges design, for example, defines design life as *'The period assumed in the design for which a structure or a structural element required to perform its intended purpose with minimal maintenance and without replacement or major structural repairs'*. The design life in AS 5100.5:2017 applicable to plain, steel-fibre reinforced, and steel-reinforced and prestressed concrete structures and members is 100 years (rather than the 50 years of AS 3600:2018 for example). It is then noted in AS 5100.5 (2017) that *'More stringent requirements may be appropriate for structures with a design life in excess of 100 years (for example, monumental structures or high-risk significant structures crossing major waterways), while some relaxation of the requirements may be acceptable for structures with a design or service life of less than 40 years (for example, temporary structures)'* (Concrete Institute of Australia, 2014).

- **Re-Birth Certificate:**
  A document, report, or technical file similar to the birth certificate for a structure, but related to the information and circumstances associated with a project for the repair/remediation/refurbishment of the structure or part thereof to extend its anticipated service life (Concrete Institute of Australia, 2014).

- **Service Life (Operational):**
  The period in which the required performance of a structure or structural element is achieved, when it is used for its intended purpose, and under the expected conditions of use. It comprises design service life and prolonged service lives (Concrete Institute of Australia, 2014).

- **Service Life (Required):**
  The stakeholders' (i.e. owners, users, contractors, society) stated period in which the required performance shall be achieved after construction (Concrete Institute of Australia, 2014).

- **Service Life (Residual):**
  The remaining period in which the required performance shall be achieved from current time until the design service life is achieved (Concrete Institute of Australia, 2014).

In terms of other service life descriptions, according to Huang et al. (2019), ACI 364.1R (2019) describes three types of service life for existing concrete structures, namely:

1. **Technical Service Life:**
   The time in service until a defined unacceptable state is reached, such as spalling of concrete, unacceptable safety level, or failure of the element. This includes when structural safety is unacceptable due to material degradation or exceeding load-carrying capacity, or severe material degradation, such as extensive corrosion of steel reinforcement or excessive deflection under service load due to decreased stiffness.

2. **Functional Service Life:**
   The time in service until the structure no longer fulfils the functional requirements or becomes obsolete due to change in functional requirements, such as when there exists a need for increased clearance, higher axle and wheel loads, or road widening, for example; or, the aesthetics become unacceptable, for example, when excessive corrosion staining appears; or the functional capacity of the structure is no longer sufficient, such as a football stadium with insufficient seating.

3. **Economic Service Life:**
The time in service until replacement of the structure or part of it is more economical than keeping it in service, such as when maintenance requirements exceed available resource limits or replacement is needed to improve economic opportunities; for example, by replacing an existing parking garage with a larger one due to increased demand.

12.4 OWNER REQUIREMENTS

If durability planning is adopted as an integral part of a project, whether it be a new build or an existing structure, then the asset owner can specify their requirements and also ensure that there is a continuous link in durability objectives from design through construction and into operation and maintenance.

Asset owner key durability planning purpose and benefits are (Concrete Institute of Australia, 2014):

- Intended design life of asset is achieved with more confidence.
- Premature damage risk is reduced during construction and design life.
- Improved confidence in the technical capability of the design and contractor team via complementary durability consultant input.
- Maintenance requirements are better taken into account during design and construction with durability related matters having greater input into operation and maintenance manuals. A proactive or reactive maintenance strategy can be adopted with economic impact shown (Figure 12.3). The proactive strategy normally requires frequent expenditure resulting in an asset that stays in good condition throughout its operational service life whereas a reactive strategy is accompanied by major rehabilitation/repair costs when poor condition is reached.
- Whole-of-life asset cost is optimised. Early project formal durability input is more cost-effective than being introduced during detailed design or construction.
- At the end of the structure or the end of design life, condition is acceptable to the asset owner and capable of prolonged service life with maintenance.

12.5 DESIGNER REQUIREMENTS

Where durability planning is an integral part of a project, whether it be new construction or service life extension of an existing structure, the Concrete Institute of Australia Durability Planning Recommended Practice Z7/01 document (Concrete Institute of Australia, 2014) indicates that designer key durability planning purpose and benefits are:

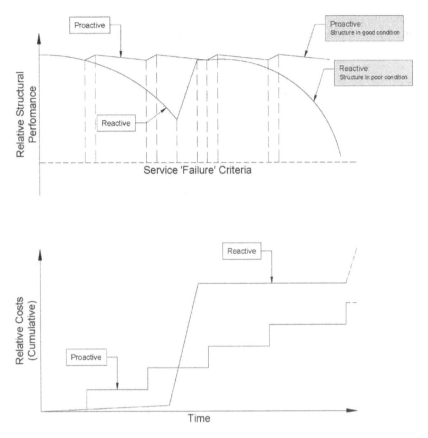

Figure 12.3 Relative costs for reactive and proactive maintenance. (Courtesy of *fib*, 2008)

- Durability input is provided that is not available from Standards and Codes, which lack durability certainty.
- Durability input is provided that is typically not available from structural and architectural design staff and is delivered in a complementary process.
- Construction materials selected by designers are more likely to achieve durability performance during design life.
- Designer durability-related mistakes or negligence are reduced.
- The design is more likely to achieve the asset owner-intended design life (as legally required in the designer agreement with the asset owner) without premature durability related damage during construction and the designer's project defects liability period.
- Design completed with first time appropriate durability input will provide cost-benefit savings for the designer (e.g. less re-design staff time and litigation claims).

A durability process that could be adopted during the design phase of a new structure or the repair of an existing structure is summarised in flowchart form at Figure 12.4 (Concrete Institute of Australia, 2014).

12.6 CONTRACTOR REQUIREMENTS

Where durability planning is an integral part of a project, whether it be new construction or service life extension of an existing structure, the Concrete Institute of Australia Durability Planning Recommended Practice Z7/01 document (Concrete Institute of Australia, 2014) indicates that contractor key durability planning purpose and benefits are:

- Durability input provided that is not available from Standards and Codes. Project specifications require construction in accordance with these Standards and Codes, which lack durability certainty.
- Durability input provided that is typically not available from structural and architectural design staff, and hence the contractor uses the durability consultant on request during construction.
- Construction materials used by the contractor are more likely to achieve durability performance during design life.
- Contractor durability-related mistakes or negligence are reduced.
- Construction is more likely to achieve the asset owner's intended design life (as legally required in the contractor agreement with the asset owner) without premature durability-related damage during construction and the contractor's project defects liability period.
- Construction completed with first-time appropriate durability input will provide cost-benefit savings for the contractor (e.g. less construction defects, project completed without delays, less defects after construction and litigation claims).

A durability process that could be adopted during the construction phase of a new structure or the repair of an existing structure is summarised in flowchart form at Figure 12.5 (Concrete Institute of Australia, 2014).

12.7 OPERATOR/MAINTAINER REQUIREMENTS

Maintenance is an integral part of ensuring durability through the life cycle of a structure, whether the structure is new or existing. A maintenance plan/asset management plan can be developed for a structure and can detail aspects such as (Concrete Institute of Australia, 2014):

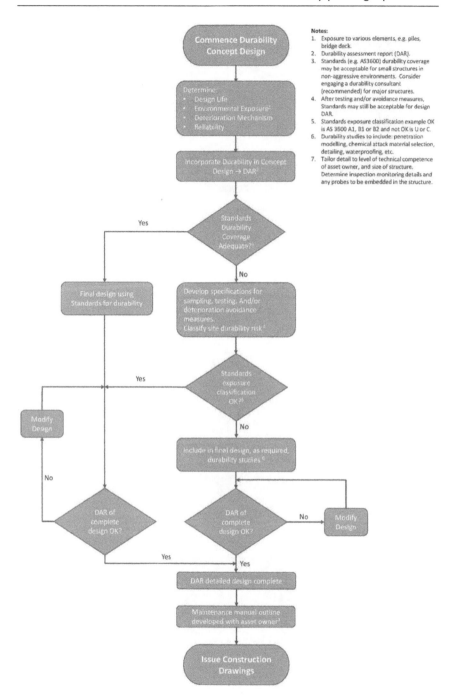

Notes:
1. Exposure to various elements, e.g. piles, bridge deck.
2. Durability assessment report (DAR).
3. Standards (e.g. AS3600) durability coverage may be acceptable for small structures in non-aggressive environments. Consider engaging a durability consultant (recommended) for major structures.
4. After testing and/or avoidance measures, Standards may still be acceptable for design DAR.
5. Standards exposure classification example OK is AS 3600 A1, B1 or B2 and not OK is U or C.
6. Durability studies to include: penetration modelling, chemical attack material selection, detailing, waterproofing, etc.
7. Tailor detail to level of technical competence of asset owner, and size of structure. Determine inspection monitoring details and any probes to be embedded in the structure.

Figure 12.4 Durability process during design. (Courtesy of Concrete Institute of Australia, 2014)

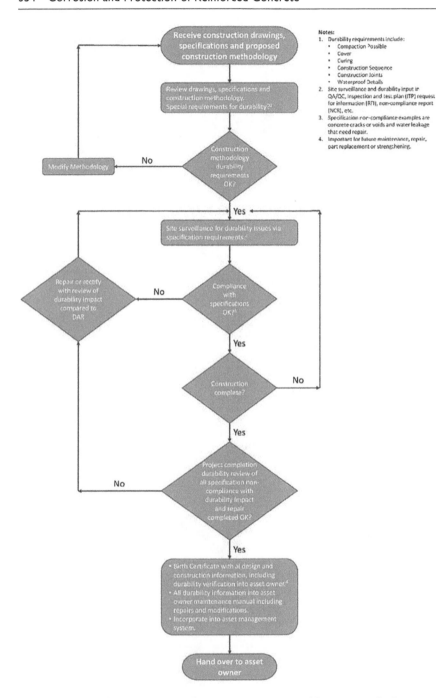

Figure 12.5 Durability process during construction. (Courtesy of Concrete Institute of Australia, 2014)

- Details of asset management inspection audits, including life cycle programme of tests to be undertaken, assessment criteria, and actions to be taken related to the inspection and test results.
- Maintenance materials and methods to be adopted throughout the service life to achieve optimum whole-of-life cost.
- Maintenance and asset management planning based on materials deterioration expectations and costs to achieve acceptable asset owner performance predicted at appropriate intervals but not exceeding 10 years.
- Identify repairs during maintenance audits to prevent further deterioration that may result in major problems.
- Minor repairs completed within required time schedule.
- System for keeping records of deterioration and repairs throughout structure life, including repair materials used.

A durability process that could be adopted during the service life (operation and maintenance) phase of a new structure or the repair of an existing structure is summarised in flowchart form at Figure 12.6 (Concrete Institute of Australia, 2014).

12.8 LIMIT STATES

A limit state is a condition of a structure beyond which it no longer fulfils the relevant design criteria. The condition may refer to a degree of loading (e.g. structural and environment) or other actions on the structure, while the criteria refer to structural integrity, fitness for use, durability, or other design requirements.

The Concrete Institute of Australia Recommended Practice document Z7/01 on Durability Planning (2014) defines three limit states, namely:

- **Durability Limit State (DLS):**
    A limit state used to define the end of the service life of a structure or structural member. The limit state may be a condition, performance, or operational limit state.

    For example, for reinforced concrete structures subject to deterioration caused by corrosion of reinforcing steel, one of more of the following DLS levels may be used to define the end of the service life:
    1. Depassivation of the reinforcing steel (or initiation of corrosion).
    2. Cracking of cover concrete.
    3. Spalling of cover concrete.

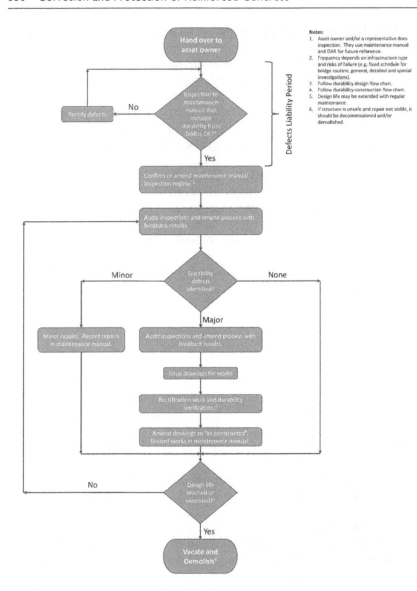

Figure 12.6 Durability process during service life. (Courtesy of Concrete Institute of Australia, 2014)

4. Loss of bond, loss of anchorage, loss of section (and reduced structural capacity).
5. Collapse of the structure.
• **Serviceability Limit State (SLS):**
A state that corresponds to conditions beyond which specified service requirements for a structure or structural member are no longer met.

- **Ultimate Limit State (ULS):**
  A state associated with collapse or with other similar forms of structural failure. Generally, the ULS corresponds to the maximum load-carrying resistance of a structure or structural member.

12.9 SERVICE LIFE DESIGN

Huang et al. (2019) advise that documents published by professional organisations define 'service life' from different perspectives and consequently there are different approaches taken to service life design.

ISO 16204 (2012), according to Huang et al. (2019), considers service life as a SLS, refer previous, and combines the principles of *fib* Bulletin 34 (2006), *fib* Bulletin 55 (2010), *fib* Bulletin 65 (2012a), and *fib* Bulletin 66 (2012b) into an international standard. A flowchart of structure service life prediction analysis specified by ISO 16204 (2012) is shown at Figure 12.7.

Huang et al. (2019) advise that *fib* Bulletin 34 (2006) predicts service life in four steps:

1. Quantify the deterioration mechanism with realistic models describing the process with sufficient accuracy.
2. Define the boundary conditions for which the structure should be designed.

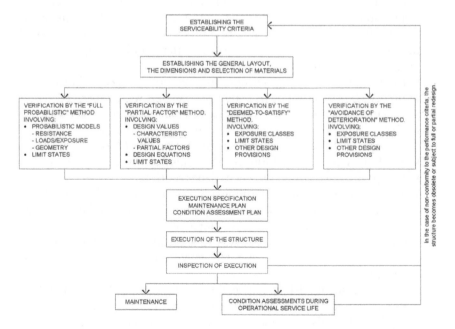

Figure 12.7 Flowchart for service life design. (Courtesy of ISO 16204, 2012)

3. Calculate the probability when the boundary limit is reached.
4. Define the type of limit state: SLS or ULS.

They go on further to say that in *fib* Bulletin 34 (2006), service life design establishes a design approach to avoid deterioration caused by environmental action comparable to load design, where a maximum load that a structure can withstand is defined quantitatively. The code uses three different levels:

- **Level 1:** a full-probabilistic approach used for exceptional structures.
- **Level 2:** a partial-factor approach considering scattered materials and environmental factors.
- **Level 3:** a deemed-to-satisfy approach based on descriptive and empirical rules.

Huang et al. (2019) advise that in *fib* Bulletin 55 (2010), the service boundary condition is defined, referring to not only safety and serviceability, but also design criteria for durability and sustainability. The document also deals with conservation strategies, condition assessments, decision-making, interventions, recording, dismantlement, removal, and recycling.

As advised at Section 5.4, full-probabilistic models for predicting the time to chloride-induced reinforcement corrosion initiation have been developed (e.g. *fib*, 2006; Kessler & Lasa, 2019; Concrete Institute of Australia, 2020). The corresponding limit state is defined as the state when a critical chloride concentration (threshold) reaches the reinforcement level. The outcome of the full-probabilistic corrosion initiation prediction is a reliability index (β) or failure probability in dependence of time that can be compared with a specific target reliability (Kessler & Lasa, 2019).

Normally, reinforcement corrosion first impairs the serviceability of the structure. A SLS corresponds to conditions beyond which specified service requirements for a structure or structural member are no longer met (EN 1990, 2002). Since reinforcement corrosion is an irreversible SLS, ISO 2394 (2015) proposes 1.5 as the target reliability (β). The *fib* Model Code for Service Life Design (*fib* 34, 2006) recommends 1.3 (β), which corresponds to a 10% probability of reinforcement depassivation.

As advised at Section 5.4, deterministic carbonation models, like chloride models, are 'stochastic', i.e. they have no unique solution as the input variables are not fixed values but are distributed around mean values. Hence a full probability analysis is the only way to ascertain the reliability at which a specific design service life will be achieved with particular concrete performance values.

Full-probabilistic modelling is available through the *fib* Bulletin 34 (2006) procedure. The recommended practice document Z7/05 (Concrete Institute of Australia, 2020) also provides full-probabilistic modelling examples.

12.10 DURABILITY ASSESSMENT – BURIED AGGRESSIVE EXPOSURE – 100-YEAR DESIGN LIFE

Detailed designs on projects are often delivered in design lots or design packages. Design reports are typically produced for each design lot (package). It is not practical to produce a durability assessment report (DAR) for each design report. However, the durability requirements for each design lot need to be incorporated in such a way that they can be readily understood, digested, and transposed into construction (Concrete Institute of Australia, 2014).

Green et al. (2009) have proposed the concept of durability checklists (matrix form) which can be incorporated directly into the design documentation and thereby also providing the required design tracking mechanism.

Durability assessment checklists (DACs) enable asset item specific durability issues to be addressed, such as:

- Element, their components, and sub-components;
- Minimum design life;
- Zone (exposure category);
- Durability issues;
- Degradation or corrosion processes;
- Risk rating (qualitative) before durability design;
- Proposed durability measures;
- Durability assessment findings;
- Inspection and maintenance;
- Remedial actions (if required); and
- Risk rating (qualitative) after durability design.

An example DAC for the reinforced concrete substructure elements of a 100-year design life bridge in a buried aggressive exposure environment is provided at Table 12.1.

12.11 DURABILITY ASSESSMENT – MARINE EXPOSURE – 100-, 150-, & 200-YEAR DESIGN LIVES

The example project considered here is a reinforced concrete (RC) bridge in a marine exposure environment. As part of the concept design for the project, a durability options analysis was to be provided. This options analysis was to consider durability planning for various design lives to determine what impact this may have on the cost of construction.

Table 12.1 Durability assessment checklist example – reinforced concrete elements buried – AS 5100 (2017) exposure classification C2

Element	Sub-Element	Minimum Design Life	Zone (Exposure Category)	Exposure Classification	Durability Issue	Degradation or Corrosion Process	Description	Risk Rating Before Durability Design	Durability Assessment Findings	Risk Rating After Durability Design
Reinforced Concrete		100 yrs	Buried	C2 (AS 5100)	Chloride exposure	Reinforcement Corrosion (chloride ion induced)	Chloride ion-induced corrosion, which could occur if chloride-contaminated groundwater penetrates external cover concrete, leading to the development of anodic sites on the reinforcement. Penetration may be by absorption, capillary sorption, wetting and drying cycles, diffusion, and/or by permeation (e.g. immersed or buried conditions). The result of this would be corrosion of reinforcement and subsequent loss of section, loss of anchorage, loss of bond and cracking, and potentially delamination and spalling of the concrete cover.	Increased	Refer Acid Attack.	Normal
					Reinforcement corrosion at cracks	Localised corrosion	The influence of cracks on the corrosion process is dependent primarily on crack geometry, whether the crack is static or active, and the orientation of reinforcement	Increased	Assurance of proper placing, compaction, and finishing.	Normal

		with respect to the crack. Anodic sites will be developed on the reinforcement at crack locations. The reinforcing bar within the un-cracked sections will become cathodic. Because of the relative areas, i.e. ratio of cathode to anode is so large, significant local corrosion could result.		Maintenance of suitable workmanship standards. Refer Acid Attack.	
Reinforcement corrosion due to local penetration	Localised corrosion (chloride ingress)	Chloride corrosion due to local penetration, which could occur if chloride-contaminated groundwater penetrates through local imperfections such as honeycombed concrete, cracks, joints, etc. Again, penetration may occur during wetting and drying cycles, capillary sorption, or under a water pressure head (permeation). Anodic sites will be developed on the reinforcement within the local imperfections. The reinforcing bar within the unexposed sections will become cathodic. Because of the relative areas, i.e. ratio of cathode to anode areas is so large, significant local corrosion could result.	Increased	Assurance of proper placing, compaction, and finishing. Maintenance of suitable workmanship standards. Refer Acid Attack.	Normal

(Continued)

Table 12.1 (Continued)

Element	Sub-Element	Minimum Design Life	Zone (Exposure Category)	Exposure Classification	Durability Issue	Degradation or Corrosion Process	Description	Risk Rating Before Durability Design	Durability Assessment Findings	Risk Rating After Durability Design
					Sulphate attack	Breakdown of cement paste/ loss of section, loss of strength	Sulphate attack of concrete is attributed to two principal chemical reactions between the sulphates and the cement leading to the formation of gypsum (calcium sulphate) and an expansive mineral, ettringite (calcium sulphoaluminate hydrate). These reactions are characterised by expansion, leading to loss of strength and stiffness, cracking, fretting of surface layers, and eventual disintegration of the concrete.	Normal	Refer Acid Attack.	Normal
					Acid sulphate soils	Breakdown of cement paste/ loss of section, loss of strength	Acid sulphate soils (ASS) is the common name given to naturally occurring soil and sediment containing pyritic sediments (iron sulphide). As long as pyritic sediment remains below the water table where it cannot be oxidised, it poses no problems. It is when pyritic sediment is exposed to air that problems occur. Exposure to air results in	Increased	Soil/Rock will only be disturbed for a very short period during construction and will stabilise. ASS/PASS (possible acid sulphate soils) to be removed	Normal

oxidation of pyrite to sulphuric acid which is highly aggressive to concrete because in addition to degrading the cement binder, it also produces an expansive reaction within the concrete matrix.

and remediated in accordance with Acid Sulphate Soil Management Plan. Refer Acid Attack.

| Acid attack | Breakdown of cement paste/ loss of section, loss of strength | Inorganic acids in general are considered to be problematic for concrete when the pH falls below 6.5 for prolonged periods. The key breakdown mechanism involves the reduction of pH due to consumption of calcium hydroxide which destabilises the calcium aluminate hydrates and the calcium silicate hydrates in the cement paste, resulting in breakdown of these cementing minerals. Ultimately the integrity of the concrete is diminished and the surface concrete gradually erodes with time. | Increased Concrete strength Normal (f'_c) = 55MPa min. Blended cement concrete 50–70% BFS or 25–40% fly ash (FA) or a ternary blended cement with 60–67% BFS and 8–10% SF or 17–20% FA and 8–10% SF. Cementitious (binder) content = 470 kg/m³ minimum. Maximum w/b ratio = 0.36. Curing continuously for at least 14 days. |

(Continued)

Table 12.1 (Continued)

Element	Sub-Element	Zone (Exposure Category)	Minimum Design Life	Exposure Classification	Durability Issue	Degradation or Corrosion Process	Description	Risk Rating Before Durability Design	Durability Assessment Findings	Risk Rating After Durability Design
					Alkali-aggregate reaction	Cracking and disruption of concrete matrix	Alkali-aggregate reaction (AAR) is a chemical process in which alkalis, present in cement combine with certain compounds in the aggregate when moisture is present. This reaction produces an alkali-silicate gel (ASR) or an alkali-carbonate gel (ACR) that can absorb water and expand to cause cracking and disruption of the concrete matrix.	Increased	Aggregate containing unstable silica minerals, such as opal, cristobalite, tridymite, or acidic glassy material must not be used. AAR assessment of aggregates (coarse & fine) in accordance with AS 5100.5-2017.	Normal
					Delayed ettringite formation	Cracking and disruption of concrete matrix	Ettringite is a complex calcium sulphoaluminate hydrate that normally forms as an early hydration product in Portland cement-based concrete pastes. Delayed ettringite formation (DEF) refers to formation of ettringite and related phases after setting and hardening of the cement paste. Formation of delayed	Increased	Concrete peak hydration temperature <70°C.	Normal

		ettringite is accompanied by the development of high pressures, which exceed the tensile strength of the concrete to cause cracking and disruption of the concrete matrix.			
Absorption, sorption & permeability effects	Various as above	Various mechanisms as above.	Normal	Refer acid attack.	Normal
1. Early thermal cracking 2. Drying shrinkage 3. Expansion and contraction cycles	Cracking, reinforcement corrosion	As per reinforcement corrosion at cracks.	Normal	Design for crack control. Detailing to minimise restraint and/or stress concentrations.	Normal
Creep	Cracking, reinforcement corrosion	Concrete components may undergo creep deformation due to load both in the short and long term. This will have no impact on the physical performance of the concrete and its resistance to chloride ion penetration or attack by other aggressive agents. However, creep deformation may result in cracking of the concrete.	Normal	Design for crack control.	Normal

(Continued)

Table 12.1 (Continued)

Element	Sub-Element	Minimum Design Life	Zone (Exposure Category)	Exposure Classification	Durability Issue	Degradation or Corrosion Process	Description	Risk Rating Before Durability Design	Durability Assessment Findings	Risk Rating After Durability Design
					Movements in response to loads or load cycles	Cracking, reinforcement corrosion	Various mechanisms as above.	Normal	Design for crack control. Limiting reinforcement stress levels.	Normal
					Differential movements	Cracking, reinforcement corrosion	Various mechanisms as above.	Normal	Detailing to minimise restraint and/or stress concentrations.	Normal
					Restraints to movements	Cracking, reinforcement corrosion	Various mechanisms as above.	Normal	Detailing to minimise restraint and/or stress concentrations.	Normal
					Fatigue	Cracking	Various mechanisms as above.	Normal	Limiting reinforcement stress levels.	Normal

Concept design durability assessment needed to outline the required concrete quality and cover requirements for select RC elements of the bridge to meet 100-, 150-, and 200-year design lives.

The main components of this bridge that were included in the concept durability assessment are as follows:

- **Piles:** assumed driven steel linings (sacrificial) with RC infill piles.
- **Pile Caps:** assumed in-situ RC.
- **Piers:** assumed in-situ RC.
- **Girders:** assumed prestressed precast concrete.

The Principal for the project was a Road Authority and any concept durability assessment needed to be on the basis of particular concrete mixes that needed to be used; see Table 12.2.

The RC elements for the bridge are required to meet the requirements of AS 5100.5 (Standards Australia, 2017) and the Road Authority structural concrete specification.

AS 5100.5 (2017) provides concrete quality and cover requirements to meet a 100-year design life but does not provide guidance on concrete requirements for more than 100 years.

The minimum concrete quality and cover requirements for the bridge concrete elements, over a 100-year design life in accordance with AS 5100.5 (2017), based on fly ash (FA) blended cement concretes only, are summarised at Table 12.3.

The concrete mixes proposed in the Road Authority structural concrete specification, refer to Table 12.2, generally meet, or exceed the above requirements.

In a coastal marine environment, the principal deterioration mechanism of the RC bridge elements considered in the concept durability options analysis was chloride-induced corrosion initiation of the reinforcement. Deterministic chloride diffusion modelling was then undertaken to determine the required reinforcement type for each of the proposed concretes and exposure environments to meet the 100-, 150-, and 200-year design lives utilising minimum covers outlined in AS 5100.5 (2017). In addition, the required minimum concrete cover to carbon steel to achieve 100-, 150-, and 200-year design lives for the main RC elements was also tabulated to allow comparison. Table 12.4 provides a summary of the findings.

The time to corrosion initiation in the modelling was also utilised as the DLS for the design life.

Table 12.2 Principal (Road Authority)-proposed concrete mixes

Element	Exposure Classification	Mix Designation	Minimum Compressive Strength (f'c)	Minimum Cementitious Content (kg/m³)	Supplementary Cementitious Content (%)	Maximum w/b Ratio
Piles (Steel lined cast in-situ)	B2 (AS 5100.5-2017) (Permanently Submerged)	VR400	40 MPa	400	25% FA	0.45
	C2 (AS 5100.5-2017) (Tidal/Splash)	VR450	50 MPa	450	30% FA	0.40
Pile Caps (Cast in-situ)	C2 (AS 5100.5-2017) (Tidal/Splash)	VR470	55 MPa	470	30% FA	0.36
	C1 (AS 5100.5-2017) (Spray)	VR450	50 MPa	450	30% FA	0.40
Pier Columns (Cast in-situ)	C2 (AS 5100.5-2017) (Tidal/Splash)	VR470	55 MPa	470	30% FA	0.36
	B2 (AS 5100.5-2017) (Coastal Atmospheric)	VR400	40 MPa	400	25% FA	0.45
Beams (Precast)	B2 (AS 5100.5-2017) (Coastal Atmospheric)	VR470	55 MPa	470	20% FA	0.45

Notes:
1. FA = Fly Ash.

Table 12.3 AS 5100.5 (2017) fly ash-blended cement concrete quality and cover requirements for a 100-year design life

Element	Exposure Classification	Minimum Compressive Strength (f'c)	Minimum Cementitious Content (kg/m³)	Supplementary Cementitious Content (%)	Maximum w/b Ratio	Minimum Cover (mm)
Piles (Steel lined cast in-situ)	B2 (AS 5100.5-2017) (Permanently Submerged)	40 MPa	400	20–30% FA	0.45	60
	C2 (AS 5100.5-2017)(Tidal/ Splash)	50 MPa	470	25–40% FA	0.36	80
Pile Caps (Cast in-situ)	C2 (AS 5100.5-2017)(Tidal/ Splash)	50 MPa	470	25–40% FA	0.36	80
Pier Columns (Cast in-situ)	C1 (AS 5100.5-2017)(Spray)	50 MPa	450	25–40% FA	0.40	70
	C2 (Tidal/Splash)	50 MPa	470	25–40% FA	0.36	80
	B2 (AS 5100.5-2017)(Coastal Atmospheric)	40 MPa	400	20–30% FA	0.45	50
Beams (Precast)	B2 (AS 5100.5-2017)(Coastal Atmospheric)	50 MPa	400	20–30% FA	0.45	40

Notes:
1. Minimum cover requirements are the same for carbon steel, galvanised steel, and stainless steel reinforcement in accordance with AS 5100.5.
2. Cover allowances do not consider additional cover requirements for curing compounds or if cast against ground.

Table 12.4 Minimum concrete and reinforcement type requirements to meet 100, 150, and 200 year design lives

Element	Exposure Classification	Minimum Compressive Strength	Minimum Cementitious Content (kg/m³)	Proposed Supplementary Cementitious Content (%)	Max. w/b Ratio	AS 5100.5 Min. Design Cover (mm)	Design Cover (mm) and Reinforcement Type Required to Meet Minimum Design Life		
							100 Years	150 Years	200 Years
Piles (Steel lined cast in-situ)[1]	B2 (Permanently Submerged)	40 MPa	400	25% FA	0.45	60	• 60 (Carbon Steel)	• 60 (Galvanised Steel) • 75 (Carbon Steel)	• 60 (Galvanised Steel) • 85 (Carbon Steel)
	C2 (Tidal/Splash)	55 MPa	470	30% FA	0.36	80	• 80 (Carbon Steel) • 80 (Carbon Steel)	• 80 (Carbon Steel) • 80 (Carbon Steel)	• 80 (Carbon Steel) • 80
Pile Caps (Cast in-situ)	C2 (Tidal/Splash)	55 MPa	470	30% FA	0.36	80		• 80 (Carbon Steel)	• 80 (Galvanised Steel) • 90 (Carbon Steel)
Pier Columns (Cast in-situ)	C1 (Spray)	50 MPa	470	30% FA	0.40	70	• 70 (Carbon Steel)	• 70 (Carbon Steel)	• 70 (Galvanised Steel) • 80 (Carbon Steel)
	C2 (Tidal/Splash)	55 MPa	470	30% FA	0.36	80	• 80 (Carbon Steel)	• 80 (Carbon Steel)	• 80 (Galvanised Steel) • 90 (Carbon Steel)

	Exposure	Strength	Cement content	FA	w/c	Cover			
	B2 (Coastal Atmospheric)	40 MPa	400	25% FA	0.45	50	• 50 (Galvanised Steel) • 80 (Carbon Steel)	• 50 (Stainless Steel) • 70 (Galvanised Steel) • 90 (Carbon Steel)	• 50 (Stainless Steel) • 80 (Galvanised Steel) • 105 (Carbon Steel)
Beams (Precast)	B2 (Coastal Atmospheric)	40 MPa	400 (500 utilised)	35% FA	0.45 (0.36 utilised)	50	• 50 (Carbon Steel)	• 50 (Galvanised Steel) • 60 (Carbon Steel)	• 50 (Galvanised Steel) • 65 (Carbon Steel)
	B2 (Coastal Atmospheric)	55 MPa	470	25% FA	0.40 (0.36 utilised)	40	• 40 (Galvanised Steel) • 60 (Carbon Steel)	• 40 (Stainless Steel) • 50 (Galvanised Steel) • 70 (Carbon Steel)	• 40 (Stainless Steel) • 60 (Galvanised Steel) • 80 (Carbon Steel)

Notes:
1. FA = Fly Ash.
2. In the tidal/splash zone the expected corrosion rate of the steel casing is 0.28mm/year. For a 16-mm casing this will perforate in approximately 50 years. Chloride exposure has therefore been assumed to begin at year 50.

REFERENCES

ACI (2019), *'Guide for Assessment of Concrete Structures Before Rehabilitation'*, ACI 364.1R-19, American Concrete Institute, Farmington, MI, USA.

Chess, P and Green, W (2020), *'Durability of Reinforced Concrete Structures'*, CRC Press, Taylor & Francis Group, Boca Raton, FL, USA.

Concrete Institute of Australia (2014), *'Z7/01 Durability Planning'*, Recommended Practice, Concrete Durability Series, ISBN 978 0 9941738 0 5, Sydney, Australia.

Concrete Institute of Australia (2020), *'Z7/05 Durability Modelling Reinforcement Corrosion in Concrete Structures'*, Recommended Practice, Concrete Durability Series, Sydney, Australia.

fib (2006), Bulletin 34: 'Model Code for Service Life Design', February, International Federation for Structural Concrete (fib), Lausanne, Switzerland.

fib (2008), Bulletin 44: 'Concrete Structure Management – Guide to ownership and good practice', February, International Federation for Structural Concrete (fib), Lausanne, Switzerland.

fib (2010), Bulletin 55: 'Model Code 2010 – First complete draft', International Federation for Structural Concrete (fib), Lausanne, Switzerland.

fib (2012a), 'Bulletin 65: Model Code 2010 – Final Draft, Volume 1', March, International Federation for Structural Concrete (fib), Lausanne, Switzerland.

fib (2012b), 'Bulletin 66: Model Code 2010 – Final Draft, Volume 2', March, International Federation for Structural Concrete (fib), Lausanne, Switzerland.

Green, W, Riordan, G, Richardson, G and Atkinson, W (2009), 'Durability assessment, design and planning – Port Botany Expansion Project', Proc. 24th Biennial Conf. Concrete Institute of Australia, Paper 65, Sydney, Australia.

Huang, I, Goodwin, F and Villani, C (2019), 'Modelling Corrosion-Related Service Life of Concrete Structures', *Materials Performance*, 58, 10, 34–39, October.

ISO 16204 (2012), *'Durability – Service Life Design of Concrete Structures'*, International Standards Organization, Geneva, Switzerland.

Kessler, S and Lasa, I (2019), 'Study on Probabilistic Service Life of Florida Bridges', *Materials Performance*, 58, 10, October, pp. 46–51.

Standards Australia (2017), AS 5100.5 'Bridge Design: Concrete', Sydney, Australia.

Standards Australia (2018), AS 3600 'Concrete structures', Sydney, Australia.

Index